Genetically Altered Viruses and the Environment

Row 1: M. A. Martin; B. N. Fields; M. D. Waters, J. R. Fowle III, A. H. Teich, R. S. Corte
D. Kamely.
Row 2: C. Y. Kawanishi, G. Khoury, A. P. Kendal, R. M. Chanock; R. C. Gallo, B. Mos
Row 3: P. M. Howley, V. Knight; J. G. Perpich, P. B. Hutt, R. S. Cortesi.
Row 4: E. D. Kilbourne; N. Hopkins; A. Helenius, R. L. Crowell.

Genetically Altered Viruses and the Environment

Edited by

BERNARD FIELDS
Harvard Medical-School

MALCOLM A. MARTIN
National Institute of Allergy and Infectious Diseases
National Institutes of Health

DAPHNE KAMELY
U.S. Environmental Protection Agency

22
Banbury Report

COLD SPRING HARBOR LABORATORY
1985

Banbury Report 22: Genetically Altered Viruses and the Environment

Printed in the United States of America
Cover and book design by Emily Harste

Library of Congress Cataloging-in-Publication Data
Main entry under title:

Genetically altered viruses and the environment.

(Banbury report, ISSN 0198-0068; 22)
Based on the Banbury Conference on Genetically Altered Viruses and the Environment held at the Banbury Center of Cold Spring Harbor Laboratory, Apr. 28–May 1, 1985.
Includes bibliographical references and indexes.
1. Viruses—Congresses. 2. Genetic engineering—Environmental aspects—Congresses. 3. Viral genetics—Congresses. I. Fields, Bernard N. II. Martin, Malcolm A. III. Kamely, Daphne. IV. Banbury Conference on Genetically Altered Viruses and the Environment (1985; Cold Spring Harbor Laboratory) V. Cold Spring Harbor Laboratory. VI. Series. [DNLM: 1. Environmental Microbiology—congresses. 2. Viruses—genetics—congresses. W3 BA19 v.22 / QW 160 G3278 1985]
QR370.G46 1985 616.9'25 85-21294
ISBN 0-87969-222-7

BANBURY REPORT SERIES

Banbury Report 1: Assessing Chemical Mutagens
Banbury Report 2: Mammalian Cell Mutagenesis
Banbury Report 3: A Safe Cigarette?
Banbury Report 4: Cancer Incidence in Defined Populations
Banbury Report 5: Ethylene Dichloride: A Potential Health Risk?
Banbury Report 6: Product Labeling and Health Risks
Banbury Report 7: Gastrointestinal Cancer: Endogenous Factors
Banbury Report 8: Hormones and Breast Cancer
Banbury Report 9: Quantification of Occupational Cancer
Banbury Report 10: Patenting of Life Forms
Banbury Report 11: Environmental Factors in Human Growth and Development
Banbury Report 12: Nitrosamines and Human Cancer
Banbury Report 13: Indicators of Genotoxic Exposure
Banbury Report 14: Recombinant DNA Applications to Human Disease
Banbury Report 15: Biological Aspects of Alzheimer's Disease
Banbury Report 16: Genetic Variability in Responses to Chemical Exposure
Banbury Report 17: Coffee and Health
Banbury Report 18: Biological Mechanisms of Dioxin Action
Banbury Report 19: Risk Quantitation and Regulatory Policy
Banbury Report 20: Genetic Manipulation of the Early Mammalian Embryo
Banbury Report 21: Viral Etiology of Cervical Cancer
Banbury Report 22: Genetically Altered Viruses and the Environment

CORPORATE SPONSORS

Agrigenetics Corporation
American Cyanamid Company
Amersham International plc
Becton Dickinson and Company
Biogen S.A.
Cetus Corporation
Ciba-Geigy Corporation
CPC International, Inc.
E.I. du Pont de Nemours & Company
Genentech, Inc.
Genetics Institute
Hoffmann-La Roche Inc.
Johnson & Johnson
Eli Lilly and Company
Mitsui Toatsu Chemicals, Inc.
Monsanto Company
Pall Corporation
Pfizer Inc.
Schering-Plough Corporation
Smith Kline & French Laboratories
The Upjohn Company

CORE SUPPORTERS

The Bristol-Myers Fund, Inc.
The Dow Chemical Company
Exxon Corporation
Grace Foundation Inc.
International Business Machines Corporation
Phillips Petroleum Foundation, Inc.
The Procter & Gamble Company
Rockwell International Corporation Trust
The Schevron Fund
Texaco Philanthropic Foundation Inc.

Participants

Rafi Ahmed, Department of Microbiology and Immunology, UCLA School of Medicine

Elizabeth L. Anderson, Office of Health and Environmental Assessment, U.S. Environmental Protection Agency

Robert M. Chanock, Laboratory of Infectious Diseases, National Institute of Allergy and Infectious Diseases, National Institutes of Health

Roger S. Cortesi, Office of Exploratory Research, U.S. Environmental Protection Agency

Richard L. Crowell, Department of Microbiology and Immunology, Hahnemann University School of Medicine

Robert L. Dixon, Office of Health Research, Research and Development, U.S. Environmental Protection Agency

Bernard N. Fields, Department of Microbiology and Molecular Genetics, Harvard Medical School

John R. Fowle III, Division of Water and Toxic Substances, Office of Health Research, U.S. Environmental Protection Agency

Robert C. Gallo, Laboratory of Tumor Cell Biology, Division of Cancer Treatment, National Cancer Institute

Ari Helenius, Department of Cell Biology, Yale University School of Medicine

Nancy Hopkins, Center for Cancer Research, Department of Biology, Massachusetts Institute of Technology

Peter M. Howley, Laboratory of Tumor Virus Biology, National Cancer Institute, National Institutes of Health

Peter Barton Hutt, Covington & Burling

Daphne Kamely, Exposure Assessment Group, U.S. Environmental Protection Agency

Clinton Y. Kawanishi, U.S. Environmental Protection Agency

Alan P. Kendal, Influenza Branch, Center for Infectious Disease, Centers for Disease Control

George Khoury, Laboratory of Molecular Virology, Division of Cancer Etiology, National Cancer Institute, National Institutes of Health

Edwin D. Kilbourne, Department of Microbiology, Mount Sinai School of Medicine of the City University of New York

Vernon Knight, Department of Microbiology and Immunology, Baylor College of Medicine

John Logan, Department of Molecular Biology, Princeton University

Malcolm A. Martin, National Institutes of Health, National Institute of Allergy and Infectious Diseases

Barbara McClintock, Cold Spring Harbor Laboratory

Thomas C. Merigan, Division of Infectious Diseases, Stanford University School of Medicine

Theodore G. Metcalf, Department of Virology and Epidemiology, Baylor College of Medicine

Henry I. Miller, Food and Drug Administration

Bernard Moss, Laboratory of Viral Diseases, National Institute of Allergy and Infectious Diseases, National Institutes of Health

Richard C. Mulligan, Whitehead Institute, Massachusetts Institute of Technology

Joseph G. Perpich, Meloy Laboratories, Inc. Revlon Health Care

Bernard Roizman, The Marjorie B. Kovler Viral Oncology Laboratories, University of Chicago

Steven Schatzow, Office of Pesticide Programs, Environmental Protection Agency

Robert E. Shope, Yale Arbovirus Research Unit, Yale University School of Medicine

Max D. Summers, Department of Entomology and The Texas Agricultural Experiment Station, Texas A&M University

Albert H. Teich, Office of Public Sector Programs, American Association for the Advancement of Science

Michael D. Waters, Health Effects Research Laboratory, Environmental Protection Agency

James D. Watson, Cold Spring Harbor Laboratory

Acknowledgments

The Banbury Center is pleased to acknowledge the role of the Office of Research and Development of the Environmental Protection Agency in facilitating and supporting this project under E.P.A. Cooperative Agreement No. CR-812187-01-0. It is also a pleasure to take this opportunity to note with gratitude the close cooperation and productive working relationship of the book's editors with the Banbury offices, the exceptionally skillful efforts of Judith Blum, the Banbury Center editor, in the production of this volume, and the continuing highly personable yet always efficient efforts of Beatrice Toliver, the Banbury Center administrative assistant.

Michael Shodell
Director
Banbury Center

Introduction

DAPHNE KAMELY AND MICHAEL D. WATERS
U.S. Environmental Protection Agency
Washington, D.C. 20460

Over the past few years, mammalian viruses have served two important functions. They provide insight into mechanisms of disease, cancer in particular. They are also used as cloning vehicles for the transfer of genes in mammalian systems. With the development of genetic engineering techniques, it is now possible to genetically alter mammalian viruses to study, in greater detail, diseases at the molecular level and to increase our understanding of gene regulation and expression. As cloning vehicles in genetic engineering experiments, genetically altered viruses provide an important tool in the development of new vaccines, therapeutic agents, and various industrial and agricultural products.

Medical and industrial applications have been increasing as has concern over the impacts of genetically altered viruses on human health and the environment. Before such organisms are released into the open environment, the survival, spread, infectivity, and potential pathogenicity of these viruses need to be characterized; yet little information is available on environmental virology. Whether naturally occurring viruses or their genetically altered counterparts are accidentally or deliberately released, their transport and fate in the environment are unknown.

For the first few years, the *NIH Guidelines for Research Involving Recombinant DNA Molecules*, published in 1976, adequately covered all government-sponsored research. Small biotechnology companies, many of which originated from university research laboratories, started manufacturing genetically engineered drugs and diagnostics. These and other bigger commercial manufacturers, which also entered the field, voluntarily agreed to abide by the NIH Guidelines. As more information became available, the NIH Recombinant DNA Advisory Committee eased the Guidelines. However, these Guidelines were not applicable to large-scale commercial production. When the number of commercial applications increased and new industrial and agricultural recombinant DNA products were developed, regulatory agencies began to examine this technology in terms of their regulatory authorities over engineered microbes. This evaluation process is still not completed.

One reason that there is less concern about the use of recombinant viruses as vaccines or in other biomedical applications may be that there are regulatory mechanisms in place to deal with such applications. The regulations for genetically engineered organisms in agriculture and in the open environment, on the other hand, are still under study. There are sufficient regulations covering toxic chemicals and chemical and biological pesticides, but regulations for genetically engineered organisms, if enacted, would break new ground.

The Environmental Protection Agency (EPA) provides a broad but not well-defined jurisdiction over biotechnological products and organisms. EPA statutes cover genetically engineered chemicals and chemical processes, pesticides and microbes for use in the environment, as well as waste products and by-products. In its concern about the effect of recombinant DNA molecules in the environment, EPA initially questioned the health consequences of deliberate or inadvertant release of genetically altered viruses.

To address these issues on genetically altered viruses in the environment, the EPA Office of Research and Development sponsored a conference at the Banbury Center of Cold Spring Harbor Laboratory in April 1985. This conference, which brought together molecular virologists, clinicians with research interests in relevant pathologies, epidemiologists, and representatives from the regulatory community was initially developed through the early planning efforts of Purnell Choppin, then of the Rockefeller Institute, Philip Sharp of MIT, Albert Teich of the AAAS, Michael Shodell from Cold Spring Harbor Laboratory, and John Fowle of the EPA, together with Bernard Fields, Malcolm Martin, and the authors of this introduction. The overall intent of the Banbury Conference on Genetically Altered Viruses and the Environment was to provide approaches in assessing potential risks that may be posed by viruses and their genetically altered counterparts released into the environment. This purpose was achieved by (1) a critical review of viral genetic mechanisms and alterations by recombinant DNA methodologies; (2) assessment of state-of-the-art knowledge of virus-cell interactions and of how genetic manipulation may affect such interactions—with emphasis on host-cell genetic regulatory mechanisms; (3) consideration of the extent to which engineered viruses present novel infectious entities; (4) assessment of possible public health and environmental hazards; and (5) clarification of possible regulatory approaches in such areas.

The conference has proved to be a great aid in delineating current research areas and the major laboratories engaged in this work, as well as in formulating possible approaches to assess potential environmental risks, both qualitative and quantitative, that may be posed by the employment of genetically altered viruses.

The impact of basic science into the legislative and regulatory processes in this area of biotechnology is important in providing a sound basis to ensure that scientific advances and technical innovation are not stifled. At the same time, we must assure ourselves that public health and the environment are sufficiently protected so that full advantage of this tremendous technology may be realized. The information contained in this volume should contribute to an understanding of the issues surrounding genetically altered viruses. It is only with such scientific knowledge that sound regulatory decisions can be made, ultimately leading to protective regulatory measures without overregulating an area of great technological and scientific promise.

Contents

Session 1: The Legislative and Regulatory Framework

Existing Regulatory Authority to Control the Processes and Products of Biotechnology

PETER BARTON HUTT
Covington & Burling
Washington, D.C.

This paper provides a broad survey of the regulatory structure within which research and development relating to biotechnology is conducted in the United States. Although this volume is focused more narrowly on genetically altered viruses, the survey undertaken in this paper applies equally to all aspects of the use of recombinant DNA technology or other techniques of the so-called new biology. Regulatory statutes are broadly directed at sources of potential hazard, not at particular types of scientific methodology.

The survey of regulatory statutes in this chapter is necessarily broad and general. No attempt is made to provide the reader with sufficient details to understand how these statutes have been applied in the past or could be applied in the future, or their detailed requirements and prohibitions. Entire volumes have been produced on many of these regulatory statutes, and their precise interpretation and application often remains a matter of uncertainty even years after enactment. The purpose of this paper is simply to identify potentially applicable regulatory statutes and to outline their general scope and purpose.

Numerous regulatory statutes enacted beginning with the first decade of this century provide comprehensive regulatory authority to the federal government to initiate whatever regulatory controls are necessary to protect the public health. The thesis of this paper is that there are no crucial legal/statutory issues presented by the prospect of federal regulation of biotechnology. The regulatory issues presented by biotechnology revolve primarily around scientific questions, not legal/statutory questions. Once scientists conclude what is and is not safe or hazardous with respect to biotechnology, regulators will experience little difficulty in using existing statutory authority to fashion appropriate regulatory controls.

The history of health and safety regulation has been the history of the development of science, not the history of the development of laws. Our regulatory laws have changed relatively little throughout history. As scientific knowledge has progressed, however, the application of those laws to provide more adequate protection of the public health has changed dramatically.

HISTORICAL BACKGROUND

The regulation of consumer products to protect the public health and safety is as old as recorded civilization.[1] The dietary laws of Moses protected consumers against unsafe food.[2] A medieval English law enacted by Parliament in 1263 prohibited the addition to staple food of any substance "not wholesome for Man's Body."[3] These are only two of many examples that could be given.

Federal regulatory statutes were enacted in the United States surprisingly early in the 19th century. Congress enacted the Vaccine Act in 1813[4] and the Import Drug Act in 1848.[5] Between 1883 and 1900, a number of laws were enacted to prevent the adulteration of imported and exported food.[6]

During the first decade of this century, four major regulatory statutes were enacted to control the marketing of important consumer products: the Biologics Act of 1902,[7] the Food and Drugs Act of 1906,[8] the Federal Meat Inspection Act of 1906[9] and 1907,[10] and the Insecticide Act of 1910.[11] These early federal statutes, administered by the Public Health Service (PHS) and the United States Department of Agriculture (USDA), and later the Food and Drug Administration (FDA), were the first to establish comprehensive regulatory structures on a national basis. All of them have since been modernized and, as amended or completely revised, remain in effect to this day.

For the next 60 years, Congress sporadically enacted new regulatory laws to protect the public health and safety, as advancing scientific knowledge demonstrated a public need. During the 1970s, three entirely new regulatory agencies—the Environmental Protection Agency (EPA), the Occupational Safety and Health Administration (OSHA), and the Consumer Product Safety Commission (CPSC)—were created to administer a host of important new health and safety statutes.

In addition to federal statutes enacted explicitly for regulatory purposes, other federal statutes have also been used for regulatory purposes even though this was not their original intent. Congress has, for example, provided authority to the

[1] P.B. Hutt and P.B. Hutt II, *A History of Government Regulation of Adulteration and Misbranding of Food,* 39 Food Drug Cosm. L.J. 2 (1984).

[2] E.g., Leviticus ch. 11.

[3] 51 Hen. III, Stat. 6 (1266), reprinted in D. Pickering, 1. *The Statutes at Large From Magna Carta to the End of the Eleventh Parliament of Great Britain* 47 (1762).

[4] 2 Stat. 806 (1813), repealed in 3 Stat. 677 (1822) after a mistake in the administration of the statute resulted in a shipment of contaminated smallpox vaccine causing an outbreak of smallpox in North Carolina. See H.R. Rep. Nos. 48 and 93 (1822).

[5] 9 Stat. 237 (1848), which remained in effect until it was superseded in 1922.

[6] 22 Stat. 451 (1883); 24 Stat. 209 (1886); 26 Stat. 414 (1890); 26 Stat. 1089 (1891); 29 Stat. 253 (1896); 29 Stat. 604 (1897); 30 Stat. 151, 210 (1897); and 30 Stat. 947, 951 (1899).

[7] 32 Stat. 728 (1902).

[8] 34 Stat. 768 (1906).

[9] 34 Stat. 669, 674 (1906).

[10] 34 Stat. 1256, 1260 (1907).

[11] 36 Stat. 331 (1910).

National Institutes of Health (NIH),[12] the National Science Foundation (NSF),[13] and USDA,[14] as well as a number of other governmental organizations, to fund scientific research. Although none of these statutes explicitly states that the federal agency may impose any condition it wishes as a requirement for those federal funds, this is in fact what NIH, NSF, and other agencies have done. Requirements for the protection of human subjects[15] and for compliance with guidelines on research with recombinant DNA molecules[16] were issued by NIH under this implicit authority.

THE CONSTITUTIONAL BASIS OF REGULATION

All government regulation is based on the premise that, in our modern technological society, individuals can no longer protect their own interests.[17] Federal and state governments may assert regulatory authority under the power of *parens patriae*[18] and the federal government may assert additional authority under its constitutional power to regulate interstate commerce.[19] Indeed, no one has ever seriously questioned the constitutional authority of federal or state governments to protect citizens from potential harm from commercial products.

Some scientists have argued that the government has no authority to interfere with basic research and inquiry because it is protected under the First Amendment right to free speech.[20] This argument misperceives the scope of the First Amendment. The First Amendment protects freedom of thought and speech (and even

---

[12] 42 U.S.C. 241(a). See *Grassetti* v. *Weinberger*, 408 F.Supp. 142 (N.D. Cal. 1976); Comment, *The Federal Conditional Spending Power: A Search for Limits*, 70 Nw L.Rev. 293 (1975); and S. Goldberg, *The Constitutional Status of American Science*, 1979 U. Ill. L.F. 1, 22–25.

[13] 42 U.S.C. 1862(a).

[14] 7 U.S.C. 2204(a) and 3101 et seq.

[15] 45 C.F.R. Pt. 46.

[16] The NIH Guidelines were first issued in 41 Fed. Reg. 27902 (July 7, 1976) and have since been revised many times but have never been codified in C.F.R.

[17] E.g., *Dalehite* v. *United States*, 346 U.S. 15, 51–52 (1953) (Jackson, J., dissenting).

[18] E.g., *Patterson* v. *Kentucky*, 97 U.S. 501, 504 (1878):

> "By the settled doctrines of this court the police power extends, at least, to the protection of the lives, the health, and the property of the community against the injurious exercise by any citizen of his own rights."

See also *Morgan's Steamship Co.* v. *Louisiana Board of Health*, 118 U.S. 455, 464–466 (1886); *Louisiana* v. *Texas*, 176 U.S. 1, 20–23 (1900); *Compagnie Francaise de Navigation a Vapeur* v. *Louisiana Board of Health*, 186 U.S. 380, 387–393 (1902); and *Jacobson* v. *Massachusetts*, 197 U.S. 11 (1905); upholding state quarantine laws.

[19] E.g., *McDermott* v. *Wisconsin*, 228 U.S. 115, 128–131 (1913); *United States* v. *Sullivan*, 332 U.S. 689, 697–698 (1948); and *United States* v. *Olsen*, 161 F.2d 669 (9th Cir. 1947).

[20] E.g., W. Szybalski, *Dangers of regulating the recombinant DNA technique*, 3 Trends in Biomedical Sciences N 243 (November 1978).

then not absolutely), not freedom of conduct and action.[21] Scientists are entirely free to hold and express their views on any aspect of biotechnology or indeed any other subject. As soon as any specific action is taken to implement those views—whether that action is characterized as basic or applied research or as product development—it is potentially subject to federal and state governmental control to protect the public health and safety. Courts have long held that any action which threatens the well-being of other people or society as a whole is subject to government regulation under both civil and criminal law.[22] The conduct of experiments on recombinant DNA in a laboratory enjoys no greater protection from government regulation than any other form of human activity that may endanger others.

THE SCOPE OF EXISTING REGULATORY STATUTES

There are two general kinds of federal regulatory statutes to protect the public health and safety: those that regulate particular products (such as food, drugs, and pesticides) and those that regulate industrial processes (such as clean air and clean water). For purposes of this paper, however, it is more useful to divide these regulatory statutes into seven different categories: (1) basic chemicals, (2) plants and animals, (3) consumer products, (4) workplace effects, (5) environmental effects, (6) transportation, and (7) general controls. This classification is, of course, entirely artificial. Many statutes could be placed in more than one category, and all of them overlap with one or more other statutes. This classification was chosen for this paper simply to impose some sense of order in the discussion that follows.[23]

Basic Chemicals

The principal statute enacted by Congress to regulate all "chemical substances" is the Toxic Substances Control Act (TSCA) of 1976,[24] administered by EPA. TSCA provides a full range of regulatory control over both old and new chemicals. EPA is authorized to impose any testing requirement or limitations upon manufacture or use that may be justified to prevent an unreasonable risk to health or the environment. TSCA covers both the chemical itself and the byproducts of the manu-

[21] E.g., *Chaplinsky* v. *New Hampshire*, 315 U.S. 568, 571–574 (1942); *Giboney* v. *Empire Storage Co.*, 336 U.S. 490, 498–504 (1949); *Yates* v. *United States*, 354 U.S. 298, 318–327 (1957); *Cox* v. *Louisiana*, 379 U.S. 536, 554–555 (1965); *United States* v. *O'Brien*, 391 U.S. 367, 376–377 (1968); *Brandenburg* v. *Ohio*, 395 U.S. 444, 447–449 (1969); and *Cohen* v. *California*, 403 U.S. 15, 18–26 (1971). See J.R. Ferguson, *Scientific Inquiry and the First Amendment*, 64 Cornell L.Rev. 639, 656–659 (1979); and S. Goldberg, note 12 *supra*, at 11–16.

[22] E.g., notes 19 and 21 *supra*.

[23] For a simple listing of federal statutes and some implementing regulations, which takes a different approach to categorization, see 49 Fed. Reg. 50856, 50859–50877 (December 31, 1984).

[24] 15 U.S.C. 2601 et seq.

facturing process. Other regulatory statutes, including those that control industrial processes, also cover the manufacturing byproducts of the chemical industry.

There was initially considerable debate as to whether the definition of a "chemical substance" in TSCA covers microorganisms and other products of biotechnology. EPA has now taken the position that the products of biotechnology are fully covered by TSCA.[25]

Plants and Animals

USDA has authority to prevent harm to plants and animals through a broad variety of statutes. The Organic Act of 1944[26] authorizes USDA to cooperate with states and other organizations to detect and control the spread of plant pests. The Federal Plant Pest Act[27] prohibits the importation or interstate movement of any plant pest except under a USDA permit. The Plant Quarantine Act[28] provides USDA with regulatory authority over the importation and interstate commerce of any plant product which may contain a plant disease or pest. The Noxious Weed Act[29] regulates the importation or interstate movement of all noxious weeds and related plant material. USDA has authority to regulate seeds under the Federal Seed Act.[30] Thus, USDA has plenary authority to regulate the development and introduction of new plants and plant pests.

Existing federal statutes also grant similar authority with respect to animals. Quarantine and related laws provide strong authority to regulate the importation and interstate movement of animals and to prevent the introduction or spread of communicable disease among animals.[31] Special authority is also given under the Federal Meat Inspection Act,[32] the Poultry Products Inspection Act,[33] and the Egg Inspection Act.[34]

The Endangered Species Act of 1973[35] authorizes the Department of the Interior, and other federal agencies as well, to protect all species of fish, wildlife, and plants listed by DOI as endangered. Under a statute authorizing DOI to regulate the importation of injurious animals and other legislative authority, the President has issued an executive order to restrict the importation and introduction of exotic

[25] 49 Fed. Reg. 50880 (December 31, 1984).
[26] 7 U.S.C. 147a.
[27] 7 U.S.C. 150bb.
[28] 7 U.S.C. 151 et seq.
[29] 7 U.S.C. 2801 et seq.
[30] 7 U.S.C. 1551 et seq.
[31] 21 U.S.C. 101 et seq.
[32] 21 U.S.C. 601 et seq.
[33] 21 U.S.C. 451 et seq.
[34] 21 U.S.C. 1031 et seq.
[35] 16 U.S.C. 1531 et seq.

species of plants or animals into natural ecosystems in the United States and abroad.[36]

Consumer Products

Specific statutes have been enacted to assure close federal regulation of a wide variety of important consumer products. Food, drugs, medical devices, cosmetics, and animal food and drugs, are regulated by FDA under the Federal Food, Drug, and Cosmetic Act of 1938.[37] Human biological drugs are regulated under the Biologics Act of 1902[38] and animal biological drugs are regulated under the Virus Serum Toxin Act of 1913.[39] All forms of pesticides and related products are regulated by EPA under the Federal Insecticide, Fungicide, and Rodenticide Act.[40] In 1972, Congress enacted the Consumer Product Safety Act to protect the public against unreasonable risks from any consumer product.[41] The CPSC was also given authority to administer the Federal Hazardous Substances Act, which requires label warnings or a ban for hazardous products intended for use in or around the household.[42] As already noted, USDA also closely regulates meat, poultry, and eggs.[43]

These product-specific statutes are also supplemented by other statutes concerned with preventing hazards in the workplace, the environment, and during transportation. These other broader statutes are discussed below.

Workplace Effects

The principal statute enacted by Congress to assure a safe workplace is the Occupational Safety and Health Act.[44] The OSH Act contains a general duty clause requiring all employers to maintain working conditions that will protect employees from harm, and authorizes OSHA to issue standards to protect workers from any significant risk of material health impairment. In addition to this broad authority, the product-specific statutes already discussed above also provide separate authority for other agencies to protect workers from hazard. OSHA has published guidelines stating that the OSH Act applies fully to all aspects of the field of biotechnology.[45]

---

[36] 18 U.S.C. 42; and Executive Order 11987, 42 Fed. Reg. 26949 (May 25, 1977).
[37] 21 U.S.C. 301 et seq.
[38] 42 U.S.C. 262.
[39] 21 U.S.C. 151 et seq.
[40] 7 U.S.C. 136 et seq.
[41] 15 U.S.C. 2051 et seq.
[42] 15 U.S.C. 1261 et seq.
[43] See notes 32–34 *supra*.
[44] 29 U.S.C. 651 et seq.
[45] 50 Fed. Reg. 14468 (April 12, 1985).

Environmental Effects

Several broad environmental statutes, all of which are administered by EPA, are available to regulate the byproducts of industrial processes in order to assure a healthy and unpolluted environment: the Clean Air Act,[46] the Water Pollution Control Act (commonly known as the Clean Water Act),[47] the Safe Water Drinking Act,[48] the Resource Conservation and Recovery Act (RCRA) (including the Solid Waste Disposal Act),[49] the Comprehensive Environmental Response, Compensation, and Liability Act (Superfund Act),[50] and the Marine Protection, Research, and Sanctuaries Act.[51] These laws authorize EPA to protect the safety of drinking water and the air, to regulate all forms of industrial waste, to control ocean dumping, and to take other action to prevent damage to the environment or to human health through the environment. As noted above, both the product-specific statutes and the OSH Act also contain authority for other agencies to take action to prevent damage to the environment or environmental hazards to human health.

Transportation

The Hazardous Materials Transportation Act, administered by the Department of Transportation, authorizes DOT to regulate the packing, labeling, and routing of the shipment of any hazardous materials, and empowers DOT to halt any shipment which is an imminent hazard.[52] Under Section 361 of the Public Health Service Act,[53] the Centers for Disease Control also regulates the shipment of all etiologic agents.

General Controls

In addition to these relatively specific regulatory statutes, additional general controls have been enacted that can be used to regulate any aspect of biotechnology. Section 361 of the Public Health Service Act[54] broadly authorizes the Public Health Service to adopt any controls over intrastate or interstate commerce to prevent the spread of any communicable disease. The National Environmental Policy Act (NEPA)[55] requires every federal agency to prepare an environmental assessment of any proposed agency action, and to compile a comprehensive environ-

[46] 42 U.S.C. 7401 et seq.
[47] 33 U.S.C. 1251 et seq.
[48] 42 U.S.C. 300f et seq.
[49] 42 U.S.C. 6901 et seq.
[50] 42 U.S.C. 9601 et seq.
[51] 33 U.S.C. 1401 et seq.
[52] 49 U.S.C. 1801 et seq.
[53] 42 U.S.C. 264.
[54] *Id.*
[55] 42 U.S.C. 4321 et seq.

mental impact statement on any major federal action significantly affecting the environment. NEPA has been used, in particular, to require that adequate attention be given to the environmental impact of research on recombinant DNA molecules.[56] The Department of Commerce also has plenary authority under the Export Administration Act[57] to restrict exportation of confidential technical information for reasons of foreign policy or national security.

Summary

Even this very brief characterization of the important regulatory statutes available to control biotechnology demonstrates their extraordinary breadth and depth. Rather than raising concern about whether these statutes are adequate to control this emerging new technology, they instead raise concern about whether existing regulatory controls may impose such restrictions that the scientific potential offered by biotechnology, for the future benefit of the public, may not be realized as quickly or efficiently as could be done without such regulation. As in all fields of technology, reasonable regulation must reflect a proper balance between the need to assure public protection and the need to permit scientific progress that promises enormous improvements in public health for the future.

THE NEED FOR NEW STATUTORY CONTROLS

The regulatory statutes described above unquestionably provide broad authority for the regulation of biotechnology. Nonetheless, two problems can occur. First, it is possible that federal agencies will not utilize the existing regulatory authority. Second, it is also possible that the particular language utilized in any given statute may yield one or more gaps, thus providing less than full regulatory control over biotechnology.

With regard to the potential failure of a federal agency to take adequate action to protect the public health under existing statutory authority, this possibility has existed for centuries under all regulatory statutes. It would continue to exist even if there were a separate regulatory statute devoted solely to biotechnology. The remedy, if one is needed, is oversight by the Executive Branch, the Congress, the judiciary, and the general public. No new statute, in itself, can compel appropriate action by appointed federal officials.

The second possibility, that gaps exist in existing legislation, is undoubtedly a reality at the present time. It is doubtful that any law has ever been written, or could ever be written, without some gaps of this nature. The questions that must be asked are whether those gaps are serious and whether they present a significant risk

---

[56] E.g., *Mack* v. *Califano,* 447 F.Supp. 668 (D.D.C. 1978); and *Foundation on Economic Trends* v. *Heckler,* No. 84-5419 (D.C. Cir., February 27, 1985).

[57] 50 U.S.C. 2401 et seq.

to public health protection. At the present time, neither of these possibilities appears to be true. Federal officials have testified before Congress that existing statutes are adequate to authorize all needed regulatory action.[58] Numerous reports by the Office of Technology Assessment[59] and others have failed to identify significant gaps that could jeopardize the public health. Thus, these possibilities provide no current basis for determining that a new statute would be advisable.

Nonetheless, for the past 10 years, a number of people have argued that a new regulatory statute, directed solely at biotechnology, should be enacted. An attempt to enact such a statute in the late 1970s[60] was forestalled when NIH adopted its guidelines on recombinant DNA research, industry agreed to abide by those guidelines, and scientists persuaded Congress that additional legislation was not needed. Further legislation to supplement the regulatory statutes already outlined above is likely to come under active consideration only if the existing laws are shown to be inadequate in some substantial way.

New legislation would, of course, reduce uncertainty about the application of existing laws to biotechnology. It might, for example, redefine the term "chemical substance" under TSCA to make clear that this definition includes microorganisms,[61] and it might also clarify other aspects of existing laws and regulations. This clarity, however, would undoubtedly be purchased at the substantial cost of far less flexibility and far greater rigidity in the law and its administration. Research and development would thus be likely to be hindered rather than helped, with no significant public benefit.

At present, as the above description of existing regulatory law demonstrates, there is no such thing as "biotechnology law." The processes and products of biotechnology are regulated under traditional statutes in the same way that the processes and products of other forms of technology are regulated. A food additive or pesticide is regulated no differently if it is produced by chemical synthesis, traditional fermentation, or recombinant DNA technology. The law, and the regulatory action taken under it, remains the same. Nor does there appear, at this time, to be any significant justification for changing this approach.

---

[58] "Biotechnology Regulation," Hearing before the Subcommittee on Oversight and Investigations of the Committee on Energy and Commerce, House of Representatives, 98th Cong., 2nd Sess. (1984). See also F.L. McChesney and R.G. Adler, *Biotechnology Released from the Lab: The Environmental Regulatory Framework*, 13 Environ. L. Rep. 10366 (November 1983).

[59] Office of Technology Assessment, *Impacts of Applied Genetics: Micro-Organisms, Plants, and Animals*, No. OTA-HR-132 (April 1981) and *Commercial Biotechnology: An International Analysis*, No. OTA-BA-218 (January 1984).

[60] J.P. Swazey et al., *Risks and Benefits, Rights and Responsibilities: A History of the Recombinant DNA Research Controversy*, 51 S. Cal. L. Rev. 1019, 1063-1076 (1978); and J.D. Watson and J. Tooze, *The DNA Story* 137-201 (1981), Cold Spring Harbor Laboratory, Cold Spring Harbor, New York.

[61] S. 3075, 98th Cong., 2d Sess. (1984).

FEDERAL/STATE RELATIONS

The potential for state and local regulation of biotechnology, in addition to federal regulation under the statutes outlined above, has been a very serious threat to biotechnology for the past decade. Just as this field was being developed, the city of Cambridge, Massachusetts, imposed its own regulatory controls, and other cities considered similar action.[62] Although those local restrictions ultimately were removed, they demonstrated the serious difficulty that could be encountered if state and local governments were to become dissatisfied with the level of control exercised by the federal government.

This threat of state and local control can be handled in several ways. First, strong federal regulatory control is likely to eliminate any need for state and local control. Second, an explicit statement by a federal agency that its regulation of biotechnology is intended to preempt state and local regulation will be highly likely to be enforced by the courts.[63] The Administrative Conference of the United States has recently recommended this approach for all federal rule-making.[64] Third, it is possible for Congress to enact specific statutory provisions preempting either the regulation of biotechnology in particular or regulatory actions taken under the existing federal statutes for all technologies.[65] Broad legislative preemption appears to be unlikely in the near future, especially in an era of increased concern about the need for state and local initiative. Fourth, the courts may, under the Supremacy and Commerce clauses, independently determine that state and local regulatory action is precluded where the federal government has asserted leadership.[66] This is uncertain and unpredictable except where state or local action directly interferes with federal regulation or interstate commerce.

The subject of biotechnology can raise very strong emotions. Future uses of biotechnology very directly raise ethical and moral issues that have already been the

[62] J.P. Swazey, note 60 *supra*, at 1053–1063; and J.D. Watson and J. Tooze, note 60 *supra*, at 91–135.

[63] *Fidelity Federal Savings & Loan Ass'n* v. *De La Cuesta*, 458 U.S. 141, 152–159 (1982); and *Conference of State Bank Supervisors v. Conover,* 710 F.2d 878, 881–883 (D.C. Cir. 1983).

[64] 49 Fed. Reg. 49838 (December 24, 1984). For examples of explicit FDA administrative preemption see 47 Fed. Reg. 50442, 50447–50448 (November 5, 1982) (tamper-resistant packaging for OTC drugs); and 47 Fed. Reg. 54750, 54756–54757 (December 3, 1982) (OTC drug pregnancy warning).

[65] Some existing federal regulatory statutes explicitly preempt additional or different state or local regulatory requirements. E.g., 15 U.S.C. 1261 (Federal Hazardous Substances Act); 15 U.S.C. 2617 (Toxic Substances Control Act); and 29 U.S.C. 667 (Occupational Safety and Health Act).

[66] E.g., *Jones* v. *Rath Packing Co.*, 430 U.S. 519 (1977); and *Cosmetic, Toiletry and Fragrance Ass'n, Inc.* v. *Minnesota*, 440 F. Supp. 1216 (D. Minn. 1977), aff'd, 575 F.2d 1256 (8th Cir. 1978).

subject of intense public debate.[67] The prospect of state and local regulation thus remains a very serious possibility, and is best forestalled by strong federal leadership.

CONCLUSION

As noted at the outset, the regulatory issues raised by the extraordinary development of biotechnology in the past decade are almost exclusively scientific rather than legal/statutory in nature. There is a plethora of statutory authority authorizing federal agencies to regulate all aspects of biotechnology. Those laws require neither overregulation nor underregulation. They are extremely broad and flexible. Their implementation must be based upon sound scientific knowledge and judgment. Thus, the real challenge is for scientists to provide a foundation on which regulation can proceed without undue interference with research and development efforts that promise such important benefits for the future.

COMMENTS

DIXON: Regarding the regulation of indoor air pollutants, there is a large concern about pollutants that come from burning wood in fireplaces. There are many independent and personal decisions involved here.

HUTT: Constitutionally, you can prohibit suicide and make that, indeed, a crime. Presumably, you can regulate almost anything. But, Bob [Dixon], you are getting right into the question of whether we need new statutes. I think we do not. We have so many statutes, we do not know what to do with them. Indeed, we do not even know where they all are and what they all mean. We have never used all of them. One can find isolated gaps, there is no doubt about it. In every statute there are always some gaps and some problems. The real question is whether they are significant enough that you need to enact a whole new regulatory mechanism.

DIXON: Could the potential hazards be regulated by the states?

HUTT: The Supreme Court has been holding in the last few years that if an agency does not say anything about the preemptive impact of its regulations,

[67] E.g., National Academy of Sciences, *Research with Recombinant DNA: An Academy Forum* (1977); President's Commission for the Study of Ethical Problems in Medicine and Biomedical and Behavioral Research, *Splicing Life: The Social and Ethical Issues of Genetic Engineering with Human Beings* (1982); H.R. 2788, 98th Cong., 1st Sess. (1983) (a bill to establish a President's Commission on the Human Applications of Genetic Engineering); and Office of Technology Assessment, *Human Gene Therapy*, No. TA-BP-BA-32 (December 1984).

it is up to the courts; and generally the state and local laws will prevail. But the Supreme Court has held that where either the statute or the agency says the intent is to preempt, even if it is just a statement by the agency, that will, in most instances—in all but rare instances—be upheld by the courts, i.e., there will be federal preemption. Recently, the Administrative Conference of the United States recommended that approach to the agencies. So when a FIFRA regulation is promulgated, it ought to be stated in the regulation as, for example, with EDB: "We intend this to be the national standard and we intend that it will preempt state and local regulations." But agencies are reluctant to do that.

DIXON: Does that mean that a local agency or a state can make any law it wants, as long as it's at least as stringent as the federal law?

HUTT: As long as the state law is not in direct conflict with the federal law. Suppose the local town council here passed an ordinance dealing with Cold Spring Harbor. As long as it just added additional requirements, but did not make it impossible to comply both with the federal and the local law—if you can comply with both—then it is usually upheld in the courts. Only if you cannot comply with both is it usually struck down by the courts.

MILLER: On the subject of additional legislation, I have a comment and then a question for you. You said a fair amount about biotechnology, but you didn't define it explicitly.

HUTT: On purpose.

MILLER: As implied in your remarks, biotechnology is not something that is discrete or homogeneous; it encompasses a spectrum from fish farming, to production of agricultural foodstuffs, to vaccine production including genetic programming of bacteria to clean up oil spills, or to produce insulin.

HUTT: It is no different under the law than nonbiotechnology. That was my point at the beginning. The reason I didn't define it is that there is no statute that is designed to deal just with biotechnology.

MILLER: I agree. It's hard to know what kind of legislation one might propose to regulate biotechnology. I think you alluded to that. But it's evident, from some of the biotechnology meetings in Washington, and from the remarks of Congressman Dingell and his committee staff, that they are drafting some kind of legislation. Would you speculate on whether it's likely to be some kind of omnibus bill or whether it will overlay existing legislation?

HUTT: It inherently will overlay other existing laws because there is no way Congress could write a bill that would include everything that I have described. You would have to rewrite all the plant and the animal authority, the food, drug, and pesticides authority, and everything else. That is crazy. My judgment is that if they really persist, I do not think they will succeed. There have been only two proposals I have seen. One is for a commission of ethics to oversee the future of gene therapy and moral and ethical issues of that kind. The second is greater control over so-called "deliberate release" which, as you know, I think is a misnomer anyway. If they ever try to define "deliberate release," I will be very interested, because my view is any time you make anything, that is a deliberate release. It is almost a nonword. I do not know how it ever entered popular usage. Senator Kennedy's bill in 1977 basically would have stopped all recombinant DNA research; it had a strong regulatory overtone to it. All of us at that time worked day and night to oppose that, and we were successful—not because of the lawyers, but because of the scientists who convinced Congress by saying, "This is the most important scientific advance in years and you are going to stop human progress."

Federal Regulatory Policies and Biotechnology Industry Product Development

JOSEPH G. PERPICH
Meloy Laboratories, Inc.
Revlon Health Care
6715 Electronic Drive
Springfield, Virginia 22151

On the tenth anniversary of the Asilomar Conference in February 1985, the National Academy of Sciences held a symposium on biotechnology. The focus of the symposium was on creating an environment for technological growth. Alexander Rich of the Massachusetts Institute of Technology noted that hundreds of thousands of recombinant DNA experiments have been performed and much of the concerns over the potential risks posed by this technology, which had been raised at the Asilomar Conference, have been laid to rest. Accordingly, the focus of public concern has shifted from biotechnology research oversight to the nature and scope of federal regulatory oversight of industrial biotechnology product development. Thus, the major focus at the NAS conference was emerging federal policies regarding the regulation of biotechnology products in the areas of health, agriculture and chemicals.

The issues and the themes that emerged for technological development at the National Academy of Sciences symposium in large part centered on pending regulations of the Environmental Protection Agency (EPA). Thus, the research presented in this volume underscores the success of the commitment by the Federal Government to support basic research—especially funding by the National Institutes of Health (NIH) that has led to the current biological revolution. It is that government support for basic research, combined with the dynamism of the private markets, that will bring biotechnology products into virtually every industrial sector. Genetic engineering companies and major companies with biotechnology divisions correspond to every major federal R&D program—health, chemicals, energy, agriculture and environment. Concomitantly, all regulatory agencies with responsibilities in health and safety will be involved in some fashion in the oversight of biotechnology product development in the various industrial sectors. It is the growth of biotechnology product development in these industrial sectors that led to the creation of a Cabinet Council Working Group on Biotechnology in the spring of 1984. The Council issued a report for public comment in December 1984 that proposed a number of regulatory changes that merit the attention of all involved in biotechnology R&D and product development (Office of Science and Technology Policy [OSTP] 1984).

FEDERAL R&D POLICIES AND THE BIOTECHNOLOGY INDUSTRY

Paul Berg at Stanford has observed that without the National Institutes of Health's (NIH) researching funding over the past 40 years in biology, he could not imagine the present development of the biotechnology industry. However, a *New York Times* editorial of January 28, 1985, asked "How much research is enough?" and challenged Congress to justify its increase of NIH's new and competing grants for 1985 from 5000 to 6500. That editorial elicited a number of letters from academic scientists in support of the NIH budget. Ronald Cape, Chairman and Chief Executive Officer of the Cetus Corporation, and I stated in a letter to the *New York Times* that the biotechnology industry vitally depends upon a vigorous basic research base primed by the NIH. (Cape and Perpich 1985).

Industry invests in research and has steadily increased its support of health-related research, spending $4.6 billion in 1984, or 39% of the national total. However, the vast majority–in excess of 60%–of total national funding for basic research in biomedical science has traditionally come from the NIH. Our current international leadership in biotechnology rests on this base. Indeed, the Congressional Office of Technology Assessment (OTA) cites federal support for basic research and training as the most important factor for U.S. international biotechnology leadership (Office of Technology Assessment 1984); and both the OTA and the National Academy of Sciences (NAS) have recommended increased federal support for biotechnology research and training–especially in plant molecular biology, microbial genetics, and bioprocess engineering.

The Commerce Department forecasts that in the 1990s the biotechnology industry is expected to be worth approximately $100 billion. Fueling the industrial advances is NIH-supported research that has helped bring forth gene splicing, monoclonal antibodies, and protein and microbial engineering. Thus, federal support for basic research is of critical importance to the biotechnology industry. With that support, the biotechnology industry can use its resources to transform basic research advances into the manufacture of a broad array of products to the benefit of all. However, getting those products into the marketplace depends on the other side of the federal ledger–namely, regulatory agency oversight of product development.

FEDERAL REGULATORY POLICIES AND THE BIOTECHNOLOGY INDUSTRY

Expanding industrial applications of biotechnology in health, agriculture, chemicals, and the environment have led Congress, the courts, and the Executive Branch to review existing federal regulatory policies for biotechnology. A Cabinet Council Working Group on Biotechnology was created last spring with George A. Keyworth II, the President's Science Advisor, as chairman. The Working Group reviewed all federal biotechnology regulatory policies and issued a report for public comment on December 31, 1984 (Office of Science and Technology Policy [OSTP] 1984). The group recommended a number of administrative and regulatory changes to accommodate the accelerating pace of industrial R&D advances. The recommendations in the report concerning the National Institutes of Health, Food and Drug Administration, and Environmental Protection Agency are discussed below.

CABINET COUNCIL WORKING GROUP REPORT RECOMMENDATIONS ON THE NIH RECOMBINANT DNA PROGRAM ADVISORY COMMITTEE (RAC)

The NIH Guidelines (developed in 1976-1978) have served to date as the standards to govern federally supported research nationally. Compliance by industry, however, has been voluntary. The system of governance includes local institutional biosafety committees whose oversight responsibilities have grown over the past 5 years. The NIH's Recombinant DNA Program Advisory Committee (RAC) is a broadly based, public advisory group that advises the NIH Director on actions and policies governing this research and initial industrial scale-up.

Noting the broad scope of applications beyond biomedical research supported by the NIH, the Cabinet Council Working Group recommended that additional recombinant advisory committees be created for other federal agencies (Agriculture, National Science Foundation, Environmental Protection Agency, and Food and Drug Administration). These committees would be overseen by a government-wide Biotechnology Science Board reporting to the Assistant Secretary for Health of the Department of Health and Human Services. Complementing the Board would be an interagency group to coordinate federal regulatory biotechnology policies. For the present, the Cabinet Council Working Group would serve that function.

Creating additional recombinant advisory committees represents a bureaucratic compromise on how best to organize recombinant DNA research oversight to meet the needs of the regulatory agencies. However, this technology cuts across scientific disciplines and the present bureaucratic organization of government research and regulatory agencies. As Dr. James Wyngaarden, the NIH Director, has observed, the traditional biomedical research disciplines are being increasingly unified scientifically by—in Dr. Arthur Kornberg's words—"the common language of chemistry" (Wyngaarden 1984). That observation reflects my own experience in industrial biotechnology. I serve on a number of task forces of the Industrial Biotechnology Association (IBA) with representatives of pharmaceutical, chemical, agricultural, and energy companies. Despite the diversity of our product interests, we share the common technology base in molecular and cell biology.

In my view, the NIH's RAC should remain as the sole body for overseeing the biotechnology research and initial industrial scale-up that precedes product development which comes under the jurisdiction of the relevant regulatory agency. The

NIH's RAC has had a splendid record in adding the necessary expertise through ad hoc groups to develop guidelines for pilot scale-up in the pharmaceutical industry, for field testing in the agricultural and chemical industries, and, most recently, for gene therapy in humans. Thus, RAC has served as the national–indeed, international–reference for biotechnology research activities, and it should continue in that unifying role. All involved in biotechnology now look to the NIH's RAC as a "supreme court," for policy development.

An alternative model for the Cabinet Council Working Group to consider is to elevate the NIH's RAC to the level of the Assistant Secretary for Health and have it serve as the Biotechnology Science Board. The concept of the NIH's RAC serving as the Biotechnology Science Board has endorsed by the Pharmaceutical Manufacturers Association (PMA) in comments filed with the Office of Science and Technology Policy (OSTP) (Pharmaceutical Manufacturers Association 1985). The PMA questioned the statutory authority, need, and practicality for the agency-based RACs to approve both research and products of biotechnology. Furthermore, a Biotechnology Science Board would have the power to develop review procedures and scientific guidelines for the agencies and to direct reviews of biotechnology proposals or classes of proposals. The PMA noted that this power provides regulatory authority to a nonregulatory body and that this might, in fact, impede the regulatory agencies' current practices. I would add that five committees and a board–all with decision-making authority–would inevitably create jurisdictional and policy conflicts that could only be resolved at the level of the Cabinet Council.

Thus, I support the PMA's position that the very substantial new procedures proposed by the Cabinet Council Working Group may be unnecessary. I have, perhaps, a vested interest in NIH's RAC for I participated in its creation. However, that committee, over the past 10 years, has withstood the test of several challenges to its development. It deserves a chance for increased responsibility under the able leadership of its chairman, Robert Mitchell, to serve as the Biotechnology Science Board at the level of the Assistant Secretary for Health.

CABINET COUNCIL WORKING GROUP REPORT RECOMMENDATIONS ON BIOTECHNOLOGY REGULATIONS

In addition to recommendations on the structure of research oversight, the Cabinet Council Report also provides policy statements by the FDA, EPA, and Agriculture (USDA) for product regulations involving health, agricultural, and chemical biotechnology applications. These applications were identified when the NIH Guidelines were developed in 1976 as long-range public policy issues beyond the scope of the NIH Guidelines. (The report, however, does not fully address an additional area that merits the attention not only of federal policymakers, but also of academic and industrial biotechnology leaders, namely, future export regulations governing biotechnology processes and products (Perpich 1985).

The PMA comments to OSTP provide an excellent framework for considering federal regulatory policies. The PMA noted that (1) products made using recombinant DNA and other biotechnology processes are not fundamentally different from products made by conventional techniques; (2) current laws and regulations for products manufactured by biotechnology are adequate for protection of the public health and safety and the environment; (3) product regulation should involve review of research, development, and manufacturing processes only to the extent required for an evaluation of safety and efficacy; and (4) proprietary information must continue to be protected during the product review process. The PMA also recommended enhanced federal support for basic research and training in the various biotechnology sectors.

CABINET COUNCIL WORKING GROUP REPORT RECOMMENDATIONS: FDA

Regulation of the first biotechnology products by FDA is an excellent model for EPA, USDA, and others to consider as they develop regulatory policies. All FDA bureaus are involved; virtually all of the biotechnology companies are involved with the FDA as products come "on line." Frank Young, the FDA Commissioner, reported in a Congressional hearing in December, 1984, that the FDA had already approved 55 devices in biotechnology, including diagnostic kits and enzyme assays, one recombinant DNA-produced drug (insulin), and three hybridoma-based diagnostics. There are 51 additional health care products and 60 veterinary products currently awaiting approval (Young 1984).

From the industrial experience to date, the FDA has been able to apply its existing regulatory authority to deal with biotechnology products without major changes in the law. The leadership of individuals, such as Irving Johnson, in the pharmaceutical industry has ensured a smooth transition from the NIH's RAC to the FDA regulatory processes.

In the Cabinet Council Working Group document, the FDA policy statement concludes that there is no need for substantive change in the current FDA laws and regulations in order to handle biotechnology products. "The use of a given biotechnological technique does not require a different administrative process. Regulation by FDA must be based on the rational scientific evaluation of products and not on a priori assumptions about certain processes" (OSTP 1984).

PMA, in its comments, endorses the FDA's statement of policy as rational, supportable, and in the national interest. PMA also commends FDA's use of "Points to Consider" documents to provide guidance on how FDA regulations apply to biotechnology products. The FDA notes that this mechanism is flexible enough to adapt to the rapidly changing scientific and product developments of biotechnology.

The FDA has issued five *Points to Consider* documents that provide general guidance to the biotechnology industry on the regulatory processes. In March 1982, FDA developed a document for the manufacture of in vitro monoclonal antibody products (FDA 1982); this document was revised in 1983. In 1983, a document was issued on the production and testing of interferon for investigational use in humans (FDA 1983a) and another document covered the manufacture of monoclonal antibody products for human use (FDA 1983b). In the fall of 1983, a document was issued covering the production and testing of new drugs and biologics produced by recombinant DNA technology (FDA 1983c) and in June 1984, the FDA issued a document on the characterization of cell lines used to produce biologicals (FDA 1984).

These documents for biotechnology companies have been extremely useful in guiding all product development efforts at Meloy. FDA scientists and staff are available for consultation on interpretations of the "Points to Consider" documents. The development of the "Points to Consider" documents has provided for public oversight and review. There are a number of questions, of course, that will need to be addressed in the future, particularly for second and third generation biotechnology products. The PMA and the Industrial Biotechnology Association (IBA) will address these issues as part of an effort to engage FDA in planning future "Points to Consider" documents.

Another important policy area in health is the application of gene splicing to human beings. The Cabinet Council Working Group Report does not address this issue, but there have been numerous Congressional hearings and reports on this subject. In addition, the NIH recently submitted a "Points to Consider" document for public comment on the design and submission of human somatic cell gene therapy protocols to the NIH. This document was prepared by a working group of the RAC, chaired by Leroy Walters of the Kennedy Institute of Ethics at Georgetown University, in anticipation of cases coming this year to the NIH for approval (National Institutes of Health [NIH] 1985). There will be considerable attention focused on these cases and attendant federal policies by the RAC in its consideration of these clinical research protocols; and there will continue to be congressional and public scrutiny of the ethical implications of gene therapy, especially prospects for germline therapy.

CABINET COUNCIL WORKING GROUP REPORT RECOMMENDATIONS: EPA

The bulk of the Cabinet Council report is devoted to the EPA policy statement on the regulation of biotechnology products under the Federal Insecticide, Fungicide, and Rodenticide Act (FIFRA) and the Toxic Substances Control Act (TSCA) (OSTP 1984). The breadth and scope of EPA's authorities under these statutes to regulate the manufacture of pesticides and chemicals bring a large segment of the biotechnology industry under EPA's product review. Furthermore, in its policy statement, EPA declares this technology may have profound consequences for the environment in terms of agricultural and chemical applications. EPA declares, "The techniques of modern molecular biology are revolutionary and they allow humans to override natural genetic constraints—new organisms produced through these techniques have genomes that do not occur in nature or have gene pools substantially altered from those that would occur through the 'natural processes' of reproduction." (OSTP 1984).

FEDERAL INSECTICIDE, FUNGICIDE, AND RODENTICIDE ACT (FIFRA)

EPA defines nonindigenous and genetically engineered microbes as the basis for FIFRA regulation. For regulation of biotechnology produced pesticides under FIFRA, EPA cites potential risks arising from a broader host range, a new toxin, enhanced virulence, greater survivability, and/or greater competitiveness than the indigenous microorganisms. According to comments submitted to the EPA by the Industrial Biotechnology Association (IBA), EPA itself has frequently said biological pest control is preferable because it reduces the reliance on chemical pesticides and presents a lower risk to the environment (Industrial Biotechnology Association 1985). EPA should keep that principle in mind in the development of these regulations because EPA-proposed regulations under FIFRA are very broad. They cover virtually all biotechnology processes, ranging from cell fusion and recombinant DNA to directed or undirected mutagenesis, conjugation, plasmid transfer, and transformation.

NIH has limited the RAC review to recombinant DNA research, and EPA should initially consider a more narrow scope for its regulations as well. The IBA in its comments also calls for a narrower scope. The IBA states that for both TSCA and FIFRA, EPA's concern should be only with those microorganisms that are truly "novel"; and IBA defines as "novel" those microorganisms produced by recombinant DNA, recombinant RNA, and cell fusion using DNA from microorganisms that could not exchange genetic material in nature. IBA recommends that if one accepts this definition, the following techniques, cited by EPA, would be excluded from regulation: (1) undirected mutagenesis; (2) directed mutagenesis; (3) microencapsulation; (4) transformation, transduction, transfection and conjugation; and (5) plasmid transfer. When these techniques involve microorganisms that could not exchange genetic material in nature, they would, by definition, be covered as recombinant DNA techniques. I believe this strategy is a reasonable one for EPA to consider in developing its regulations.

The regulation of field testing raises EPA authority over research and development. Having served on several IBA task forces, I believe that field testing is research and should generally be exempt from regulation. Indeed, the EPA notes that under the present FIFRA rules, small-scale field testing (under 10 acres) is exempt. However, because a higher risk may be present with direct release into the environment, EPA could require an environmental use permit (EUP) for all small-scale field testing of biotechnology products. Instead, EPA is suggesting that an applicant notify EPA before the company conducts small-scale field studies. If EPA is satisfied within 90 days, the field test can proceed; if not, a permit would be required.

EPA's approach is reasonable, but I would urge that if the NIH's RAC is reconstituted as the Biotechnology Science Board, it have the authority to conduct the initial review for small-scale field testing (under 10 acres). Just as the NIH's RAC is advisory to the NIH Director, here the Committee would serve in a comparable role for the EPA Administrator. The EPA could make a decision based on that review as to whether a permit would be required.

TOXIC SUBSTANCES CONTROL ACT (TSCA)

EPA addresses biotechnology regulations under the Toxic Substances Control Act (TSCA), and the proposed TSCA policies are analogous to those of FIFRA cited above. To establish its regulatory authority under TSCA, EPA defines DNA and living organisms to be chemical substances. When recombinant DNA legislation was being considered in 1976–1977 by a federal interagency committee, there were questions raised as to whether EPA could establish TSCA jurisdiction by this means. No doubt, this issue of TSCA jurisdiction will be ultimately decided by the Congress or the courts.

EPA defines its regulatory boundaries under TSCA by declaring that the Departments of Agriculture and Interior control plants and animals and that FDA jurisdiction extends to food, food additives, drugs, and cosmetics. Both PMA and IBA endorse this statement in their comments on the EPA proposal. EPA does, however, note that microorganisms used to produce these products could be considered "chemical intermediates" subject to TSCA, but EPA defers to FDA's regulatory authority.

For purposes of TSCA, EPA states that its requirements could be based strictly on the concept of a natural gene pool. However, EPA declares it is simply not possible to describe the limits for the pool and to define natural exchange boundaries. Thus, the most appropriate way to distinguish between "new" and "naturally occurring" microorganisms (which are exempt from TSCA) is by the methods or processes by which they are produced and the level of human intervention involved. The regulatory scope the EPA proposes is thus very broad, encompassing, once again, all of the processes used by the biotechnology industry.

The IBA, in its comments, states that EPA's definitions of "new" versus "naturally occurring" substances are useful in clarifying those products of biotechnology that may be subject to TSCA oversight. IBA notes, however, that the EPA should not be quick to dismiss the concept of a "natural" gene pool. Genetic exchange of material in nature is well documented for many organisms. Recombinant RNA, recombinant DNA, or cell fusion techniques that combine genetic material from microorganisms capable of exchanging DNA by natural means are not "new" and should not be subject to EPA requirements. Further, microorganisms resulting from natural processes (such as undirected mutagenesis) should not be considered "new," for these techniques have been used for years to alter and select microbes.

Currently, chemical R&D is exempt from TSCA regulation. The EPA now proposes to bring biotechnology R&D under TSCA and thereby eliminate the R&D exemption currently in place for chemicals. The IBA believes that the exemption for chemical R&D is equally valid for biotechnology R&D whether it is conducted in a laboratory, a greenhouse, a growth chamber, a small field plant, a pilot facility, or other facilities provided that there is appropriate containment and adequate supervision by technically trained people.

For R&D involving field testing of microorganisms, IBA and PMA both recommend that the EPA and NIH work through an interagency mechanism to make certain oversight in this area is consistent both for industrial and academic research. The EPA states that further discussions will be held with the NIH and the Cabinet Council Working Group. If the NIH's RAC is reconstituted as the Biotechnology Science Board, I would recommend that it have the authority for conducting the initial review of biotechnology R&D, including field testing. EPA can then determine, based on that RAC review, whether additional review is required.

The EPA, via a "grandfather" clause, declares that companies need not go through the premanufacture notice (PMN) process when genetically engineered substances are already in commerce. To expedite the PMN review, data requirements, risk assessment, and other criteria will be developed as will guidelines for placing genetically engineered microorganisms on the TSCA chemical substance inventory list once EPA approval has been granted. EPA also notes it has the authority to exclude chemical substances from TSCA requirements if they are found not to present an unreasonable risk. PMA and IBA both recommend that the classes of experiments exempt now from the NIH Guidelines, such as *E. coli K-12, B. subtilis*, and *Saccharomyces*, be so defined for purposes of TSCA.

In sum, EPA has been actively soliciting the advice of industrial, public, and environmental organizations over the past 2 years. Based on that process, the agency developed the policy statement found in the Cabinet Council Working Group Report. Industry, in large part, believes that the framework EPA has proposed for consideration is a promising beginning for developing regulations to cover agricultural and chemical product development. But much work remains to be done on the scope and implementation of those regulations to ensure rapid and effective review of biotechnology products.

CABINET COUNCIL WORKING GROUP REPORT RECOMMENDATIONS: USDA

The U.S. Department of Agriculture (USDA) also published a policy statement in the Cabinet Council Working Group Report. (OSTP 1984). The USDA, like the FDA, reports that agricultural and forestry products developed by modern biotechnology do not differ fundamentally from conventional products. Therefore, the present USDA statutory and regulatory authorities for veterinary and biological products, plants and plant products, seeds, meat and poultry products apply equally well to biotechnology products in these areas.

CONCLUSIONS

In biotechnology today, we have a vital and growing partnership—among government, university, and industry—priming the flow of basic science discoveries and applications in virtually every industrial sector. As Dr. Lewis Thomas notes in an essay that concludes a *Technology in Society* series on biotechnology, "Today's biological revolution is unquestionably the greatest upheaval in the history of biology and medicine" (Thomas 1984). So far, our system for science support and industrial applications has served us well. The NIH and the National Science Foundation (NSF), in their support of investigator-initiated research grants through a rigorous peer review system, have fostered excellence in science, fueling advances such as today's biotechnological revolution.

The Cabinet Council Working Group Report provides a valuable guide to the future organization of regulatory activities for our industry's product development efforts. Yet, in evaluating the comments received from the public, industry, and the academic community, the Working Group may be persuaded that the NIH's RAC should serve as the sole source for research oversight by becoming the Biotechnology Science Board. Over the next several months, the FDA and EPA will be meeting with the trade associations, public organizations, and other groups to consider their comments on future regulatory policies.

I conclude with observations made by David Padwa, former Chairman of Agrigenetics, an agricultural biotechnology company (Padwa 1984). Padwa notes that regulation of the risks posed by technologies are often based on fear of the consequences of the new technology itself. Science can provide an assessment of a risk, but society must ultimately decide the acceptability of that risk and what to do about it. Padwa states that we must work through our institutions but that the organization of our institutions must be flexible enough to manage risk in light of the never-ending debate over how much risk is acceptable. His concluding statement is that institutions cannot solve all of these political problems through reason alone and, therefore, a commitment to muddling through is the rational course.

To honor that commitment and to preserve the public goodwill on which biotechnology depends, David Bazelon, in his article in the *Technology in Society* series, reminds all involved in biotechnology that society must be informed about "what is known, what is feared, what is hoped, and what is yet to be learned" (Bazelon 1983). The flexibility of our institutions and the responsibility of those involved in biotechnology to ensure public awareness and oversight of this technology have been the hallmarks of the process established by the NIH and its RAC. As the policy arena shifts to the regulatory agencies, the spirit of openness and public engagement continues to be well served by the Cabinet Council Working Group's efforts in developing its report. With cooperation between the research community and the regulatory agencies, as exemplified by the conference on which this volume is based, regulatory agencies like EPA will be able to develop a sound data base for risk assessment; as the data base grows, speedy and effective regulatory review of biotechnology products should permit the promise of industrial biotechnology to be realized.

REFERENCES

Bazelon, D.L. 1983. Governing technology: Values, choices and scientific progress. *Technol. Soc.* **5**: 15.

Cape, R. and J. Perpich. 1985. How we can stay ahead in biotechnology. Letter in the *New York Times*, April 23, 1985, p. A-26.

Food and Drug Administration (FDA), Office of Biologies Research and Review, Center for Drugs and Biologics. 1982. *Points to consider in the manufacture of in vitro monoclonal antibody products subject to licensure.*

——. 1983a. *Points to consider in the production and testing of interferon for investigational use in humans.*

——. 1983b. *Points to consider in the manufacture of injectable monoclonal antibody products intended for human use in vivo.*

——. 1983c. *Points to consider in the production and testing of new drugs and biologicals produced by recombinant DNA technology.*

——. 1984. *Points to consider in the characterization of cell lines used to produce biologicals.*

Industrial Biotechnology Association (IBA). 1985. *Comments on the Environmental Protection Agency (EPA) policy statement in the OSTP proposal for a coordinated framework for regulation of biotechnology* March 25, 1985.

National Institutes of Health (NIH), Department of Health and Human Services. 1985. Request for public comment. Points to consider in the design and submission of human somatic-cell gene therapy protocols. *Federal Register* **50**: 2940.

Office of Science and Technology Policy (OSTP). 1984. Proposal for a coordinated framework for regulation of biotechnology: Notice. *Federal Register* **49**: 50856. (For the statement on the proposed Biotechnology Science Board, see

p. 50904 and for policy statements by the regulatory agencies, *see* p. 50878 [Food and Drug Administration, FDA], p. 50880 [Environmental Protection Agency, EPA] and p. 50897 [Department of Agriculture, USDA]).

Office of Technology Assessment (OTA). 1984. *Commercial biotechnology–An international analysis* p. 13. U.S. Congress, Washington, D.C.

Padwa, D. 1984. Lysenko and others. Biotech '84, USA, an International Conference on Biotechnology, Washington, D.C. September 10, 1984.

Perpich, J. 1985. The last word–Export controls on biotechnology. *Bio/Technology* **3**: 384.

Pharmaceutical Manufacturers Association (PMA). 1985. Statement to the Deputy Director, Office of Science and Technology Policy (OSTP), Executive Office of the President, on the OSTP Proposal for a Coordinated Framework for Regulation of Biotechnology April 11, 1985.

Thomas, L.S. 1984. Oswald Avery and the cascade of surprises. *Technol. Soc.* **6**: 37.

Wyngaarden, J.B. 1984. Nurturing the scientific enterprise. *Science* **223**: 361.

Young, F.E. 1984. Statement by Commissioner, Food and Drug Administration, Department of Health and Human Services, before the Subcommittee on Oversight and Investigations, Committee on Energy and Commerce, U.S. House of Representatives December 11, 1984.

COMMENTS

FOWLE: I find a paradox in your statement about compliance with NEPA. You mentioned earlier that in the mid-1970s it was very appropriate for NIH to comply with NEPA, and, although it was a very tedious exercise, it was very worthwhile. With the increasing experience of the NIH, through the establishment of RAC and the risk assessment studies, you learned a great deal. You came up with the modified *E. coli* strains for laboratory experiments, and also for industrial fermentation and your experience resulted in no problems. In large measure, the level of safety was achieved based on containment, either biological containment or physical containment. With respect to release to the environment, you now want organisms that are going to survive long enough to carry out their function. Therefore, you will not have physical containment, and although you may be able to build in some sort of biological containment, the organisms are going to exist in the environment–at least long enough to do their job. Rifkin and others have pointed this out rather forcefully, and there is also public concern. I would argue that it is very important for governmental organizations such as NIH to examine such issues carefully. They should comply with NEPA and complete environmental impact statements because only if they act in a responsible manner will the public's trust be gained. Regarding the environmental release of genetically engineered organisms, I believe that if we follow a responsible course of action, our experience will show that there will be no problem. I think that in the next ten years, we'll be in the same situation, with respect to environmental release of a variety of organisms, as we are now with respect to the laboratory and industrial applications using *E. coli* K-12. However, I find an incredible paradox in your statement, given your past experience at NIH, in addressing the public's concerns.

PERPICH: The NIH discontinued formal assessments and relied on the RAC's review of proposed experiments as constituting environmental assessments. However, in the proposed field testing of genetically engineered bacteria to prevent frost damage, Judge Sirica found that the NIH must conduct formal environmental impact assessments and the NIH will now do so. An assessment will be done to determine whether the proposed experiment may pose a significant environmental impact. If the conclusion is that it does, then one must prepare an Environmental Impact Statement (EIS). However, there is no reason a priori to assume that one must file an EIS on all proposed field testing experiments. An assessment can be done and published in a timely fashion; an EIS requires a more exhaustive inquiry and far greater public scrutiny because of the possibility of a significant environmental impact. For example, assessments done by the NIH are usually completed in 3-6 months. However, the EIS that NIH published on the NIH Guidelines was begun in the spring of 1976, with a draft EIS submitted for public comment in September and a final statement published in October 1977. Peter Hutt participated in the NIH public hearings on the development of the NIH Guidelines. He strongly supported NIH's decision to file an EIS on the Guidelines and the subsequent assessments that were done. I have no doubt he would have urged the NIH not to discontinue those assessments because, when challenged in court, the NIH would be found not to have met the procedural requirements of the NEPA.

HUTT: Let me second what you have said. I think that where the federal government most often makes a mistake is procedurally, rather than substantively. If you do not follow the procedural requirements of doing the assessment, you are going to get "cut off at the knees," exactly the way NIH was, preferably by the courts. The courts took a look at what they did and said, "We are not saying the experiment is unsafe or you should not do it, but you have not followed the procedure that the law requires"—and, in fact, they had not done so. When I was at FDA, I decided to test whether the NEPA applied to regulatory activity. I took the position that when FDA approved a food additive, it was not required to file a NEPA and do an assessment because the Food and Drug Act was not amended and NEPA is only a procedural statute. I set this up as a test case. The district court held against me and said that when FDA approves a product or when EPA approves a pesticide, you have to do an environmental impact assessment, a preliminary assessment, to determine whether there will be a major impact on the environment. If so, you do a statement; if not, you don't do a statement. In my judgment, that has now been clearly held, and any government agency that doesn't follow it is crazy. I agree with you.

SCHATZOW: We've been sued under every environmental statute that we implement.

HUTT: Wait until they hit you with NEPA under FIFRA.

KHOURY: The message about adequate legislation is perfectly clear with these overlapping authorities, and it seems to me that if scientists can define an obvious and present danger and back it up with data, one of these organizations is going to look into the situation and take some action. I also like the idea of one RAC, rather than five RACs. It seems to me the problem is how you staff the RAC. You told us initially that there is a problem, as cited by Dr. Bernadine Healy, in getting the NIH and RAC to take the responsibility, which they consider beyond their ability. Can you staff one RAC with enough people with broad enough interests, authority, and understanding to cover the responsibilities that are being assigned to these five committees? It's very hard to get scientists, who are perhaps the best-informed, to take time away from their work to carry out such additional responsibilities. You really haven't addressed this issue.

PERPICH: The RAC has obtained relevant expertise by creating task forces with outside experts joining the RAC committee members. These task forces have developed standards and procedures in such areas as industrial scale-up in the pharmaceutical industry, field testing of genetically engineered microbes and, most recently, gene therapy in humans. The RAC, by these means, is able to handle the relevant review of the proposed research. When the focus shifts to product development, then oversight becomes the responsibility of the relevant federal regulatory agency. The members of the RAC, over the past 10 years, have done an extraordinary public service in ensuring appropriate oversight. I agree with you, Jack [Khoury], that it would be difficult to duplicate that commitment from the scientific community with five RACs and a Biotechnology Board as proposed in the OSTP document.

HUTT: I would like to discuss one element that has troubled me. I think it is time for the RAC to bow out. Once you get to the area of a product that is being regulated by one of the substantive agencies, such as FDA or EPA, there is no need for the RAC. For example, suppose you have an NIH-funded drug made through some sort of recombinant DNA mechanism. That is required to be approved by an institutional review board and by the FDA through an IND. Should NIH and the RAC be involved in that at all? My answer is no. Yet my understanding is that the RAC persists in being involved, and you therefore have yet another irrelevant review.

PERPICH: My point is that the RAC's role should continue to include oversight of all biotechnology research, including industrial applications. Biotechnology product review obviously is the responsibility of the relevant federal regulatory agency.

HUTT: But when you go to humans, you have an IND and you have full FDA control. Why do you need the RAC?

MILLER: Generally you do not need the RAC early in the process.

SCHATZOW: Are you suggesting that someone other than EPA, with whatever consultation it has, should have the authority to control the commercial licensing of a microbial pesticide?

HUTT: The RAC should be involved until EPA exerts jurisdiction, then the RAC should bow out.

SCHATZOW: It seems clear to me that under the statute EPA has the authority. Section 5 of FIFRA talks about Experimental Use Permits. The question is how far back you go, at what point do you draw a line between research to determine the pesticidal properties and research to develop the pesticidal properties further? I don't know where you draw that line. But to say that the RAC should begin somehow by playing a semi-regulatory role, and then at some "magic" point shift over to EPA, suggests the existence of a line. I don't think that line exists.

PERPICH: But it did. It worked with the NIH and FDA whereby the RAC provided the review for initial scale-up in the pharmaceutical industry.

HUTT: I do not agree with you, Joe [Perpich]. I think that the way it ought to work is precisely the way Steve [Schatzow] suggests: The moment that the substantive agency's, (EPA or FDA) jurisdiction attaches, through an experimental use permit or an IND, the RAC ought to say, "We give up here. We will turn it over to FDA and EPA." Why does the RAC want to stay in and have two regulatory agencies doing the same job?

PERPICH: Historically, that has been the responsibility of the RAC.

HUTT: I did not say it had not done it; I said it should not do it.

PERPICH: The RAC was created to oversee all recombinant DNA research and thereby ensure a uniformity of standards and procedures for all involved in this research. We have had an international set of standards and procedures based largely on the RAC model. As industrial applications come to the fore, the role of the RAC has grown to provide the oversight for initial industrial applications. In my view, EPA should follow FDA's lead and permit the RAC to review small-scale field testing. The RAC has served as the national and international reference for biotechnology research activities and it should continue in that unifying role.

HUTT: I do not agree.

Risk Assessments/Risk Management for Environmental Uses of Biological Agents

ELIZABETH L. ANDERSON,* ROY E. ALBERT,† AND DAPHNE KAMELY*
*Office of Health and Environmental Assessment
U.S. Environmental Protection Agency
401 M Street, SW
Washington, D.C.
†New York University Medical Center
440 First Avenue
New York, New York 10016

OVERVIEW

Through recent advances in molecular biology, especially biotechnology, regulatory agencies are faced with evaluating a range of biological agents for potential risks they pose to human health and the environment. Although biological risks are of relatively recent concern, this has not been the case for chemicals. In fact, most environmental regulatory actions in the United States focus on hazardous agents of chemical nature. For this reason, over the past several years, the Environmental Protection Agency (EPA) has developed methodologies for assessing health risks associated with chemical pollutants; the most experience has been with chemical carcinogens. In anticipation of regulating biological agents, including genetically engineered microorganisms, the Agency must stimulate the generation of a solid scientific basis in order to adapt its present, chemically based risk assessment approaches to biological agents. A review of the risk assessment approaches that have been adapted for chemical carcinogens may be useful to focus attention on what is needed to develop similar approaches for biological agents. Assessing biological risk poses unique problems owing to the diversity (hormones, enzymes, bacteria, fungi, viruses, plasmids) and the ability of live molecules to replicate, transform, transduce, and infect. These potential risks need to be better understood so that they can be weighed against benefits to adequately protect public health and the environment and to ensure that regulatory actions do not inhibit progress in this rapidly developing field.

INTRODUCTION

Risk assessment is a process for assembling and evaluating scientific information to determine the likelihood that risk may exist under defined circumstances and to predict the potential magnitude of this risk should the event occur. These risk methods have been widely applied to a variety of data bases to estimate potential risk, e.g., safety assessment for engineering failures, failures such as nuclear power plants, transportation equipment, economic forecasting, and insurance risk.

Major regulatory agencies in the United States use risk assessment methods to estimate risk associated with a variety of pollutant concerns including radiation exposures, air pollutant concentrations, food contaminant concentrations, pesticide residues, water contaminants, worker and consumer exposure to many agents, hazardous waste contaminant concentrations, and hazardous spills. At the EPA, these risk assessments are necessary to define the nature of environmental problems as a basis for regulating toxic pollutants. This regulatory decision process is separated from the risk assessment process and generally includes considerations of social and economic factors together with the results of the risk assessment to decide how much risk management is necessary (National Research Council 1983). A typical risk assessment consists of four parts: (1) an assessment of all biomedical data to determine what health effects might be associated with exposure to the agent, (2) characterization of the dose-response relationship, (3) an assessment of exposures to various populations, (4) a quantitative estimate of the current and anticipated health effects and related dose-response information evaluated in the first steps, including acute toxicity, mutagenicity, carcinogenicity, and reproductive effects. Exposure assessment evaluates parameters that result from exposure to hazardous chemicals such as dose, persistence, and route of exposure (Environmental Protection Agency 1984a).

The EPA has developed a procedure for risk assessment of carcinogens which has also been applied and extended to other health effects (EPA 1976; 1984b,c,d; Albert et al. 1977). This paper will focus on EPA's carcinogen assessment approach, as well as the problems and uncertainties associated with this process. The paper will then elaborate on concerns and potential risks posed by biological agents to human health and the environment from the point of view of the various EPA programs, consisting of air, water, toxic substances, pesticides, and hazardous waste. Emphasis will be placed on applications of genetically engineered microbes and viruses and the need for focused research to support risk assessment for these agents.

CARCINOGEN RISK ASSESSMENT

In 1976, EPA adopted guidelines for carcinogen risk assessment which provided for a two-step approach to the evaluation of carcinogenesis data (EPA 1976). The first step, the qualitative assessment (hazard assessment) takes a weight-of-evidence approach to the likelihood that an agent is a human carcinogen. The second step, the quantitative assessment, which coupled parts two and three of risk assessment—the dose-response relationship of the chemical and the assessment of environmental exposures—provides quantitative estimates of increased, individual lifetime risk to subpopulation groups and nationwide impacts on an annual basis (EPA 1976;

Albert et al. 1977). Finally, steps one and two are expressed together as the final part four of risk assessment. These guidelines have recently been updated and incorporate more detailed guidance than the earlier guidelines (EPA 1984b). These guidelines are fully consistent with background documents on the scientific basis for carcinogen risk assessment that have been published by federal interagency committees (Interagency Regulatory Liason Group 1979; Office of Science and Technology Policy 1985).

Qualitative Assessment (Hazard Assessment)

Since 1976, EPA has been evaluating data indicating carcinogenic potential using a weight-of-evidence approach. The strongest evidence comes from human observations backed up by animal studies. Most often such human studies are not available; the next best evidence then is derived from lifetime animal bioassay studies. Short-term in vitro and in vivo studies are useful as supporting evidence for the determination of carcinogenic potential for humans.

In assessing human information, it is usually possible to establish the length of time between the first exposure and the end of the follow-up period. As the type of population at risk is also known, such factors as age and type of exposed populations can be factored into the assessment. The more difficult to determine parameters are duration and intensity of the exposure as well as behavior patterns of the population exposed. It is important to know whether a given population was exposed to other carcinogens, including cigarettes, and whether everybody in a subgroup was exposed to the same extent during a period of time. It should be emphasized that the variation in the available human data leads to a high degree of uncertainty which should be treated with caution.

In contrast, animal studies can be designed and conducted under controlled laboratory conditions. Nevertheless, the interpretation of animal bioassay results must consider the weight to be given to a host of factors including the interpretation of positive results in the face of negative studies, the significance of benign tumors, interpretation of tumor outcomes in test animals where the controls have a lower background of tumors than historical controls, the significance of the mouse liver tumor system, interpretation of tumor outcomes in the face of organ toxicity or reduced sensitivity because of animal death in the treated groups, mechanisms of carcinogenic response under the conditions of the test, lack of dose-response relationships, and treatment of the statistical significance of excess tumors of types in the aggregate where no single site is significant.

In summary, qualitative assessment consists of literature searches and evaluation of all the biomedical data, including both negative and positive studies, pharmacokinetic studies and information from short-term tests to determine carcinogenic potential. The recently proposed EPA guidelines lay the ground rules for interpreting the qualitative weight-of-evidence and further propose a system for stratifying this weight-of-evidence according to criteria for these categories (EPA 1984b).

Quantitative Assessment

The second step provides quantitative estimates of public health impacts. Assuming that the agent is a human carcinogen, risks are bracketed between a lower bound approaching zero and a plausible upper bound based on the linear, nonthreshold cancer model (EPA 1984b). The upper-bound risks are expressed both in terms of the individual, increased lifetime cancer risks in exposed population subgroups (i.e., one in 10^{-5} risk of cancer) and the nationwide impact in terms of increased number of cancer cases per year. This second, quantitative step gives regulators a feel for the potency of a suspect carcinogen, as well as some quantitative information on public health data. In the absence of biological understanding of the mechanisms of cancer, it is impossible to describe risk precisely; yet, it is important to provide some quantitative measure of risk. The linear nonthreshold model that is used by EPA, places plausible upper bounds on risk; it is usually not possible to quantify the risk more precisely (Anderson et al. 1983).

In the absence of definitive scientific data, the Agency makes the following assumptions (EPA 1984b):

A. Lifetime incidence in humans is the same as in animals receiving an equivalent dose.
B. Dose (in mg/surface area) is equivalent between species.
C. Humans may be as sensitive as the most sensitive animal species.
D. A linear, nonthreshold model places a plausible upper-bound on the risk at low doses for single chemical exposures, unless there is mechanistic information to the contrary.
E. Lifetime incidence is proportional to the total lifetime dose received, averaged on a daily basis.
F. If the experiment is terminated early, lifetime incidence is estimated assuming that cumulative incidence increases by the third power of age.
G. The linearized multistage model is appropriate for extrapolation and the upper 95% confidence limit of the linear term is appropriate for expressing the upper-bound of potency.
H. The upper-bound risk is probably less plausible if there is no evidence of mutagenicity.
I. Where data permit, human data are used in preference to animal data as a basis for risk extrapolation.
J. Where exposure data are adequate, negative human data may be used in preference to animal data as the basis for placing an upper-bound on the dose-response slope.
K. For human data, the method of analysis is tailored to the completeness and quality of data available. A model that is linear at low doses is used for extrapolation.

L. For risk extrapolation, animals with one or more tumor sites showing significant elevated levels of tumors are pooled.
M. Generally, benign and malignant tumors are combined unless the benign tumors are not considered to have the potential to progress to the associated malignancies of the same morphologic type.

The EPA risk assessment approach is still developing and there are many uncertainties associated with the current methods. Nevertheless, risk assessment is necessary to provide a conceptual basis for balancing risks against social and economic concerns and for setting priorities for Agency regulatory actions. The following discussion provides examples to illustrate the use of risk assessment in risk management decisions.

THE USE OF RISK ASSESSMENT AND RISK MANAGEMENT

Carcinogen risk assessment has provided the scientific basis for a range of policy decisions by federal agencies. Quantitative risk estimates, generally expressed as upper-bound estimates, coupled with the qualitative evaluation of the weight of biological evidence, provide policymakers with rough estimates of risk which serve as a basis for balancing risks and benefits including social and economic factors. In this process of risk management, Agency decisionmakers decide upon target levels of risk, set priorities, decide the acceptability of residual risk after the application of the best available technology, or decide the urgency of the public health situations where inadvertent exposures occur. These decisions are based not only on measurements of risk but also on judgment of their acceptability.

In the absence of risk assessments, acceptable levels of risk/exposure would have to be decided by aiming as low as possible but establishing the cutoff based on some measure of achievability. This approach could lead to underregulation of important health problems or regulation of trivial risk at high cost. It should be emphasized that the acceptable risk/exposure is a matter of regulatory policy and that risk is acceptable only in light of benefits.

For example, recently the EPA made a final decision to regulate two sources of benzene emissions to the ambient air but not to regulate three other source categories (Table 1). This decision was based on risk assessment information which concluded that the weight-of-evidence for benzene is clear-cut and strong, i.e., that benzene causes leukemia in humans. The distinction between sources was based on estimates of exposure, before and after the application of best available technology, to populations living near sources of benzene emissions in combination with dose-response relationships for benzene-induced leukemia in workers (Table 2). Incidentally, animal bioassay studies appear to roughly duplicate the human potency estimates (Table 3). The cost of control for the first year capital investment and

Table 1
Summary of Five Source Categories[a]

Source category	Number of existing facilities	Emissions (megagrams/year) Before	After	Total stationary source emissions
	Intent to regulate under section 112			
Benzene fugitive	229	7,900	2500	18%
Coke by-product	55	29,000	3500	53%
	Intent to propose withdrawal of proposed standards			
Maleic anhydride	3	960	120	2%
Ethylbenzene/styrene	13	210	68	0.4%
Benzene storage	126	620	400	1%

[a]Background information for national emission standards for hazardous air pollutants: Benzene. EPA press statement released December 15, 1983.

Table 2
Carcinogenic Potency of Benzene Calculated on the Basis of Nonlymphatic Leukemia Mortality Rates Observed in Three Epidemiologic Studies[a]

Data base	Estimated relative risk	Estimated continuous exposure	Lifetime risk per ppm	Lifetime risk per $\mu g/m^3$
Infante et al. (1977)	6.4	2.73	1.33×10^{-2}/ppm	$4.09 \times 10^{-5}/\mu g/m^3$
Aksoy (1974, 1976, 1977)	9.96	2.22	1.82×10^{-2}/ppm	$5.60 \times 10^{-6}/\mu g/m^3$
Ott et al. (1977)	3.75	0.17	4.64×10^{-2}/ppm	$1.43 \times 10^{-6}/\mu g/m^3$
Geometric mean			2.23×10^{-2}/ppm	$7.08 \times 10^{-6}/\mu g/m^3$

[a]Reprinted from EPA (1984).

Table 1 *(continued)*

Maximum lifetime individual risk		Annual cancer cases			Cost (millions)	
Before	After	Before	After	Differences	Capital	Annual
		Intent to regulate under section 112				
15/10,000	4.5/10,000	0.45	0.14	0.31	5.5	0.4
83/10,000	3.5/10,000	2.60	0.23	2.37	30.9	(1.3)
		Intent to propose withdrawal of proposed standards				
76/million	5.3/million	0.029	0.016	0.013	6.4	2.8
140/million	9.0/million	0.0057	0.00058	0.00051	2.7	0.97
3.6/100,000	2.3/100,000	0.043	0.028	0.015	7.3	1.3

Table 3
Carcinogenic Potency of Benzene Calculated on the Basis of Animal Data

Data base	Lifetime risk per ppm	Lifetime risk per $\mu g/m^3$
Female rats (Maltoni et al. 1982)[a,c]	3.4×10^{-2}	1.1×10^{-5}
Male rats (NTP 1984)[a,d]	2.0×10^{-2}	6.0×10^{-6}
Female rats (NTP 1984)[a,d]	3.3×10^{-2}	1.0×10^{-5}
Male mice (Snyder et al. 1980)[b,c]	1.4×10^{-2}	4.3×10^{-6}
Geometric mean	2.4×10^{-2}	7.3×10^{-6}

[a]Zymbal gland carcinomas (gavage study)
[b]Hematopoietic neoplasms (inhalation study)
[c]Carcinogen Assessment Group (1983)
[d]C. Chen, pers. comm.

continued maintenance cost was compared to the incremental gain in public health from regulation (Table 1). The risk assessment information was a primary basis for the risk management decision.

Although the distinction among these sources of benzene is a complex matter, fraught with scientific uncertainty, decisions must be made. This is most certainly going to be the case for the many important biological agents now under development for use in the environment. This technology dictates the need for scientists involved in these developing technologies to also develop the scientific data on which to base sound environmental risk management decisions.

BIOLOGICAL RISK ASSESSMENT

The risk assessment/risk management process has become considerably complex as EPA has been faced with the evaluation and clearance of biological agents. Risk assessment approaches need to be developed for biological agents. In the past few years, EPA has been asked to evaluate biological agents such as enzymes, hormones, plasmids, viruses, and various bacteria, and fungi. Recently, genetically modified products have been submitted to the Toxics and Pesticides Office at EPA for clearance and registration. The number of genetically engineered microorganisms and products intended for commercial use is increasing as is EPA's role. Several EPA statutes can be used to regulate new biological agents and genetically altered organisms which may pose new risks to human health and the environment. Specifically, in addition to the conventional risk assessment parameters evaluated on chemical pollutants, these genetically engineered organisms replicate, infect, transduce, transform, and spread in the environment. They may affect ecosystems in an advantageous manner. Their release into the environment may also cause adverse effects that are irreversible.

As this volume focuses on genetically altered viruses, EPA's concerns in assessing biological risks are best illustrated by addressing potential problems with genetically altered viruses in each of EPA's major programs.

Toxic Substances and Pesticides

The Toxic Substances Control Act (TSCA) mandates the regulation of chemical and biological products that are manufactured for commercial use. Although genetically engineered products can be initially assessed as chemicals, the purity of the product poses a potential problem. Since the biological product is cloned in a vector and a host, it is conceivable that during the isolation procedure, part of the vector may become incorporated into the final product. If the vector is a virus, it cannot be ruled out with certainty that a deleterious gene, e.g., a cancer gene, is not contained as a residue in the final product. Alternatively, a control element in the virus can activate a cellular gene resulting in expression of cancer as another harmful trait.

TSCA also covers the manufacturing process of chemicals. In this case, exposure to the process of synthesizing a product via genetic engineering techniques, the leakage or escape of genetic material into the environment, and subsequent health and environmental effects, become an integral part of the risk assessment procedure.

Biological pesticides and fertilizer pose even greater assessment challenges for the Agency and can be regulated under the Federal Insecticide, Fungicide and Rodenticide Act (FIFRA). Although the manufacturing and use of chemical products can be contained in a laboratory, pesticides and fertilizers are intended for use in the environment.

EPA has learned of viruses and bacteria that are intended to improve crops and make them pest and frost-resistant. There is an effort to produce nitrogen-fixing bacteria that may be applied to plants so that plants can fix their own nitrogen and will no longer require fertilizers. There are enormous benefits to the agricultural industry in developing and using these genetically engineered microbes. Yet the risks to ecosystems and ultimately to animals and humans who consume these crops remain unknown. By releasing these new genetically altered microorganisms and their components into the environment, some irreversible processes may take place, all of which need to be assessed in advance. The new organisms may transform, replicate, survive, and grow beyond control under certain advantageous conditions in a new environment and could cause harmful effects.

Risks associated with the release of these molecules need to be assessed before application. Alternatively, industry needs to build safeguards into the plasmids or hosts so that they may be self-destructive once they have fulfilled their intended function.

Under both TSCA and FIFRA, if the new biological product is identical to the one manufactured by conventional techniques, it still needs to be assessed as an unknown entity because of the new manufacturing process. EPA has to ensure that additional risks are not added on by new genetically engineered techniques.

Air and Water Programs

Although the Office of Pesticides and Toxic Substances is in the process of assessing biotechnology products, the Office of Air and Radiation Programs and the Office of Water Programs at EPA have taken a less active role in the risk assessment of genetically engineered microorganisms. Nevertheless, the risks of survival and spread of modified organisms in air and water are beginning to be evaluated under the Clean Air Act and the Clean Water Act. In fact, an immediate problem is posed by various kinds of viruses in drinking water. Certain kinds of viruses survive the chlorine treatment of municipal water systems causing various enteric diseases in the population (Spendlove et al. 1985; C. Gerba, pers. comm.). Genetically altered viruses, especially cancer-causing viruses, could survive equally well in drinking water. They can also enter the waters and streams and eventually cause harmful effects in the food chain. Both the air and water programs need to follow the exposures, fate and transport of modified microbes and viruses in their respective media.

Hazardous Waste Program

Biological activity in waste seems to be an emerging problem. Not all biological products, whether manufactured through conventional or genetic engineering methods, are disposed of properly. Furthermore, current disposal methods may not be adequate to inactivate all microbes and viruses in industrial or research use. Should it be determined that biological activity may cause adverse health effects to the nearby population, the EPA will be called upon to investigate the waste sites in question. Risk assessments are necessary at such sites to assess the direct and indirect health effects on the populations and ecosystems exposed.

EPA regulates waste disposal under the Resource, Conservation, and Recovery Act (RCRA). The Agency regulates spills under the Comprehensive Environmental Response, Compensation, and Liability Act (CERCLA). As the biotechnology field grows and becomes more diverse, the biological waste issues, including biotechnology, will need to be addressed.

CONCLUSION

Despite the careful attention to the development of a step-wise approach for carcinogen risk assessment, considerable uncertainty is associated with these evaluations. Most likely, we must expect a higher level of uncertainty for risk assessments of biological agents, particularly genetically altered microorganisms, their products and by-products. It is critical that scientists involved in these developing techniques also focus particular attention on developing the data and methods necessary to answer the risk assessment questions: (1) How likely are these agents to be threats to human health and the environment? (2) If they might be, what is the magnitude of the possible impact? Tables 4 and 5 summarize the components of risk assessment for chemical agents and biologically altered agents. This is also the appropriate time for regulatory agencies to prepare guidance and, so far as possible, assess the potential risks to human health and the environment posed by this technology. With the development of reliable procedures and sound safety measures, the controversy over the health and the social-ethical issues surrounding biotechnology will subside. As knowledge of the risks posed by genetic manipulation increases, society may be able to proceed cautiously and benefit from the tremendous possibilities of biotechnology.

Table 4
Risk Assessment for Chemical Agents

Steps in assessment	
Protocol for laboratory tests	Whole animal life study tests Acute studies Chronic studies In vitro tests, e.g., such as cell transformation Health effects tests, e.g., mutagenicity, carcinogenicity
Hazard assessment (qualitative assessment)	Epidemiology Evaluation of data Identification of potential hazards associated with suspect agents
Dose-response	Mechanisms of action Low dose and threshold effects
Exposure	Evaluation of parameters, such as fate, transport, transformation, persistence, bioaccumulation, degradation, and route of exposure
Quantitative risk	Evaluate the potential of harm to human health and the environment Estimate the magnitude of risk

Table 5
Risk Assessment for Biologically Altered Organisms

Steps in assessment	
Protocol for laboratory tests	Whole animal tests Test for pathogenicity Test for infectivity Short-term tests for mutagenicity/cell transformation Chronic tests health effects, e.g., cancer
Hazard assessment (qualitative assessment)	Epidemiology data, related organisms Assessment of data on health effects
Dose-response	Unknown
Exposure	Evaluation of parameters, such as survival, spread, fate, transport, transformation, persistence, accumulation, degradation, route of exposure, and metabolic variables
Quantitative risk	Evaluate the potential of harm to human health and the environment Estimate the magnitude of risk

REFERENCES

Aksoy, M. 1977. Testimony of Mazaffer Aksoy, M.D. to Occupational Safety and Health Administration, U.S. Department of Labor. July 19–20, 1977.

Aksoy, M., S. Erdem, and G. Dincol. 1974. Leukemia in shoe-workers exposed chronically to benzene. *Blood* **44(6)**: 837.

——. 1976. Types of leukemia in chronic benzene poisoning: A study in thirty-four patients. *Acta Haematol.* **55**: 65.

Albert, R.E., R.E. Train, and E. Anderson. 1977. Rationale developed by the Environmental Protection Agency for the assessment of carcinogenic risk. *J. Natl. Cancer Inst.* **58**: 1537.

Anderson, E.L., et al. 1983. Quantitative approaches in use to assess cancer risk. *Risk Analysis* **3**: 277.

Carcinogen Assessment Group. 1983. Evaluation of the carcinogenicity of benzene. Revised August 30, 1983. *Internal Review Draft* C-10.

Environmental Protection Agency (EPA). 1976. Interim procedures and guidelines for health risks and economic impact assessments of suspected carcinogens. May 25, 1976. *Federal Register* **41**: 21402.

——. 1984a. Proposed guidelines for exposure assessment. Request for comment. November 23, 1984. *Federal Register* **49**: 46304.

——. 1984b. Proposed guidelines for carcinogen risk assessment. Request for comment. November 23, 1984. *Federal Register* **49**: 46294.

——. 1984c. Proposed guidelines for mutagenicity risk assessment. Request for comment. November 23, 1984. *Federal Register* **49**: 46314.

——. 1984d. Proposed guidelines for the health assessment of suspect developmental toxicants. Request for comment. November 23, 1984. *Federal Register* **49**: 46324.

——. 1984e. Response to public comments on EPA's listing of benzene under section 112. *EPA Report No. 450* 5-82-003:62.

——. 1985. Proposed guidelines for the health risk assessment of chemical mixtures. Request for comment. January 9, 1985. *Federal Register* **50**: 1170.

Infante, P., R. Rinsky, J. Wagoner, and R. Young. 1977. Leukemia in benzene workers. *Lancet* **ii**: 76.

Interagency Regulatory Liaison Group (IRLG). 1979. Scientific basis for the identification of potential carcinogens and estimation of risks. *J. Natl. Cancer Inst.* **63**: 243.

Maltoni, C., G. Cotti, L. Valgimigli, and A. Mandrioli. 1982. Zymbal gland carcinomas in rats following exposure to benzene by inhalation. *Amer. J. Ind. Med.* **3**: 11.

National Research Council. 1983. *Risk assessment in the federal government: Managing the process.* Prepared by the Committee on the Institutional Means for Assessment of Risk to Public Health, Commission on Life Sciences. National Academy Press, Washington, D.C.

National Toxicology Program (NTP) Technical Report of the Toxicology and Carcinogenesis Studies of Benzene. 1984. Draft NTP TR 289, NIH publication No. 84-2545, NTP-84-072.

Office of Science and Technology Policy (OSTP). 1985. Chemical carcinogens; A review of the science and its associated principles. March 14, 1985. *Federal Register* **50**: 10372.

Ott, G., J. Townsend, W. Fishbeck, and R. Langner. 1977. *Mortality among individuals occupationally exposed to benzene.* Dow Chemical Company, Midland, Michigan.

Snyder, C.A., B. Goldstein, A. Sellakumar, I. Bromberg, S. Laskin, and R.E. Albert. 1980. The inhalation toxicology of benzene: Incidence of hematopoietic neoplasms and hematotoxicity in AKR/J and C57B1/6J mice. *Toxicol. Appl. Pharmacol.* **54**: 323.

Spendlove, R.S., et al. 1985. Reoviruses in water pollution testing. *EPA Report No. 600*/51-84-022.

COMMENTS

MARTIN: I have several questions. There's a deja vu about some of the things you said that reminds me of 1974 and 1975, the "dark period" of recombinant DNA research. Many of the scenarios of doom that were proposed then, such as epidemics of cancer and rampant infectious diseases, were considered quite likely to occur. People weren't using the words, hypothetical or conjectural risks, when they discussed possible dangers connected with recombinant DNA experimentation. What type of dangers are you concerned with in the risk assessment program you propose? Are they real or conjectural? I don't think we're talking about carcinogens or radiation or even infectious viruses when we discuss recombinant DNA. For that reason, I am not convinced a toxic chemical model is relevant here. I don't know how to measure hypothetical risks, but I do know that during the last 10 or 15 years the most revealing information has not come from planned risk assessment experiments but from basic research areas–from basic bacteriology, basic virology, molecular biology.

ANDERSON: Absolutely. We are certainly very much aware of this. This is an important issue because there is an enormous fear of toxic chemicals. We take environmental agents and look at these hypothetical risks, and we establish an upper-bound risk. We consider quite low risks of one in 1,000,000 or one in 100,000. We stress the hypothetical nature of the work we perform–and compare it to cigarette smoke or to the use of seat belts. We also try to have public discussions about things that are not hypothetical, that we accept every day at much higher risk.

MARTIN: But toxic chemicals, by definition, are toxic. There is a measured biological effect. I am not aware of any recombinant DNA molecule that exhibits unexpected hazard.

ANDERSON: When you start to talk to some of the people who have a real stake in some of these so-called toxic chemicals, you will find that they do not believe that these chemicals are even toxic. That's when we get into weight-of-evidence—whether some of these chemicals are actually toxic to humans or not. I am not sure the discussions for some of these chemical agents are all that different from what you are driving at. Obviously, the primary purpose of our work with scientific data on chemical carcinogens is to provide policy-makers with the tools with which to distinguish an urgent situation from one that is not urgent. We recognize that there is not a good way to convey this information to the public. In fact, we often see the public take positions that are quite inconsistent with the risk insofar as it can be described by knowledgeable scientists. If you take a poll, people working in risk assessment will rank nuclear power near the bottom; people who are not working in risk assessment will rank it at the top. There is a great inconsistency about the way people perceive and understand possible hypothetical risk in a whole variety of areas.

MARTIN: Do you know of a particular biological product on which EPA is considering doing some risk assessment that might be a good example for this group to discuss in terms of hypothetical risk?

ANDERSON: We actually thought of taking something like genetically altered bacteria that would give us the frost-free protection of crops—obviously something very important to many people—and trying to track it through this process. We decided not to do it because my colleagues thought it would be a very dangerous thing to do. But I think that that's the kind of thinking we ought to do, namely take some agents over which there had been public furor and try to answer the questions: "Are you worried that it would cause a hazard? If there might be some hypothetical potential that it would, would it be a very bad hazard? Are we talking about much of a quantitative hazard at all?"

SUMMERS: Toxicity is one thing, but you are dealing with biologically replicating entities. How do you factor that into your risk assessment? When you release these agents, they are not like a chemical and are not going to be a static agent in the environment. Your statistics may not apply—in fact, I don't think they will.

ANDERSON: Well, that is exposure. If they are replicating in the environment and are being altered into a different product, we have a different exposure situation to take into account as compared with chemical agents. Thus we may deal with other products or by-products which we have to place at the top of the risk assessment process and see if we think those are hazardous. What is the dose-response relationship and the exposure patterns for those? What would be their potential effect on the ecological systems and human effects?

SUMMERS: But, theoretically, all you need is one virus and one bacterium for injection of an organism. I have difficulty in grasping your dose-response in that respect.

ANDERSON: I'm not suggesting that I have any answers to this. I am saying if we are going to release genetically altered bacteria into the environment to control frost, the decisionmakers may be asking the scientists: "Do you think this agent poses a hazard? Is it going to change in the environment in a way that may cause other agents to pose a risk? Is it going to become incorporated in any kind of mammalian cells? If you think that any of the above may happen, can you define a dose-response relationship so that small amounts may cause a risk, or could you get a fair amount out there before you begin to worry? Finally, if you look at the potential for exposure, is it a short-term exposure resulting in the death of the microbes? Are they going to persist a long time? Are they going to multiply? Are they going to produce some other kinds of organisms? Are they going to give off by-products that we have to be concerned about? That's why I think the assessment of risk for these particular agents is a very difficult matter. It involves not one agent, but a number of different interactions. Of course, with chemicals we can get transformation in the environment, but it is not the primary concern. With biological agents, transformation will be the primary concern. I do not underestimate the complicated nature of the questions we are asking.

KILBOURNE: Is there a danger of divorcing assessment of risk from benefits?

ANDERSON: No.

KILBOURNE: I think this inevitably must be factored in. You have to look at your target. I'm thinking specifically of vaccines – this is all too often forgotten these days – that there is tangible and perceptible risk with all such biologic agents. But, it seems to me, the reasonable factoring has to involve the benefits that may accrue from reduction in mortality and reduction in morbidity. I think what Max [Summers] was saying is that there are quantitative aspects to this risk depending on the population confronted with either the vaccine or the disease, and that there is a certain danger in equating biologic agents with toxic substances.

ANDERSON: It's not so much trying to equate them. I was trying to focus the questions posed to the scientists. When policymakers make a judgment as to whether to permit the release of an agent in the environment for some beneficial purpose, they are going to ask the kinds of questions that I posed. Whether or not we can answer them is another matter. I think there is going to be a long period of development before we can answer these questions with any degree of certainty. But, if these questions are going to be answered, we cannot divorce the idea of trying to answer these questions from the benefits of using the products. The second step, which we term risk management, is used by the National Academy of Sciences and certainly intends factoring the two together. Because of too many scientific uncertainties, we defined risk management as a separate step. It is therefore conceivable that a risk manager who would like to get a certain result could simply tinker with the uncertainties in different ways for different agents to obtain the desired answer in order to make a policy decision in one direction or another. The reason the two are presented separately, then, is to see if we can organize the science, first, about what we know about the potential risk, and then factor in the benefits that accrue from the use. If you think of it as the risk being the push not to want to do something, and then all of the benefits and the social and economic factors pushing against it, this is what policymakers, such as Steve [Schatzow], cope with every day in making judgments about various agents.

KILBOURNE: I realize you have to make an assumption on the status quo in "plugging these things in," but that's not reality. That illusion and everything else is obviously going on, and the risk is going to probably vary as well as the benefits in any particular situation. There is a time dimension in this also.

MILLER: You have alluded a number of times to altered organisms, but you have studiously avoided defining what you mean by that. There is a tendency by EPA that is notable in the *Federal Register* statements in general to discriminate, in the form of more stringent or more Draconian regulation, against organisms that are produced with the newer, better understood, more meticulous techniques. I wonder how you can defend that.

ANDERSON: I was not trying to distinguish any biological agent from any other one, but rather to present an idea and a concept, that when we think about biological agents, many of the questions that we ask of chemical agents are going to be asked of biological agents. Whether we like it or not, we need to get organized to answer them. We have many unaltered biological agents right now that we are equally concerned with, such as viruses in sludge. There is a concern about these agents. I think the ones that are altered frighten people more because they may think this is something different. Therefore, I was not trying to make a distinction, but, rather, to raise the same kinds of thought-provoking questions that are asked of scientists when policymakers have to deal with judgments about environmental agents–whether they are chemical agents or radiation–or whether they are going to be biological agents in the future.

The Role of the Environmental Protection Agency

STEVEN SCHATZOW
Director, Office of Pesticide Programs
Environmental Protection Agency
401 M Street, SW
Washington, D.C. 20460

Biotechnology research is an exciting field with the potential to provide substantial benefits to society in terms of increased food production, public health, and pollution control, to name a few. Federal regulation of biotechnology has been a major focus at the Environmental Protection Agency (EPA) for the past several months since it became apparent that a variety of commercial products of biotechnology would soon emerge. EPA regulates products of biotechnology under two statutes: genetically altered pesticides under the Federal Insecticide, Fungicide, and Rodenticide Act (FIFRA) and genetically altered products for general industrial purposes under the Toxic Substances Control Act (TSCA). The primary focus of this paper will be on regulating pesticide products of biotechnology under FIFRA. TSCA activities will be discussed only briefly.

In general, EPA's regulatory approach is to strike a proper balance between the risks and benefits of using products of biotechnology and not to stifle innovative research. However, because of the novel aspects of biotechnology and the many unanswered questions about how genetically altered materials will react in the open environment, we are proceeding cautiously. Our basic approach, at this point, is to evaluate on a case-by-case basis.

EPA's regulatory approach was described in two *Federal Register* notices published in October and December 1984. The first set forth an interim policy under FIFRA to require notification to EPA prior to any small-scale field testing of genetically altered pesticides. The second much broader notice published by the White House Office of Science and Technology Policy (OSTP) described a unified federal approach to regulating products of biotechnology. It included a matrix of all federal statutes which apply to biotechnology; proposed statements of policy by EPA, the Food and Drug Administration (FDA), and United States Department of Agriculture (USDA); and an interagency proposal for a science advisory system for reviewing biotechnology products.

Since publication of our interim policy under FIFRA, EPA has received five notifications to field test genetically altered microbial pesticides in the environment. These include only bacteria and fungi thus far, but we believe commercialization of genetically altered viruses for pesticidal use is on the near horizon.

There are four topics I want to discuss. These include (1) EPA's regulatory framework, (2) the review process for genetically altered microorganisms, (3) development of guiding principles for sound risk/benefit assessment, and (4) public reaction to our proposed regulatory approach.

FIFRA is a licensing statute under which EPA registers pesticides on a no "unreasonable adverse effects" standard. That is, a pesticide is registered only when data supplied by the manufacturer demonstrates that the product will not pose unreasonable adverse effects to humans or the environment when used according to common and widespread practice. Pesticide decisionmaking involves consideration of both risks and benefits and striking a balance between the two.

There are two broad classes of pesticides regulated by EPA under FIFRA—conventional chemical pesticides and biochemical and microbial pest control agents. Biochemical and microbial pesticides are distinguished from conventional chemical pesticides by unique modes of action, low use volume, target species specificity, or natural occurrence. Microbial pesticides include bacteria, fungi, viruses, and protozoa.

EPA has a long history of evaluating naturally occurring microbial pesticides. The first was registered in 1948; and today there are 14 microbial pest control agents registered in over 100 products for use in agriculture, forestry, mosquito control, and home owner situations. The majority of these products were registered within the past 6-8 years. These include four nuclear polyhedrosis viruses used against various insects during their larval stage. In general, the microbials are very insect-specific, and demonstrate low toxicity to humans and other mammals.

Although naturally occurring microbial pesticides have been used without adverse effects, they are living entities capable of survival, reproduction, and pathogenicity and infectivity to nontarget hosts. These risk concerns apply as well to genetically altered microbes.

EPA drew heavily upon experience with regulating naturally occurring microbes in developing its regulatory approach for genetically altered microbes. For example, Part 158, the data requirements for pesticide registration published in the summer of 1984, apply to both types of microbes. Since assessment techniques and data requirements are not static, Part 158 explicitly gives the Agency the flexibility to require additional data to assess special concerns, such as those associated with genetically altered microbials. Additional information might include characteristics and origin of the product organism, genetic manipulation techniques used, identity of the inserted gene segment, a description of the "new" traits or characteristics expressed, and tests to evaluate potential for genetic transfer and assess environmental fate.

Likewise, the recommended protocols for testing contained in EPA's Pesticide Assessment Guidelines have been developed, to the extent possible, to be applicable to both naturally occurring and genetically altered microbials. The guidelines for microbials (Subdivision M) were developed carefully in conjunction with EPA's Office of Research and Development and the American Institute of Biological Sciences. Subdivision M is now being reviewed to determine where refinements or updated protocols are needed.

TSCA gives EPA broad authority to review and regulate a wide range of chemical substances (including living organisms) in industrial and commercial applications. The Act specifically excludes drugs, food, and pesticides. Like FIFRA, TSCA's basic regulatory objectives are achieved through a balancing of risks and benefits. The types of genetically altered products likely to be subject to regulation under TSCA include both chemicals produced by microorganisms, such as industrial enzymes, and microorganisms themselves when used for purposes such as producing commercial chemicals, degrading pollutants, and enhancing oil recovery.

The Agency has proposed to use TSCA's premanufacture notification (PMN) provisions to regulate genetically altered substances. The provisions require companies to notify EPA before producing "new chemical substances," which the Agency has interpreted to include products of recombinant DNA, recombinant RNA, and cell fusion. The Agency's Office of Toxic Substances has not yet received a PMN for a genetically altered product, but based on discussions with several companies, expects to receive PMNs later this year.

Returning to FIFRA, I will discuss our experience to date in reviewing genetically altered microbials under the interim notification policy. This policy requires notification to EPA before small-scale field tests that involve release to the environment may be carried out. Small-scale field tests typically involve application of a substance on under 10 acres of land. The interim policy specifies what type of information must be submitted to EPA for review. EPA has 90 days to review the information and let the applicant know its decision. If EPA decides that the proposed project raises health or environmental concerns, an Experimental Use Permit (EUP) may be required before testing can begin. The decision to require an EUP ensures that there will be sufficient information available to evaluate risks adequately, that the results of testing are reported to EPA, and that EPA has oversight over the testing. The interim policy does not apply to studies conducted when there is no direct release to the environment, such as in growth chambers, contained facilities, and greenhouses, so as not to infringe on research and development.

To date, EPA has received five notifications under the interim policy. Two proposed to test genetically altered, naturally occurring, frost-promoting bacteria. The manipulation involved deletion of the naturally occurring bacteria's frost-promoting gene. The altered bacteria may be able to compete with and displace their natural counterparts. The third notification involved insertion of a gene into a common type of soil bacterium which causes the bacterium to make an insect toxin, that is, it produces its own insecticide. The inserted gene comes from the common naturally occurring microbial, *Bacillus Thuringiensis*. The fourth notification proposed to test strains of a fungus altered by undirected mutagenesis using ultraviolet irradiation and chemical treatment to provide improved control of the target fungi. (The notification period on this proposal was suspended by the applicant pending submission of more complete information.) Finally, the fifth notification was withdrawn also pending development of additional data.

In the first three cases, EPA decided that an EUP would be necessary to address critical questions about the ability of the microorganisms to replicate and spread beyond the application site. This does not necessarily mean that we believe the products pose a high risk, but rather that additional information is needed to evaluate the risks fully.

What sort of review mechanism does EPA have in place to evaluate such proposals? The Agency has developed a procedure which ensures that a thorough and accurate assessment takes place. Basically, internal review within the Office of Pesticide Programs (OPP) is accompanied by review by other EPA program offices to achieve regulatory consistency. This is followed by external peer review conducted under the auspices of EPA's legislatively mandated Scientific Advisory Panel (SAP). The SAP peer reviewers are handpicked by the Agency to ensure they have appropriate expertise and familiarity with the issues at hand. This ensures that we receive the best possible feedback. The Agency may also consult with the applicant to obtain more detailed information. Finally, other federal agencies may also be consulted if there is a potential jurisdictional problem or if another agency has particular expertise to contribute. This intense review process is frequently not easy to complete within only 90 days.

Although not put to the test, the Office of Toxic Substances expects to use a similar process, with emphasis on external scientific review and extensive discussion with applicants prior to submission.

To ensure that the review process is consistent Agency-wide, EPA is considering forming a single body to review applications for products of biotechnology submitted under both TSCA and FIFRA.

Although there is a sound foundation for assessing risks of genetically altered microbials, there are several risk concerns that are unique to such substances and require the case-by-case evaluation process we use at present. These generally include evaluation of dissemination and containment, stability of genetic elements, ecological issues, and availability of test methods and monitoring capabilities. For viruses, there are more complex uncertainties including the following:

1. Host range. Since viruses can only reproduce in living cells, establishing host range is essential. What is a representative selection of potential hosts? Which should be tested? Can host range be predicted from in vitro tests?
2. Dissemination from a test plot may be difficult to prevent, especially if there are many mobile host vectors or if the virus is very persistent outside the host organism.
3. Transfer of genes from one organism to another is a concern with bacteria. Can genes engineered into viruses be transferred to unintended hosts and express harmful properties in the environment?
4. It may be difficult to predict the environmental fate of engineered viruses, that is, persistence and multiplication. Thus, accurate nontarget host range determination is critical.

5. Are currently available laboratory monitoring methods able to follow these viruses in the environment? Will genetic probes work with viruses in the environment?

Subdivision M contains virus testing protocols, and they are being updated as necessary. EPA's Office of Research and Development is sponsoring research to further develop and improve them. In the meantime, EPA strongly urges anyone contemplating the submission of a viral product to contact the Agency before initiating any testing. Also, if the research can shed any light on the above mentioned risk assessment concerns, researchers should contact EPA.

Moving away from scientific concerns to my last topic, the public and the regulated industry commented thoughtfully and candidly on EPA's proposed regulatory approach. The Agency received nearly 75 comments from the pesticide industry, research companies, academia, and environmental and public interest groups. Although most comments generally supported the need for a coherent federal approach and the particular regulatory scheme proposed by EPA, several questions were raised. Concerning FIFRA, academia and other research-oriented entities expressed serious concern that the policy on small-scale field testing would stifle research and that the proposed regulatory scheme would be burdensome. The pesticide industry generally supported EPA's regulatory approach, stressing that we should continue to be flexible and evaluate on a case-by-case basis. Specific issues raised included confidentiality, clarity of definitions, and the scope of regulation.

Comments on TSCA issues included what should be considered "new" under the PMN provisions; should small-scale experimental releases be considered strictly research and development and exempted; and what level of information is necessary to adequately assess the potential risks of a "new" microbe.

The Agency is now evaluating these comments. Clearly, one of the major issues we will have to deal with is where testing that is purely research and development ends and where testing leading to commercialization begins. It is frequently difficult to draw the line. The second issue is determining what effects genetically altered substances will have on the environment's ecological balance. The Agency's ability to assess human risk has become more sophisticated in recent years. However, our ability to evaluate successfully ecological risks has not emerged as quickly. There is a strong need to develop this capability soon. Biotechnology research is forcing the issue and will help us to regulate overall the environmental impacts of pesticides. We look forward to working with researchers and the regulated industry as we deal with these challenges in the months to come.

COMMENTS

MILLER: There is a fallacy in your presentation, namely that the initial uses of these products in field trials are intended to provide the first attempts to demonstrate that they are efficacious in an outdoor setting and that there are no untoward effects. I think that generally you need to be much more lenient with such limited, isolated applications. This is the first step towards obtaining valuable data; it is how industry determines whether to move toward commercialization, the documentation for which is extremely expensive and time-consuming. FDA has the same dilemma with determining requirements for initial clinical trials of drug products for marketing approval. The differences are vast. We require much less documentation for initial clinial trials, invariably conducted in small numbers of patients. By analogy then, I do not think that yours is a useful rationale for justifying EUPs.

SCHATZOW: With biological agents, obviously, there is a potential for replication that is not present for the chemical drugs that FDA has traditionally regulated. Also, marketing lead time is different. Most biological agents are typically exempted from the requirements to conduct long-term chronic testing which may take 4 years to complete. My point, Henry [Miller], is that the ability to market products more rapidly is going to be greater with biological agents than it is with chemicals. While chronic testing is being conducted, manufacturers also conduct major experimental use for chemical agents on thousands and thousands of acres as a premarket entry strategy. Otherwise, you may be able to gather, on a much smaller-scale testing, the necessary information to allow for commercialization. I think the fact that companies go through 3 or 4 years of testing a corn herbicide all over the country on 25,000 acres is not because they necessarily need that information, but because they are trying to make the chemical known to the farmers. So, I don't agree with you fully.

MARTIN: What kind of risk assessment did you carry out on the five biological pesticides?

SCHATZOW: Very qualitative. Again, we were trying to answer some questions. The major issue that we have taken back to the companies is: What can you tell us about the fate of this organism in the environment? Is it going to spread? Where is it going to spread? Is it going to last over the winter? What kind of potential risk does that present? In the case of the Monsanto toxin, we are interested in how many different insects it may affect. Is it going to be specific to the one target insect, and how do you know its specificity? These are the kinds of questions that we are asking.

MARTIN: How were the licensed microbes evaluated before you approved them?

SCHATZOW: The 14 naturally occurring microbials were registered before my time. The Agency has never had an explicit risk-benefit assessment for initial licensing. We have carried out much more explicit risk-benefit assessment on existing chemicals that we wanted to take off the market.

FOWLE: Steve [Schatzow], don't you require producers of microbial pesticide control agents to generate data to support risk assessment under Subpart M of the pesticide guidelines, including infectivity tests in a number of different species?

SCHATZOW: Yes, we have a whole series of tests most recently enunciated in our Part 158 Data Requirements for Pesticide Registration. I did not mean to suggest that we license chemicals on an ad hoc basis. We have a battery of tests to be conducted and the results reported to EPA before a major chemical can be registered. My point was that we do not have a formal, step-wise procedure to analyze that data.

TEICH: You mentioned that a number of microbial pesticides have already been licensed. When were they licensed? Was it prior to the establishment of the current set of procedures?

SCHATZOW: It was prior to the formal establishment of data requirements, although many of the tests now required were conducted as the microbials were licensed—and again, the more recent ones were scrutinized more fully.

MERIGAN: How about confidentiality questions?

SCHATZOW: That is a problem.

MERIGAN: How do you handle that?

SCHATZOW: Not very well.

MERIGAN: The FDA is really good at that, in the sense that an NDA and an IND are real, private documents; the committees are reviewed carefully to decide whether there is conflict of interest. I wonder how you handle confidentiality.

SCHATZOW: There is protection under FIFRA for confidential business information.

MERIGAN: Is all of the science of proprietary interest? How do you handle it? Do you allow public disclosure of such information?

SCHATZOW: No, we do not publicly disclose information. In fact, we have had the committee meet in closed session because of the claims of confidential business information (CBI). I'm uncomfortable with it because the claims are so broad that we cannot communicate with the public. For instance, Monsanto had given us a notification and had press conferences and press releases about this notification. When we responded to Monsanto's submission, Monsanto said that our letter to them was confidential and that we could not release that letter. It had nothing to do with the content of the letter; we just could not release any information at all on that product.

PERPICH: I would like to comment on that point. When the Federal Interagency Recombinant DNA Committee recommended legislation for 1977 for all recombinant DNA activities, there was no provision for the protection of proprietary information. The Committee could not agree on the scope of protection because of EPA and CEQ concerns on the need for public disclosure of potential hazards. That issue may be a sensitive one for EPA in fashioning and implementing its biotechnology regulations.

SCHATZOW: We deal with the general CBI issue under FIFRA. There are no particular problems with that. We have enormous mechanisms to protect confidential business information. My concerns are the other side of the issue. When you have so much public concern and desire to have public education and to inform the public in this area, how can you do it when a company says, "Everything I have given you is confidential business information." All you can do is put a notice in the *Federal Register* that says, "We are having a meeting. It will be a closed meeting. We can't tell you what it's about, but we're having it."

KAMELY: Steve [Schatzow], I wonder whether you can comment briefly about how your program plans to handle the importation of genetically altered pesticides. For example, if a company such as Monsanto goes outside the country, develops a pesticide, or any other genetically altered microbe, and then imports it into the United States, how would EPA handle the imported microbe?

SCHATZOW: It would be no different than the way we would handle a chemical. If it is intended for sale in the U.S., it has to be registered in this country.

KAMELY: Have you faced that for recombinant DNA products, as well?

SCHATZOW: I'm not aware of any situation like that.

MOSS: If a live genetically altered virus is intended for veterinary or medical use, is that handled solely by FDA or by the Department of Agriculture?

SCHATZOW: That's right. I think we've divided everything up fairly well. There is some overlap with animal drugs and whether they are internal or external. There are certain external animal formulations that are considered pesticidal. But EPA has worked out these problems with FDA, so I don't see any potential for duplication.

MERIGAN: Then the new human vaccines will not be your regulatory province?

SCHATZOW: I wouldn't think so. Henry [Miller], are they?

MILLER: They are not. Whatever is regulated under the Federal Drug and Cosmetic Act or HS or Virus, Serum, and Toxin Act is exempt.

SCHATZOW: Again, FIFRA is just pesticides. TSCA is everything except pesticides, food, and drugs. So, FDA has foods and drugs; we have pesticides; the Office of Toxic Substances has at least everything else under TSCA that is unlike either the Federal Food, Drug, and Cosmetic Act and unlike FIFRA, both of which are licensing provisions. TSCA, in essence, contains notification provisions and is not a licensing statute.

HOWLEY: You mentioned that one of your applications deals with undirected mutagenesis. Would EPA have to become involved with directed selection not involving mutagenesis that a company just selected naturally for a variant of the bacteria to make a certain product?

SCHATZOW: Again, if it were the development of a pesticide, then it would have to be licensed by EPA. If it were a new "chemical," I suppose, that was not a pesticide or a food or drug, then there should be notification to TSCA for premanufacturing review.

MILLER: That's a tricky one because the *Federal Register* notice expresses uncertainty about what will be considered a novel microorganism for purposes of TSCA, so that naturally occurring organisms, spontaneous mutants, and those obtained by undirected chemical mutagenesis are not considered novel. There is uncertainty in the notice about those that are manipulated by conventional genetic engineering techniques–like transduction and transfection. Those which use the modern techniques, recombinant DNA and cell fusion, would be considered novel. The issues are therefore complicated and worthy of comment.

FOWLE: EPA asked for public comment on what the regulatory restrictions should be. That's an open question.

Session 2:
Environmental Virology

Viral Spread Between Hosts

ROBERT ELLIS SHOPE
Yale Arbovirus Research Unit
Yale University School of Medicine
New Haven, Connecticut 06510

OVERVIEW

Virus spread between hosts is a mechanical event. The degree and efficiency of spread depend on multiple factors including the amount of virus excreted by the donor host, the site and means of exit from the host, the stability of the virus, the accessibility of fomites or other vehicles of spread, the distance between hosts, and the availability of receptors for entry into the recipient host. Viruses are commonly spread directly host-to-host by respiratory droplets or by the fecal-oral route. However direct spread may also occur through ingestion of infected food, sexual contact, needle, hand-to-mouth or mouth-to-mouth, or hand-to-nose or conjunctiva. Arthropods may be a vehicle either by mechanical transmission or biological transmission. Transovarial and transstadial maintenance of biologically transmitted virus may occur in the arthropod. Some viruses are transmitted animal-to-animal by droplet or by bite, with human beings serving as accidental hosts. The choice of a virus to be genetically altered and used as a vector should take into consideration the mechanism and efficiency of spread of the parent virus. The altered virus should then be extensively tested to detect and document changes in its ability to be transmitted. For most applications the ideal vector will be one which is not readily spread from host-to-host; however, a rapid and efficiently transmitted genetically altered virus, shown to be safe, may be of enhanced usefulness for some environmental applications.

INTRODUCTION

Jacob Henle (Henle 1840), writing at the age of 31 just after he assumed the position of Professor of Anatomy in Zurich, described most of the concepts of spread of viruses between hosts that we espouse today. His essay was called *Von den Miasmen und Kontagien* (*On Miasmata and Contagia*). Although he did not know about viruses, he proposed the concepts of a living pathogen, of epidemics, and of endemics where case after case occurred locally and in persons from distant districts who entered the endemic zone. He described two types of infections. One required direct contact with the patient or contaminated fomites, was present in blood and body secretions, and could be transmitted by inoculation. Another was transmitted through the atmosphere, originated in the lungs or in skin lesions, and was communicated to the mucous membranes. Henle recognized as contagious many of the major viral diseases of people and domestic animals known today:

measles, rubella, smallpox, certain types of cold, influenza, rabies, rinderpest, and sheep pox.

He puzzled over yellow fever but did not understand how it spread. The transmission of yellow fever by *Aedes aegypti* with attendant long extrinsic incubation in the mosquito was not to be determined until the work of Carlos Finlay of Cuba and the Walter Reed Commission 60 years later (Warren 1951). Neither did Henle understand the concept of inapparent infection with or without shedding of an infectious agent; but he otherwise laid the foundation for virtually all of our understanding of spread of viruses today.

Following Henle's lead, this paper will review the factors which determine the degree and efficiency of viral spread, and will discuss the mechanisms of spread as we understand them today. It will also illustrate the methods for determining the extent and modes of spread, using examples of live attenuated viral vaccines and live viruses employed as biological control agents.

RESULTS

Virus spread between hosts is a mechanical event. The degree and efficiency of spread depend on the following factors: amount of virus excreted by the donor host, the site and means of exit from the host, the stability of the virus, the accessibility of fomites or other vehicles of spread, the distance between hosts, and the availability and avidity of receptors for attachment and entry into the recipient host. The recipient host must of course also be non-immune and susceptible to infection.

What are our presently accepted concepts governing the mechanisms of spread of viruses? These principles are outlined in Table 1 as presented by Benenson (1985) and are not too different from those proposed by Henle in 1840. Viruses may spread by direct contact. Touching—either hand to mouth, hand to eye, or hand to nose—is commonly implicated among others in the spread of poliovirus, rotavirus, hepatitis A, adenovirus, and herpesvirus. Rabies virus is spread by the bite of dogs or other animals. Kissing, in developed countries, is a major spreader of Epstein-Barr virus, the cause of infectious mononucleosis, especially among college populations. Sexual intercourse has been implicated in the spread of herpes virus type 2, and intimate contact between male homosexuals spreads hepatitis B virus and the virus causing acquired immunodeficiency syndrome. These latter 2 viruses as well as any virus which causes viremia may also be spread by blood transfusion and surgical procedures.

Virus is also spread directly by the acts of sneezing, coughing, spitting, singing, and talking when the donor host is within one meter or less of the recipient host. Polio, rhino, influenza, and other viruses which colonize the upper respiratory tract are spread this way. Close contact is required because the droplets containing virus are relatively large and are removed rapidly from air by force of gravity.

Table 1
Mechanisms of Spread of Viruses Between Hosts[a]

Direct
Direct contact
Touching
Biting
Kissing
Sexual intercourse
Transfusion
Surgery
Droplet spread to conjuctiva or mucous membrane, usually limited to 1 meter
Sneezing
Coughing
Spitting
Singing
Talking
Indirect
Vehicle-borne
Fomites such as handkerchiefs, bedding, eating utensils
Ingested such as water, food, milk
Tissues such as blood, scabs, organs
Vector-borne including arthropods and rodents
Mechanical
Biological
Airborne. The dissemination of aerosol suspension of virus particles 1–5 microns in air, usually to the respiratory tract
Droplet nuclei: Residue from droplets, remain suspended for long periods
Dust: Particles arise from soil, clothes, bedding

[a]Modified from Benenson 1985

Fomites such as handkerchiefs, bedding, clothes, drinking glasses, and other eating utensils serve as vehicles for spread of viruses. For spread to be successful, the viruses must retain viability on these surfaces. Viruses such as polio and hepatitis A can be spread by drinking water and food. Shellfish consumed raw are efficient spreaders because they concentrate and hold viruses for long periods of time. Viruses are also spread indirectly by dried blood, scabs, or other body tissues. These mechanisms of transmission apply not only to viruses of vertebrate animals, but also to viruses of insects and plants; for instance, the baculoviruses of insects are transmitted when the larvae eat contaminated plant material.

Arthropods such as mosquitoes, ticks, sand flies, and midges transmit over 400 different viruses, known as arboviruses. These include yellow fever, dengue, and equine encephalitides. The arboviruses replicate in the arthropod midgut, pass to the salivary glands, and are transmitted to the recipient host when the arthropod takes a blood meal. In some cases the virus also is maintained transovarially and transtadially in the arthropod. Arthropods also transmit viruses such as myxoma (see below) mechanically, without replication of the virus. In this mechanism, the virus persists sometimes for long periods sequestered in the mouthparts and is transferred to the recipient host when probing occurs. Rodents may also spread viruses to people. Viruses such as Lassa, Machupo (the cause of Bolivian hemorrhagic fever), and Hantaan (the cause of Korean hemorrhagic fever) are carried by rodents which excrete the virus in urine and respiratory secretions. The spread to people is by aerosol or contact with contaminated fomites.

Small droplets between 1 and 5 microns in size remain in the air and reach the alveoli of the lungs when inhaled. Virus in these aerosols remains suspended in air for long periods and if the virus is stable, it is efficiently spread. Aerosols may be formed from droplet nuclei or from dust particles arising from soil, personal articles, or bedding.

If we understand the mechanism of spread of a given virus, we can then intervene to diminish spread or, if we wish, to enhance spread. Handwashing and vector control can be very successful intervention measures. Other measures such as refraining from fondling children and from kissing are neither deemed practical nor accepted socially.

We lack either field or laboratory data on spread of genetically altered viruses between hosts. Nor do we have for most viruses the information base on genetic control of factors such as the amount of virus excreted by a donor host, the stability in the environment, or the receptor avidity. We do, however, have some experience with spread of live viruses which have intentionally been liberated in natural populations, either as vaccines or as biological control agents. There have so far been no natural disasters as a consequence, and some lessons have been learned. Two vaccines, the Sabin oral poliovirus and vaccinia, and two biological control agents, myxoma virus and baculoviruses, serve as examples. Keep in mind that in some cases, such as poliovirus vaccine, spread between hosts was the desired outcome and in other cases, such as vaccinia, natural spread was not desired. Presumably in the case of genetically altered viruses also, some will be designed to spread and others not to spread.

Poliovirus Vaccine

Wild-type polioviruses are excreted for weeks in the stool and for shorter periods in the nasopharynx. The virus spreads to a high percentage of non-immune contacts of naturally infected individuals irregardless of the socioeconomic status. With the development of live attenuated vaccines, it was expected that they would be shed and would spread also. Spread between hosts would be desirable because it would result in infection of community contacts who might have been missed in vaccination campaigns and would result in a higher degree of herd immunity. If, however, the vaccine virus should revert to neurovirulence, spread would be undesirable.

A series of experiments undertaken by Fox and his colleagues in 1958 in southern Louisiana are illustrative (Fox and Hall 1980). They chose families which had at least three susceptible members. They vaccinated an index family member and carried out surveillance for poliovirus excretion and seroconversion in the other members. Non-immune children from neighboring families were also exposed in the homes of the index cases, sometimes for periods as short as 2 hours. Most vaccinees shed the vaccine virus. There was more transmission in lower socioeconomic status families and their outside contacts than in families of higher socioeconomic status. There was no spread from vaccinated adults, only from index children. As shown in Table 2, children fed a relatively high dose of poliovirus, type 3, infected siblings and many of the extrahousehold contacts, even those with short contact periods. It was not possible to determine whether pharyngeal or fecal excretion of virus was the more important source for virus spread; pharyngeal excretion appeared to be important, but some contacts were infected when the index case was shedding only in feces. Virus was recovered from swabs taken from the buttocks of children, from the floors, and from the hands on multiple occasions (Table 3). Although primary contacts were regularly infected, at no time did vaccine virus spread widely or cause community epidemics.

Vaccinia Virus

Vaccinia is a virus of uncertain ancestry. Experience for over a century indicates that, although many animal species are susceptible, spread with establishment of continuing transmission does not occur. Cases in a children's dermatology ward have been reported (Sommerville et al. 1951), and 8 of 36 strains from cows in Holland had characteristics of vaccinia (Dekking 1950); but it was neither proven that these were vaccinia nor that spread was from human vaccinees.

Table 2
Infection of Family and Extrahousehold (EHH) Contacts of Families in Which One Child Was Fed 7.1 log ID_{50} Sabin Type 3 Poliovirus Vaccine[a]

Unit	Index child socioeconomic status	Age	Fecal virus excreted at contact	Susceptible child family contacts	Susceptible child EHH contacts[b]
45	White lower-economic	4	Yes	4/4[c]	2/3
49	Black	3	Yes	3/3	1/5
46	White lower-economic	4	Yes	4/4	3/9

[a]Data from Fox and Hall (1980)
[b]Most of EHH contacts were for 2 hrs only
[c]Number of infected/number of exposed

Table 3
Recovery of Enteric Viruses from External Sites in Households

Unit	Socio-economic status[a]	Type fed	Number of susceptibles infected	Viruses recovered	
				Type	Site
28	N	P2	3	P2	Buttock of index child
39	WL	P3	4	P3	Kitchen floor
40	N	P3	3	P3	Buttock of sibling
49	N	P3	4	P3	Hands (four occasions) and buttock of index child
15	N	P1	1[b]	P2	Living room floor

[a]N = Negro; WL = white lower-economic.
[b]In addition, three susceptibles were infected with type 2 poliovirus.
Reprinted, with permission, from Fox (1980).

The use of vaccinia as a vector has been proposed partly on the basis that it will not be readily spread in the environment. There is no credible evidence, based on past experience, that environmental spread will pose a problem.

Myxoma Virus

Myxoma is a poxvirus of rabbits that in its native habitat of South America causes a mild disease. When transmitted to the European rabbit *Oryctolagus cuniculus*, however, the infection results in mucinous tumors and is usually lethal. This virus was used in a grand experiment in Australia to control the massive populations of introduced European rabbits which were competing with sheep for forage. The virus was not genetically altered, but the experiment serves as an example for this conference since myxoma virus had not occurred naturally in Australia.

In 1950, after the virus had undergone several years of preliminary laboratory study in Australia and the Americas, it was released in the Murray Valley of Australia with the intent of controlling rabbits (Fenner and Ratcliffe 1965). The virus appeared initially to have disappeared, however in December 1950 it reappeared in several sites in the Murray Valley and spread with surprising speed over much of southeastern Australia. It established itself as an enzootic infection and still persists.

The mechanism of transmission was suspected of involving arthropods prior to release of the virus in 1950. Subsequent studies described in detail by Fenner and Ratcliffe (1965) confirmed that many biting arthropods, including mosquitoes, ticks, fleas, mites, blackflies, and lice, transmitted the virus, but that *Culex annulirostris* and *Anopheles annulipes* mosquitoes were the significant vectors. The studies also showed that the virus was mechanically transmitted, being sequestered on the mouthparts of mosquitoes during the first week after the mosquito probed the skin tumors (Table 4). Some mosquitoes were still infectious for rabbits during

Table 4
Persistence of Myxoma Virus on Mouth Parts of *Aedes aegypti* as Measured by Transmission to Rabbits

Day: 0–7	8–14	15–21	22–28
249/770[a]	17/136	1/22	1/14

[a]Number of mosquitoes transmitting/number of attempts
Data from Fenner and Ratcliffe (1965)

the fourth week after probing. Epizootics were most prominent during years and in areas of heavy rainfall. Most infected rabbits died, although with time, genetically resistant rabbits appeared. Also, the ability to kill rabbits was examined for field isolates of myxoma virus. Isolates with both decreased and increased virulence were found, but the most commonly encountered strains killed about 90% of inoculated rabbits.

The lessons from the myxomatosis experience are not necessarily predictive of results to be had from releasing a genetically altered poxvirus into the environment. The experiment is the biggest and the best we have, however, and it showed that both the virus and the vertebrate host adapted and that no catastrophic events (other than death of rabbits) were found.

Baculoviruses

Baculoviruses are double-stranded DNA viruses which infect and kill insects. There are now four baculoviruses registered by the U.S. Environmental Protection Agency as pesticides for agricultural and forest application. The viruses are not genetically altered, but their mechanism of spread between hosts and their potential as vectors of recombinant DNA (Summers and Smith, this volume) command our attention in the context of this conference.

Table 5 lists the characteristics of baculoviruses which favor spread between hosts. The virus kills the insect and in so doing produces large numbers of polyhedral inclusion bodies which contain infective virions encapsulated in polyhedral

Table 5
Characteristics of Baculoviruses Favoring Spread Between Hosts

Virus kills and is liberated from dead insects
Viability protected from environmental degradation by polyhedral protein
Infectivity inactivated by ultra-violet radiation
Virus persists in soil, undersides of leaves, surface of insect eggs
Virus disseminated by vertebrates, but does not infect them
Polyhedral protein is solubilized at alkaline pH in insect gut to liberate infective virions
Virus infects insect gut

protein. The insect integument ruptures spilling the polyhedrons into the atmosphere. The polyhedral protein protects the virion from environmental degradation, although ultraviolet irradiation from sunlight inactivates infectivity. For this reason, inclusion bodies in sites such as soil, under leaves, and on the surface of eggs have the best survival and may persist for several years in the environment. The virus infects only insects, although vertebrates ingest polyhedrons and disseminate virus. Larvae ingest the polyhedrons on the surface of natural foods. The alkaline pH of the larval gut dissolves the polyhedral protein and the liberated virions infect the insect gut.

Experience to date with the registered baculovirus pesticides in the United States indicate their safe use, both with respect to vertebrates and to nontarget insects in the environment.

DISCUSSION AND CONCLUSIONS

The mechanisms of spread of viruses between hosts have been understood in a descriptive sense since 1840. The genetic basis governing efficiency of spread, however, is still not understood today. The experience with attenuated viral vaccines and with viruses used as biological control agents, both in zones where the virus has previously existed and in new geographic areas, has been salutary.

The immediate challenge with genetically altered viruses is to find or develop strains that are genetically stable and do not spread between hosts. This should be relatively easy, although the testing of viruses to be released may be expensive and tedious.

The more interesting and difficult challenge is to find or construct stable viruses for use as vectors, which are efficiently transmitted to specific hosts in a controlled manner, and better yet whose transmission could be stopped by immunization, anti-viral compounds or other mechanisms in the event of an unforeseen adverse effect on the environment.

REFERENCES

Benenson, A.S. (ed.). 1985. *Control of communicable diseases in man.* 14th ed., p. 457. American Public Health Association, Washington, D.C.

Dekking, F. 1950. Koepokken en vaccinia. *T. Diergeneesk* **75**: 248. In *Viral and rickettsial infections of man* (ed. F.L. Horsfall, Jr. and I. Tamm), 4th ed., p. 953. Lippincott, Philadelphia.

Fenner, F., and F.N. Ratcliffe (ed.). 1965. *Myxomatosis* Cambridge University Press, Cambridge.

Fox, J.P. and C.E. Hall. 1980. Experimental studies with vaccine strains of polioviruses. In *Viruses in families*, p. 152. PSG Publishing Company, Littleton, Massachusetts.

Henle, J. 1840. *On miasmata and contagia* (translated by G. Rosen, 1938). Johns Hopkins Press, Baltimore.

Sommerville, J., W. Napier, and A. Dick. 1951. Kaposi's varicelliform eruption: Record of an outbreak. *Br. J. Dermatol.* **63**: 203.

Warren, A.J. 1951. Landmarks in the conquest of yellow fever. In *Yellow fever* (ed. G.K. Strode), p. 1. McGraw-Hill, New York.

COMMENTS

DIXON: Your references to the Sabin vaccine reminded me of a *McNeil-Lehrer Report* that you may have seen last month. The point was that there is a small incidence of paralytic disease among people who are treated with the Sabin vaccine, and the virus can apparently even go from the person who's treated to a member of the family. Have there been careful risk assessment studies of this? What are those numbers?

SHOPE: If I remember correctly it is somewhere in the neighborhood of five or six cases per year in the United States. Bob [Chanock], is that about right?

CHANOCK: It's about once per 10,000,000 vaccinations.

SHOPE: It is a relatively rare event. I think when the vaccine was originally developed, it was an event which would probably not have been, on the basis of chance, encountered in the original vaccine trials. However, it is a risk which the Public Health Service considers worth taking because of the benefit of the vaccine.

DIXON: Recognizing the fact that it's a very rare event, I wonder if there is a feeling as to whether it is a host susceptibility or that it is something that attentuates in the virus? Realizing it's as rare as it is, does anyone have a guess as to why it happens?

CHANOCK: I'm going to discuss that in my presentation. The data show that 15% of the vaccine-associated cases occur in children who are immunodeficient.

DIXON: I think the program left an uninformed viewer with a level of concern greater than what might be deserved.

CHANOCK: I think these isssues were discussed very clearly because John Fox, who performed the studies on familial spread alluded to previously, summarized the benefits and the risks of immunization with oral poliovirus vaccine. The figure of one adverse reaction in 7,000,000 vaccinations was mentioned. The point was also made that you are much more likely to be involved in a fatal automobile accident or be hit by a car while riding a bicycle than be adversely affected by oral poliovirus vaccine. If you consider the risk in this perspective, it becomes considerably less significant.

MILLER: I'm surprised at the discussion of whether this polio vaccine would be licensed today. I think that the risk-benefit characteristics are clearly appropriate for licensing; and, in fact, the proof of that is that we license other vaccines, such as pertussis, whose risk-benefit characteristics are less favorable than the polio virus vaccine we were discussing.

FIELDS: One of the issues here is that if you genetically manipulate a virus or introduce a recombinant DNA molecule, what would the impact be on the virus, both in the host, which relates to the vaccine issue and also to the environment? You cited four viruses where there are different genetic alterations or manipulation. I have two questions: First, do you know of any instances where a genetic alteration of a virus by in vitro means increases or changes the capacity of the virus to remain in the environment; and particularly are you aware of any in vitro manipulations that don't decrease the capacity of a virus to survive in the environment? Second, is there any insight into any specific genetic manipulations of which you're aware that affects the capacity of a virus to persist in the environment?

SHOPE: I think the second question may be easier than the first one—for me, anyway—and it's hypothetical. But, if you genetically alter a virus to express a protein which is a surface protein of another virus, that surface protein certainly has the possibility that it might incorporate itself into the assembled virus to form a pseudotype or something like a pseudotype. That, indeed, might alter its ability to attach to a receptor site or some other site. However, I don't know whether that is an issue at this point in time.

FIELDS: I'm not asking the theoretical question. I'm really asking whether you know of any experiments that have been carried out concerning how specific viral mutations affect the environmental stability, spread, or other viral properties. There is much more information concerning the impact of viral mutations on the host. Part of the issue is, in fact, these other environmentally-related issues—stability, spread, mortality, and survival. Is there anything known or is there anyone who would like to comment?

CHANOCK: I think that there is quite a bit known. John Fox did some other studies that Bob [Shope] didn't have time to discuss. Fox compared the interfamilial spread of the vaccine strains and then compared the spread to that observed several years before in the same sort of family when wild-type viruses came in. He found that in the lower economic families there really wasn't much difference; however in the upper economic families very little spread occurred with the vaccine virus, but it was almost complete with the wild-type. Next he did a community study and found that the vaccine virus actually dies out after about two cycles; it doesn't go through more that two cycles. So, one of the reasons why there is so much transmission in the general population is that the virus is being pulsed in all the time; every infant at the age of 6 or 8 weeks receives a polio vaccine, therefore the virus is constantly being introduced. However, if virus is pulsed into a population, which occurred on three separate occasions—once in the United States, once in Mexico, and once in Czechoslovakia—involving hundreds of thousands of individuals, after 2 months the vaccine virus could not be recovered in the general population. So, the virus is clearly not as transmissible. It dies out and is not able to sustain itself. Vaccinia also cannot sustain itself because once the general vaccination campaign stopped, the virus disappeared. Vaccinia has not been isolated from humans since vaccination stopped.

KILBOURNE: This is also true, but it's not an engineered virus.

CHANOCK: It's an altered virus.

KILBOURNE: It is an empirically selected mutant.

CHANOCK: But that doesn't make any difference. It's a genetically altered virus. We don't know who did it or how it happened. The point is that vaccinia cannot sustain itself.

SHOPE: This is also true of baculoviruses which are also not altered viruses. When you put them back into a population of insects, they may, for several generations, maintain an increased prevalence, but they eventually go back down to normal levels.

Airborne Transmission of Virus Infections

VERNON KNIGHT, BRIAN E. GILBERT, AND SAMUEL Z. WILSON
Department of Microbiology and Immunology
Baylor College of Medicine
Houston, Texas 77030

OVERVIEW

In order for airborne transmission of virus infection to occur, virus-containing particles must fulfill two criteria: (1) They must be small enough (< 10 μm) to remain suspended long enough to be carried beyond the immediate environment of their source (i.e., infected person or animal); and (2) they must retain viability long enough to initiate infection when deposited within a susceptible host. This paper presents evidence that some natural and experimental viral infections of man and animals are spread by the airborne route. The paper also examines the quantitive aspects of generation of airborne virus, its dissemination, and the dosages required to produce infection, as well as information concerning the survival of viruses in aerosol.

INTRODUCTION

The problem of environmental contamination with airborne viruses resolves itself into a study of viruses contained in particles small enough that their settling velocities due to gravity are low. For example, a 10 μm spherical particle of unit density in still air will fall the height of a room (3 meters) in 17 minutes, while smaller particles, particularly those less than 5 μm in diameter, may remain suspended in air almost indefinitely, especially if the air is turbulent. Virus containing small particles derived from the infected respiratory tract of man can be disseminated widely, and when inhaled in sufficient concentration, will cause infection. Based on evidence of transmission of infection in animal care facilities, it appears that small particles containing virus may also be generated by such activities as snuffling and coughing by infected animals and by changing contaminated bedding (McGarrity and Dion 1978). We will give examples of long distance spread of hoof and mouth disease virus.

In contrast to the small particle aerosols, large particles sediment rapidly due to gravity, and transmission by these larger particles must, therefore, largely occur by direct spray from coughs from infected persons to susceptibles, or by hand or other contact transmission of deposited particles in the immediate environment of the infected person or animal. Thus, transmission of large particles is not distinguishable from transmission by direct contact of infected secretions, urine, or feces without involvement of the airborne route.

In this paper we will consider the present state of knowledge of airborne transmission of virus infection. It will focus on the production of infection by deposition of virus-containing aerosols within the respiratory tract. This is the most important site of infection with airborne viruses and most studies have been directed to this problem. The quantitative aspects of generation of virus-containing aerosols, factors which influence their survival, and infectious doses required to produce infection by inhalation will be covered along with examples of experimental and natural airborne transmission of viral infection.

RESULTS

Sites of Deposition of Aerosols in the Respiratory Tract

Figure 1 shows the deposition patterns of inhaled particles in the respiratory tract according to particle size and hygroscopicity. The differences between hygroscopic and nonhygroscopic particles is that hygroscopic particles, such as those generated

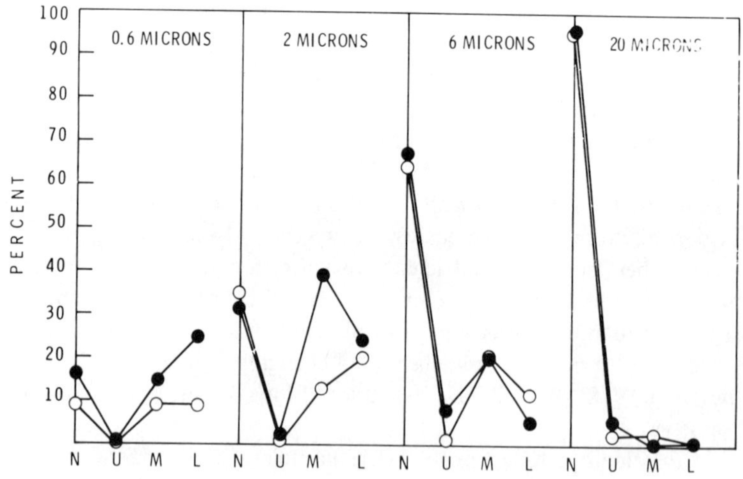

Figure 1
The effect of hygroscopicity and consequent increase in particle size on the site of deposition and percentage retention of the particles. (*N*) nose breathing, 6% of Tidal Air; (*U*) upper respiratory tract, pharynx to and including bronchi, 10% Tidal Air, (*M*) middle respiratory tract, bronchioles, 20% Tidal Air; (*L*) lower respiratory tract including alveolar ducts, 63% TA. (●—●) Hygroscopic; (○—○) nonhygroscopic. Data from Landahl (1972). Reprinted, with permission, from Knight (1973).

from aqueous solutions, equilibrate rapidly with changes in humidity. Thus, at the time of formation they will have a size corresponding to the nearly 100% humidity within the nebulizer flask, but within milliseconds after entering the lower relative humidity of ambient air will shrink due to evaporation of water to a smaller size. When inhaled, the particles will increase in size in the high humidity of the airway by accreting water. Figure 1 shows that larger sizes of either type of particle deposit in high proportion in the nose. This is because inertia determines the site of deposition of larger particles and the nose is the first site encountered by inhaled particles. Differences between deposition of hygroscopic and nonhygroscopic particles are greatest at particle diameters of 0.6 μm and 2.0 μm when the increasing size of the hygroscopic particles due to addition of water leads to their greater deposition in the middle and lower lung areas. Naturally occurring airborne viral particles generated from respiratory secretions will be hygroscopic in nature.

The overall deposition of inhaled particles is determined by particle size and this is shown in Table 1. The hygroscopic particle of subject K (1.3 μm) has the deposition characteristic of the larger nonhygroscopic particles although there is considerable variation of deposition fractions among individuals. This has to do with variation in the size and conformation of airways. Disease will also alter dimensions of small airways and cause differences in deposition fractions.

Another factor that affects particles introduced into the respiratory tract is the clearance by muco-ciliary action after the initial deposition. Figure 2 shows the retention of iron oxide particles over a period of 12 hours following inhalation. The differences in retention are related to differences among the volunteer subjects; however, particles larger than 4 μm in diameter are cleared at rates increasing with particle size. Presumably virus-containing particles would be handled similarly although attachment and penetration of infectious virus that occurs within a shorter time period (an hour or so) would substantially obviate the effect of clearance activities.

Table 1
Deposition Efficiency at Controlled Respiration Frequences and Tidal Volumes for Various Particle Diameters (Nose In-mouth Out Breathing)

		Particle diameter, microns			
		0.7	1.2	1.30	1.80
	Tidal vol	Resp/min	Resp/min	Resp/min	Resp/min
Subject	(cm^3)	12	12	14	12
K[a]	730	(Ribavirin aerosol)		64.0[c]	
5[b]	750	24.0	34.0		67.0
11[b]	750	23.0	29.5		38.0
15[b]	750	27.0	43.0		65.0

[a]Ribavirin in distilled water (hygroscopic) (V. Knight et al., unpubl.)
[b]Cornuba wax particles (nonhygroscopic) (Giacomelli-Maltoni et al. 1972)
[c]Percent

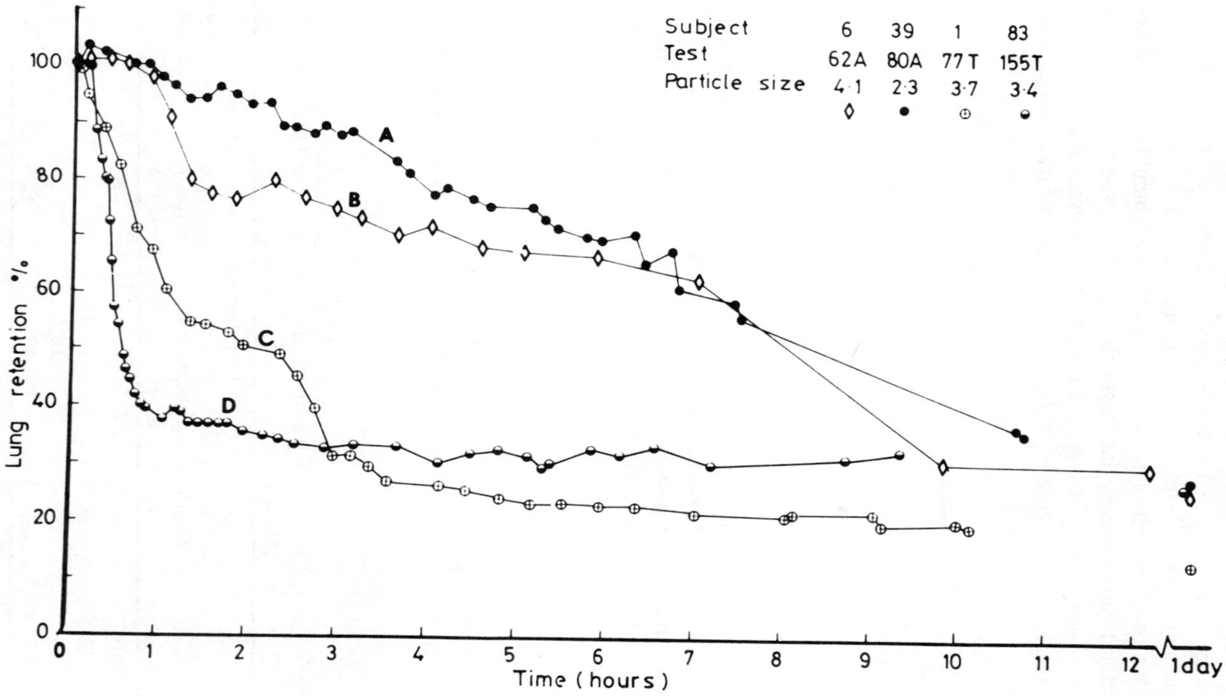

Figure 2
Retention of γ-tagged, monodisperse ferric oxide microspheres as a function of time after 1 min inhalation exposure via a mouthpiece for four healthy nonsmoking men. Reprinted, with permission, from Lippmann et al. (1980).

Lung Cancer and Site of Deposition of Inhaled Particles

Lippmann et al. (1980) have shown a close correlation between the fractional deposition of inhaled particles in different anatomical regions of the lung and the occurrence of pulmonary cancer at these sites (Fig. 3). These findings are highly suggestive of an action of inhaled small particles, presumably carcinogens, in the initiation of neoplastic change. Inhaled small particles containing viruses would distribute in this way and if they possessed transforming capacity could cause neoplastic change.

Human Infectious Dose for Four Respiratory Viruses Administered by Small Particle Aerosol or by Nasal Drops

Table 2 shows the infectious dose by aerosol or nasal drops for four respiratory viruses (Knight 1973). Aerosol was administered to volunteers through a face mask from a flowing aerosol within seconds of its nebulization. Subjects inhaled measured volumes of aerosol and the dosage was estimated by concurrent sampling of the aerosol with an all glass impinger and assaying for virus with an appropriate tissue culture system. Three viruses, rhinovirus type 15, adenovirus type 14, and influenza A virus produce illness at dosages indistinguishable from one infectious particle when given by aerosol. Infectious doses by nasal drops for these viruses are variable, being 70-fold larger for adenovirus, but being more than 20-fold less for rhinovirus. Coxsackievirus A type 21 is slightly more infectious in nasal drops than aerosol whereas influenza virus is slightly more infectious as an aerosol. In all cases, very small dosages of virus were found to be infectious by both routes of inoculation.

Most of the subjects who became infected with these viruses developed mild respiratory illness. The most severe illness patterns were seen in the subjects who received aerosol inoculation with adenovirus type 4. Figure 4 shows the course of illness in a volunteer who developed typical acute respiratory disease syndrome beginning 5 days after inoculation with seven 50% tissue culture infectious doses ($TCID_{50}$) of virus. The highest titers of virus, exceeding 10^6 per ml, were recovered from sputum and oral secretions. Virus shedding was first detected on day 3 in sputum, oral secretions, and anal swabs. Virus largely disappeared by day 18 about 1 week following recovery from illness. This volunteer had no measurable antibody to the virus at the time of inoculation, but neutralizing antibody reached a titer of 1:256 at day 27 after inoculation.

The Size of Particles Produced in Coughs and Sneezes

Figure 5 shows the distribution of particles according to size produced by coughs and sneezes of uninfected volunteers. Particles less than 4 μm were measured in a particle size analyzer (light scattering technique); the larger particles were measured

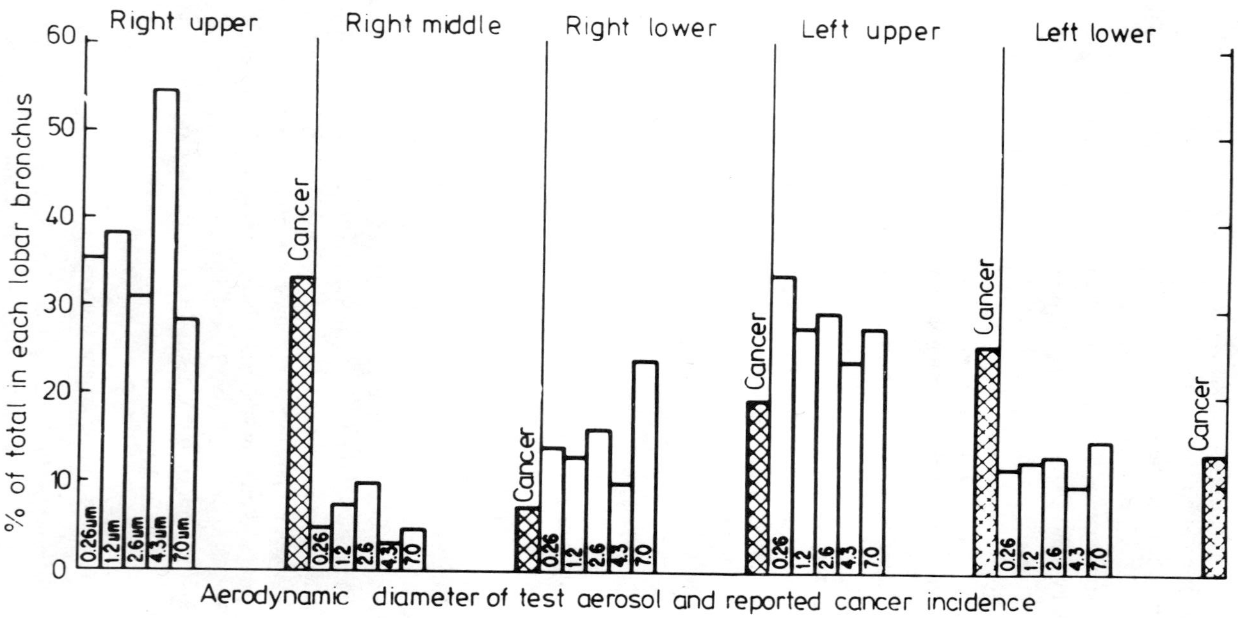

Figure 3
Percentage of total particle deposition within hollow airway casts made from human lungs. Bifurcations accumulate a disproportionate share of tracheobronchial deposits. There appears to be close correspondence between the density of the airway deposition and the incidence of lung cancer sites. The largest number of human cancers are found in lobar bronchi. The largest number occur in the right upper bronchus and the lowest in the right middle bronchus. Data from Schlesinger and Lippmann (1972). Reprinted, with permission, from Lippmann et al. (1980).

Table 2
Infectious Dose (50%) for Four Viruses Given in Small Particle Aerosol of Aerodynamic Mass Median Diameter (AMMD) 1.5 μm and by Nasal Drops

Virus	Aerosol (TCID$_{50}$)	Nasal drops (TCID$_{50}$)
Rhinovirus type 15	0.68 (0.2–2.0)[a]	0.03 (SD = 0)
Coxsackievirus A type 21	28.00 (15–49)	6.00 (3–13)
Adenovirus type 4	0.50 (0.2–1.4)	35.00 (8–157)
Influenza A/Bethesda/63(H2N2)	3.00[b]	15–30[b]

[a]95% confidence limits
[b]Estimated

by visual examination on an oiled slide in the Bourdillon slit sampler. Also shown in the figure is the volume of particles in $\log_{10}$ cubic μm. The largest number of particles are in the less than 1 μm group and decline with increasing size. Despite decreasing numbers of particles with increasing size, the volume of 8–16 μm diameter particles is several orders of magnitude greater than that of small particles. There are about tenfold more particles produced in sneezes than in coughs. The

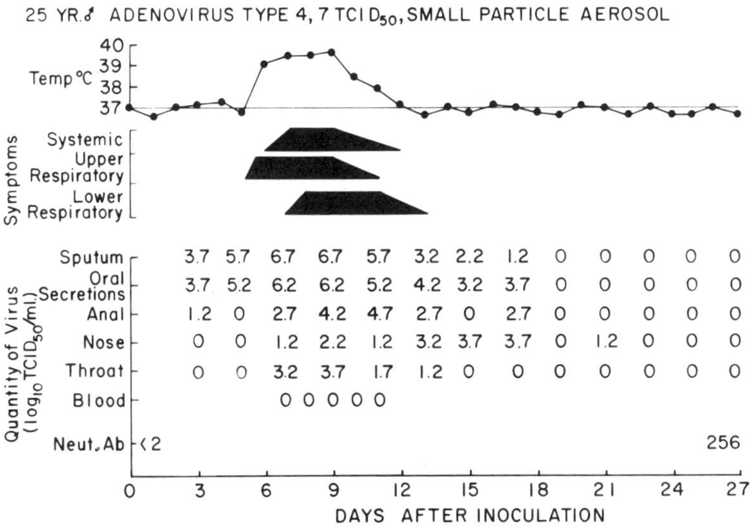

Figure 4
Illness in volunteer after inoculation with adenovirus type 4 by 1.5 μm diameter aerosol. (Case of Dr. R.B. Couch.) Reprinted, with permission, from Knight (1973).

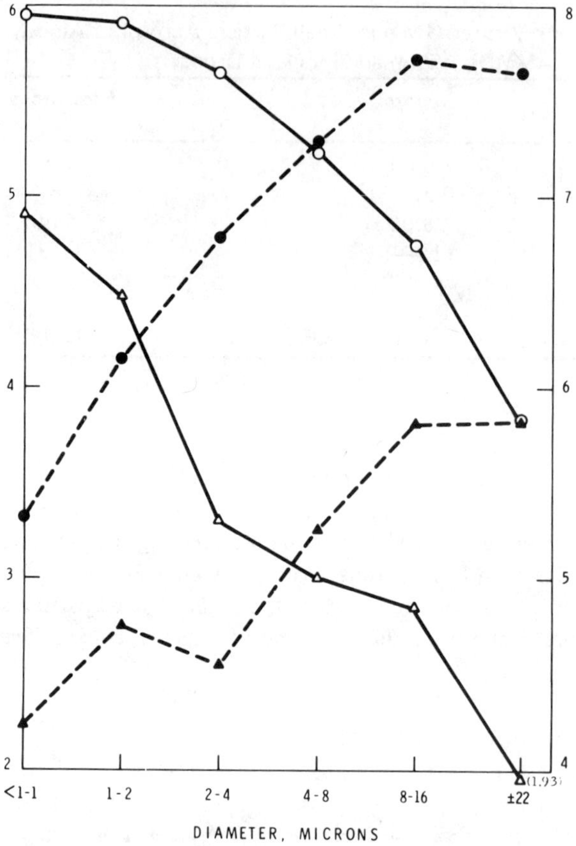

Figure 5
Distribution of particles in coughs and sneezes. Number of particles $\log_{10}$: (○—○) sneeze; (△—△) cough. Volume μ^3, $\log_{10}$: (●- - -●) sneeze; (▲- - -▲) cough. Data from Duguid (1946), and from Gerone et al. (1966). Reprinted, with permission, from Knight (1973).

numbers of particles are variable and are greatly influenced by the amount of secretions in the nasal passage. Because of their capacity to remain airborne, the small particles are the most important in airborne transmission of infection.

Recovery of Airborne Coxsackievirus A Type 21 from Coughs and Sneezes

Gerone et al. (1966) recovered coxsackievirus A type 21 from weather balloons used to collect discharges from coughs and sneezes of infected volunteers. Virus was present in washings from the wall of the balloon as well as in sampled air. The wall samples would contain material from large, rapidly sedimenting particles while air samples would contain the more slowly sedimenting smaller particles. In 52 samplings of coughs, 30% of air samples were positive while 12% of wall samples were positive. Air and wall samples of sneezes were both about 25% positive. The titers of virus were generally higher in wall samples as might be expected.

Recovery of Coxsackievirus A Type 21 from Rooms Housing Infected Volunteers

A large volume air sampler was operated at 10,000 1 pm to sample 82% of a 70,000 liter room (Gerone et al. 1966). The samples were taken following a 2-4-hour period in which ventilation was shut off. From hospital rooms occupied by 1-3 volunteers who were discharging virus in their respiratory secretions, 5 of 16 room air samples yielded low titers of virus. In a control test coxsackievirus A type 21 and fluorescein dye were sprayed into a room with an atomizer producing a large population of small particles. The average recovery of viable coxsackievirus A type 21 was 10% contrasted to 67% of the fluorescein. Thus, the room air sampling procedure detected only a small fraction of the airborne infectious virus.

Transmission of Coxsackievirus A Type 21 in Volunteers by the Airborne Route

In subsequent experiments Couch et al. (1970) housed volunteers infected with coxsackievirus A type 21 on one side of a barrack separated from the other side by a double wire screen (depth 4.5 feet) (Fig. 6). It was shown that most of the experimentally infected volunteers discharged virus in coughs and sneezes over a several day period, during which all of the 19 volunteers housed across the barrier developed coxsackievirus A type 21 infection, indicating beyond reasonable doubt that airborne transmission of infection had occurred.

Natural Outbreaks of Viral Infection Due to Airborne Infection

Influenza is the most important respiratory viral disease of man in terms of its morbidity and mortality on a worldwide basis. Evidence has been presented above which indicated that very small airborne inocula of influenza virus will cause infection and illness. Opportunities to demonstrate the natural occurrence of this phenomenon are infrequent despite a general feeling that this mechanism is the basis for the spread of influenza epidemics. This does not exclude, however, a contribution of contact and large droplet transmission among patients and susceptibles in near proximity. Because of the rapid inactivation of influenza virus at ambient temperatures, environmental contamination would seem to be less important.

Moser et al. (1979) studied an outbreak of influenza on an airliner in 1977, which is best explained as an example of small particle dissemination of influenza.

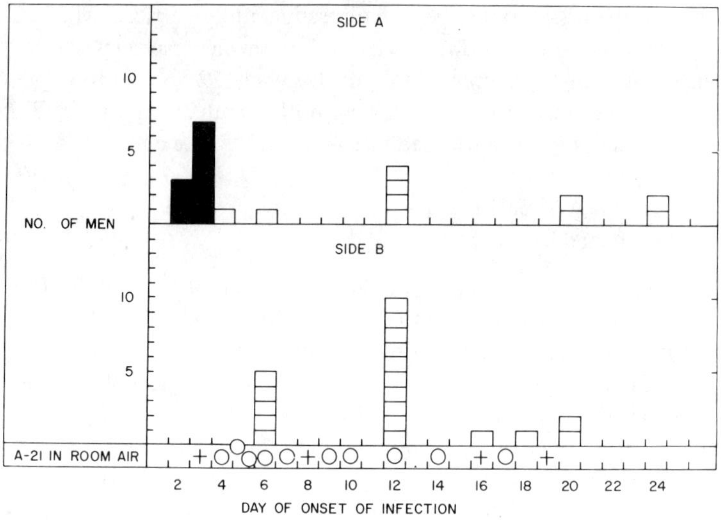

Figure 6
Number and location of volunteers with time of onset of shedding of coxsackievirus A type 21 following exposure to volunteers inoculated by aerosol. (+) Air samples positive for virus; (○) negative; (■) inoculated by aerosol; (□) inoculated by exposure. Reprinted, with permission, from Couch et al. (1970).

A jet airliner with 54 passengers was delayed on the ground in Alaska for 3 hours because of engine failure. During that time the air conditioning was shut off although most of the passengers remained aboard, including one woman who had developed acute febrile systemic illness resembling influenza 15 minutes after boarding at the previous stop. She remained aboard the plane, lying down during the 4.5 hours the plane was stopped. She continued to have severe cough. Following this event 72% of the 54 passengers became ill on the third to seventh day after exposure with syndromes of illness resembling influenza. The disease was determined to be due to influenza A/Texas/1/77 by virus isolation from several patients. Virus was not isolated from the index patient but a serologic rise to influenza A/Texas was demonstrated. The frequency of illness was greatest in persons who remained longest on the aircraft during the time it was stopped. The environment of the crowded plane was comfortably warm and by inference the humidity was low. These circumstances are all consistent with dispersion of airborne particles of influenza virus from the index case and with an extremely high attack rate with a short incubation period thereafter.

Another example of almost certain airborne transmission was the occurrence of measles in susceptible children who visited the office of a pediatrician over a several hour period following a 1½ hour visit there by a child who was acutely ill with measles and coughing frequently (Remington et al. 1985). The four secondary cases did not have contact with the index case. Analysis of airflow in the office suggested that the index case was producing 144 infectious doses per minute while in the office. At the time of departure of the index case from the office, there were an estimated 24 infectious doses per cubic meter of air. At this concentration of virus only a few minutes would be required for a susceptible person to inhale an infectious dose.

Foot and mouth disease (FMD) is a virus infection of cloven-hoofed animals caused by an RNA virus of 23 nm diameter (Falk and Hunt 1980). It is acid labile, but otherwise resistant to inactivation. There is little cross immunity among the 53 serotypes of FMD virus. Domestic animals, deer, goats, and antelopes may be infected. Animals are usually infected by the oropharyngeal route and the disease is manifested by epithelial vesicles developing on the tongue. Viremia may follow, but the mortality is usually low. The disease is widespread. Outbreaks occurred in the U.S. in 1929 leading to expensive slaughter programs, and it occurred in Mexico in 1946 where thousands of cattle were destroyed to control the infection.

Major outbreaks occurred in England in 1967-68 that became the basis for several critical epidemiologic studies. A feature of these outbreaks was the apparent long distance spread of the virus. Review of the circumstances of one series of outbreaks produced a hypothesis concerning the mechanism of transmission (Tinline 1970). It was conceived that small airborne particles (less than 10 μm in diameter) (Sellers et al. 1973) containing foot and mouth disease virus would be carried aloft by wind currents designated "lee waves," with the characteristic that viral particulates would be carried as a bolus, undiluted, to ground sites every 18-20 km, when a downward thrust of these air currents would occur (Fig. 7) and spread virus over animals and pastures. Because of horizontal movement of air, virus could penetrate into the interior of barns. Outbreaks of foot and mouth disease that coincided with the downward movement of these waves are shown on the figure. Another explanation not requiring the "lee wave" hypothesis is simply that a strong wind would raise small particles from infected animals and environmental sites and carry them for long distances (Hugh-Jones and Wright 1970). Diffusion of particles in air is a function of time rather than distance, so a strong wind could carry virus without much dilution over a considerable distance. Deposition would be aided by rain which clears the air of particles and deposits them on the ground surface where ground level aerosols may be produced to contaminate crops, fodder, and animals. The virus appears to be rather stable under cool, wet weather conditions since new infections have occurred up to 4 months after removal of infected animals from a site. During this series of epidemics in England, 443,000 domestic animals were slaughtered in a 120-day period in an effort to control the disease (Falk and Hunt 1980).

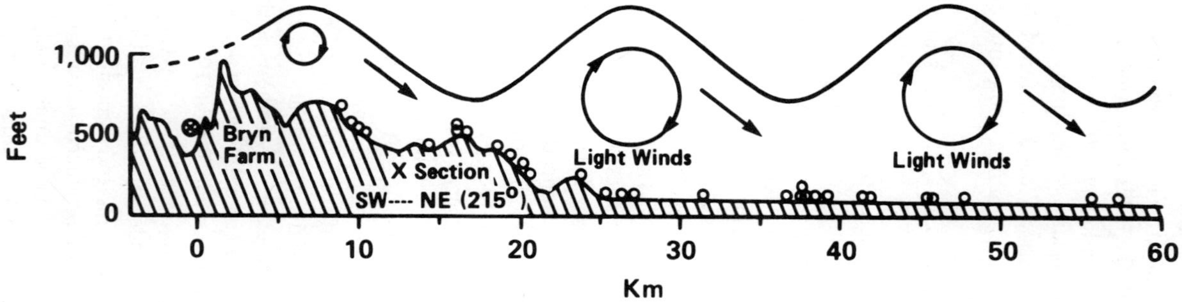

Figure 7
Profile through Bryn farm illustrating the lee wave hypothesis. Redrawn from Tinline (1970).

Spread of Virus Infection by Aerosol in Animal Care Facilities

Infection has always been a problem of animal and human health in animal care facilities. The airborne route is important as a mechanism of transmission of virus, and other infections in these locations as new agents and new hazards of infection periodically appear. Some estimate of the significance of airborne transmission in animal care facilities is revealed in the following studies.

McGarrity et al. (1976) and McGarrity and Dion (1978) studied airborne transmission of polyoma virus among mice in an animal care facility. Infected mice discharge large amounts of polyoma virus in urine, feces, and saliva. These viruses are also relatively resistant to dehydration and other environmental influences. In experiments in which mice were exposed to airborne infection and exposure by contact was excluded, acquisition of airborne infection was proportional to airborne exposure. McGarrity and coworkers determined that the 50% tissue culture infectious dose was about one, and that 50% of animals exposed developed infection after 9 hours of exposure, or after inhaling about 13 liters of air. Air sampling showed 1 $TCID_{50}$ per 22 liters of room air indicating about 50% recovery of infectious virus, a reasonable degree of agreement. The authors considered that the infectious particles were those less than 9.5 μm in diameter. Of interest was the finding that no seroconversion to polyoma virus was detected in 138 serum specimens from 44 persons involved in these studies.

McKissick et al. (1970) found that 39.5% of mice exposed to a small particle aerosol dose of 2675 $TCID_{50}$ of Rauscher murine leukemia virus developed leukemia during 25 months following aerosol exposure. The aerosol was produced in the Collison generator which probably produced particles averaging 1.5 μm in diameter. Only about 9% of animals given 27 $TCID_{50}$ developed leukemia during 25 months of observation. The doses of virus required to produce leukemia seem larger than doses of other viruses required to produce infection by the aerosol route.

Virus Survival in Aerosols

When viruses are nebulized into aerosols, the size of the particle is determined by the amount of energy utilized in the process. The relative humidity (RH) in the aerosol nebulizer is nearly 100% and particles when first produced by the Collison nebulizer are about 4 μm in diameter. When they enter ambient air with a relative humidity of 60-70%, their size is reduced in milliseconds to about 1.5 μm (95% with a diameter of less than 5 μm). After inhalation, they accrete water as they pass downward in the respiratory tract, and by the time they have reached the trachea, they will have again increased in size to about 4 μm in diameter in the high relative humidity of the lower respiratory tract.

Virus nebulized into aerosol undergoes reduction in concentration in proportion to the degree of dilution of the suspending medium. This dilution can be expressed as the reciprocal of the concentration of virus per liter in aerosol divided by the concentration of virus per liter in nebulizer fluid (spray factor). This factor usually exceeds 10^5. Dissolved salts in nebulizer fluids will increase in concentration as well when the particle equilibrates to a lower relative humidity, and decreases after inhalation when rehumidification occurs.

Viruses can be distinguished by their survival in different relative humidities according to the presence or absence of a lipid envelope. deJong et al. (1973) summarized data from several sources which show that lipid-free RNA and DNA viruses undergo a major loss of viability in 1 hour in an aerosol with low relative humidity (< 50%), but are relatively stable at high humidity (> 70%) (Fig. 8). Lipid enveloped RNA and DNA viruses are more stable in aerosol than nonenveloped viruses, but lose some viability at high relative humidity (> 50%). There are exceptions to the generalization due at least in part to the effects of solutes present in the nebulizer fluid, which increase in concentration when the aerosol particles undergo

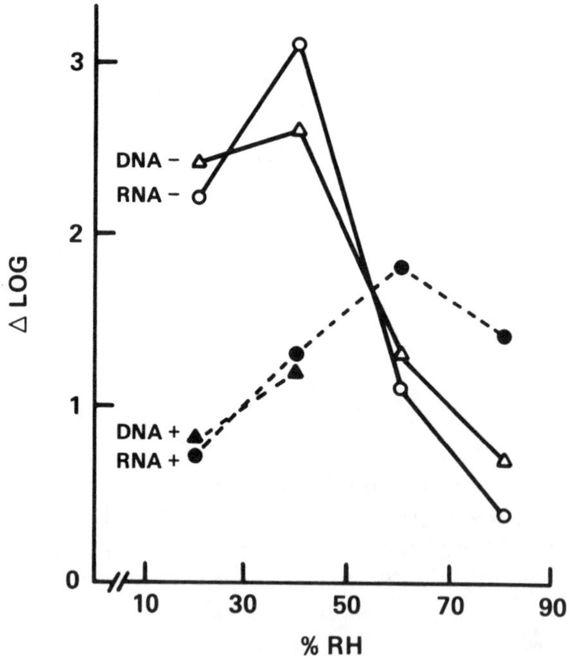

Figure 8
Inactivation of viruses in aerosols (pooled literature data). Viruses sprayed from unpurified suspensions. (Δ log t) The difference between log titre calculated and log titre found 1 hr after atomization; (DNA –) DNA-containing, lipid-free viruses; (RNA –) RNA-containing, lipid-free viruses; (DNA +) DNA-containing, lipid-containing viruses; (RNA +) RNA-containing, lipid-containing viruses. Redrawn from deJong et al. (1973).

evaporation in lower relative humidities. Benbough (1971) found that increasing the sodium chloride concentration up to 0.5% in the nebulizer fluid resulted in rapid loss of viability of Langat virus (a group B enveloped RNA virus) in aerosol within 5 minutes. Serum albumin, 0.1%, added to the nebulizer fluid prevented this loss. The polyhydroxy compounds inositol, sorbitol, and glucose stabilized the virus held in aerosol at 50% relative humidity for periods up to 1 hour, in contrast to otherwise nearly complete inactivation without these substances. Webb et al. (1963) tested the effect of inositol on survival of Rous sarcoma virus in aerosol (Fig. 9). This virus was unstable in aerosol nebulized from water at low humidity, but was quite stable at humidities above 70%. Citrate buffer (0.005 M, pH7) improved stability at low humidity while inositol (6.0% in nebulizer fluid) stabilized the virus in aerosol through the full range of relative humidities.

Adenovirus, type 12, exhibits a response to changing humidity in aerosols typical of nonenveloped viruses (Davis et al. 1971). At 89% RH, $10^{6.7}$ $TCID_{50}$ of virus was recovered from aerosol, compared to $10^{6.0}$ and $10^{4.3}$ at 51% and 32% RH, respectively. Newborn hamsters were exposed to aerosols of the virus at these humidities, and assays of the lungs showed smaller but proportionate amounts of virus.

Most workers have presumed that the changing survival of viruses resulting from nebulization and storage in aerosol at various relative humidities is due to gain or loss of water molecules from the surface of the virus particles. The other major target of the effect of aerosolization would be nucleic acids within the virus. deJong and Winkler (1968) compared the inactivation of poliovirus type 1 and its nucleic acid at intervals of up to 1 hour exposure in aerosol (Fig. 10). Preparations were

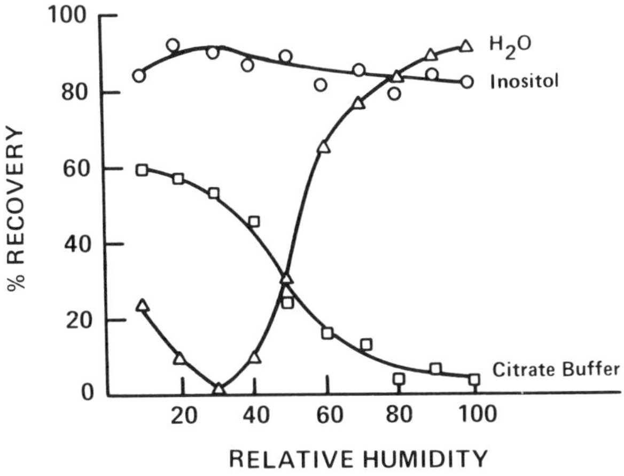

Figure 9
The effect of relative humidity and chemicals on the 5-hr survival of Rous sarcoma virus in an aerosol. Redrawn from Webb et al. (1963).

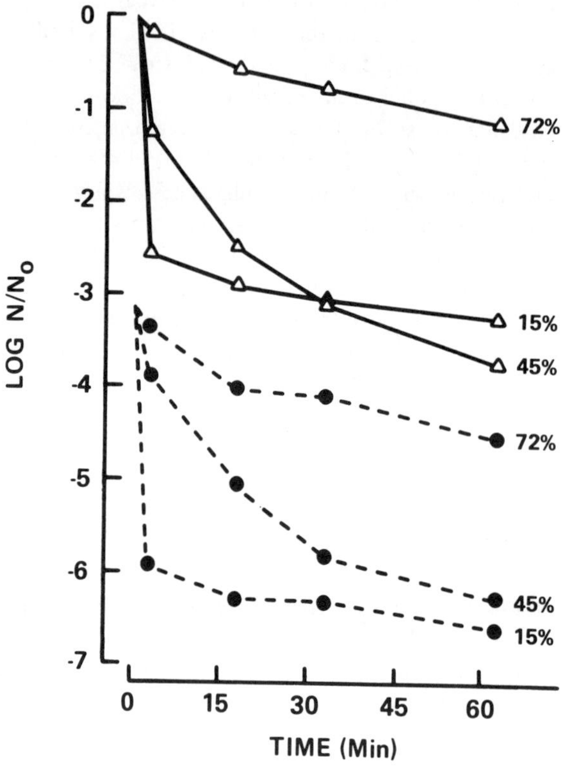

Figure 10
Comparison of the decay of poliovirus in air with the inactivation of the RNA in the virus particles. The mean titres of four experiments are plotted against time. (△) Virus; (●) iRNA. Redrawn from deJong and Winkler (1968).

tested at relative humidities of 72%, 45%, and 12%. Consistent with the principle that nonenveloped viruses survive better at high relative humidity in aerosol, the least decrease in viability occurred at 72 RH, and was progressively greater with decreasing RH in the aerosol aging chamber. Viral nucleic acid infectivity was reduced parallel to loss of virus viability. However, deJong et al. (1973) studied another small, lipid-free RNA virus, encephalomyocarditis virus (Fig. 11), and found that virus viability and hemagglutinating activity diminished 100-fold within a few minutes after nebulization, whereas infectious RNA, sprayed into aerosol in purified form or separated from whole virus after exposure in aerosol, retained all of its infectivity. Akers and Hatch (1968) also found that encephalomyocarditis virus RNA in aerosol was not inactivated by relative humidities that inactivated intact virus.

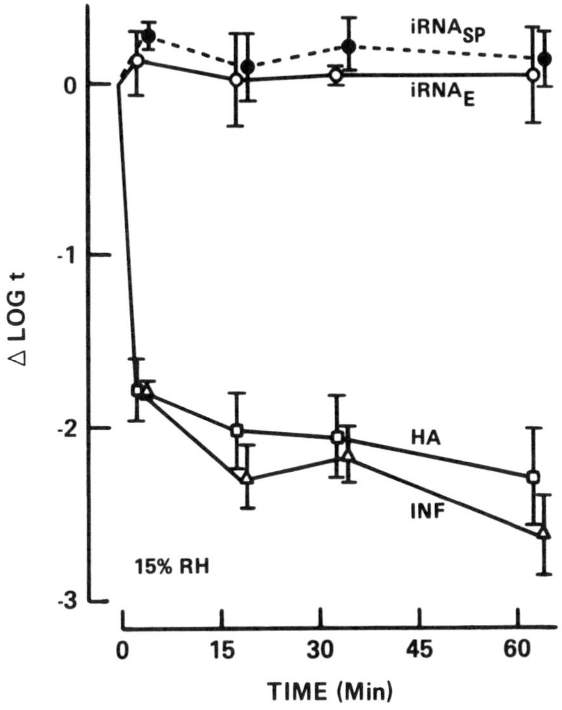

Figure 11
Inactivation of EMC-virus in aerosols. (Δ log T) The difference between log titre calculated (being about 8 for virus and 5 for iRNA) and log titre found; (INF) infectivity of complete virus [PFU]; (HA) hemagglutinating activity [sheep erythrocytes]; (IRNA$_E$) infectious RNA extracted by phenol from the impinger fluid; (IRNA$_{SP}$) infectious RNA sprayed in the free state. Redrawn from deJong et al. (1973).

deJong et al. (1976) in another study examined survival of Semliki Forest virus (an enveloped RNA virus) and its nucleic acid in aerosol (Fig. 12). In contrast to the result with poliovirus, they found that nucleic acid activity remained virtually unchanged during an exposure period of 1 hour at 50% RH and that extracted infectious RNA also retained its activity during such exposure. However, there was a steady decline in viability and hemagglutinating activity of whole virus during exposure. Dubovi (1971) found no loss of activity of the nucleic acids of the DNA-containing coliphage ϕX-174 and the RNA-containing coliphage MS-2 in aerosol. Thus, in general, inactivation of virus in aerosols appears to be an effect of humidity on the surface structures of the virion and not on the internal nucleic acids.

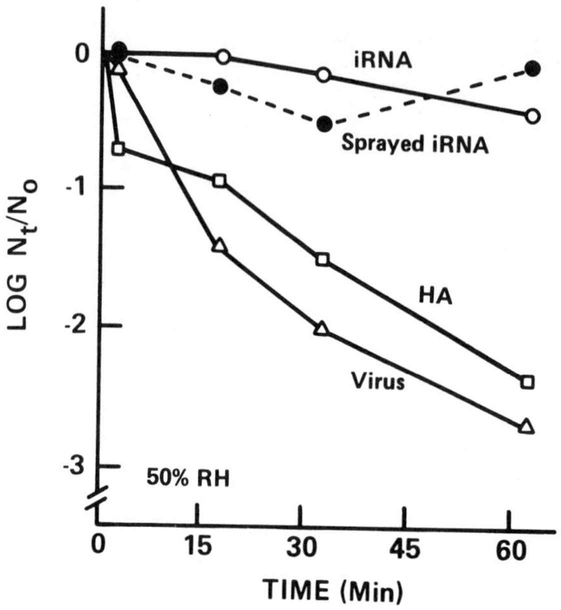

Figure 12
Inactivation of SFV, SFV-HA, and SFV-RNA in aerosols at 50% RH. After spraying purified SFV, aerosol samples were taken and titrated on virus infectivity, on HA, and, after extraction with phenol, on iRNA (○). After spraying of free SFV-iRNA, samples were assayed for iRNA (●). Redrawn from deJong et al. (1976).

DISCUSSION AND SUMMARY

We have described several outbreaks of viral infection in man and animals that, based on epidemiologic evidence, appear to have been transmitted by airborne virus. Experimental studies in man established that coughs and sneezes produce large populations of particles small enough to remain airborne for long periods. Studies with normal volunteers revealed the presence of virus in coughs and sneezes, and airborne transmission was demonstrated in an experiment with volunteers.

Studies of virus survival in aerosol showed that lipid enveloped viruses survive best in low relative humidity; nonenveloped viruses survive best in high relative humidity. Exceptions to this generalization appear to result from effects of salts and other substances in nebulizer fluid that are concentrated when particles reduce in size by evaporation in ambient humidity. Inositol and other polyhydric compounds protected viability of viruses in aerosols. Viral nucleic acids tolerated nebulization and storage in aerosol with little loss of activity.

There is a massive occurrence of airborne infections, but limitations of current experimental methods have prevented regular detection of the small amounts of airborne virus that must be responsible for the infections that occur. The problem can be seen more in context when it is appreciated that the lung surface of an adult comprises an area of about 70 square meters (Weibel 1963). In this area and in the upper respiratory tract, about 70% of small particles in the 900 liters per hour of air inhaled by the average adult human will be deposited. Given the high infectivity of respiratory viruses, one can see that the respiratory tract is a sensitive sampling device for airborne viruses. However, with large volume samplers that are currently used to collect aerosol, a large proportion of infectivity of airborne viruses may be destroyed. Thus, alternative procedures should be sought, such as recovery of virus-specific nucleic acids from large volume air samples. There could be many possible benefits of improved knowledge of the quantitative aspects of airborne transmission of viral infection.

REFERENCES

Akers, T.D. and M.T. Hatch. 1968. Survival of a picornavirus and its infectious ribonucleic acid after aerosolization. *Appl. Microbiol.* **16**: 1811.

Benbough, J.E. 1971. Some factors affecting the survival of airborne viruses. *J. Gen. Virol.* **10**: 209.

Couch, R.B., R.G. Douglas, Jr., K.M. Lindgren, P.J. Gerone, and V. Knight. 1970. Airborne transmission of respiratory infection with coxsackievirus A type 21. *Am. J. Epidemiol.* **91**: 78.

Davis, G.W., R.A. Griesemer, J.A. Shadduck, and R.L. Farrell. 1971. Effect of relative humidity on dynamic aerosols of adenovirus 12. *Appl. Microbiol.* **21**: 676.

de Jong, J.C. and K.C. Winkler. 1968. The inactivation of poliovirus in aerosols. *J. Hyg.* **66**: 557.

de Jong, J.C., T. Trouwborst, and K.C. Winkler. 1973. Mechanisms of inactivation of viruses and macromolecules in air. In *Airborne transmission and airborne infection. VIth international symposium on aerobiology.* (ed. J.F.Ph. Hers and K.C. Winkler), p. 126. Oosthoek Publishing, Utrecht, The Netherlands.

de Jong, J.C., M. Harmsen, A.D. Plantinga, and T. Trouwborst. 1976. Inactivation of Semliki Forest virus in aerosols. *Appl. Environ. Microbiol.* **32**: 315.

Dubovi, E.J. 1971. Biological activity of the nucleic acids extracted from two aerosolized bacterial viruses. *Appl. Microbiol.* **21**: 761.

Duguid, J.P. 1946. The size and the duration of air carriage of respiratory droplets and droplet nuclei. *J. Hyg.* **44**: 471.

Falk, L.A., Jr. and R.D. Hunt. 1980. Overview of airborne contagion in animals. Airborne contagion. *Ann. N.Y. Acad. Sci.* **353**: 174.

Gerone, P.J., R.B. Couch, G.V. Keefer, R.G. Douglas, E.B. Derrenbacher, and V. Knight. 1966. Assessment of experimental and natural viral aerosols. *Bacteriol. Rev.* **30**: 576.

Giacomelli-Maltoni, G., C. Melandri, V. Prodi, and G. Tarroni. 1972. Deposition efficiency of monodisperse particles in human respiratory tract. *Am. Ind. Hyg. Assoc. J.* **33**: 603.

Hugh-Jones, M.E. and P.B. Wright. 1970. Studies on the 1967–8 foot and mouth disease epidemic. *J. Hyg.* **68**: 253.

Knight, V. 1973. *Viral and mycoplasmal infections of the respiratory tract*, p. 1. Lea and Febiger, Philadelphia, Pennsylvania.

Landahl, H. 1972. The effect of gravity, hygroscopicity and particle size on the amount and site of deposition of inhaled particles, with particular reference to hazard due to airborne viruses. In *Assessment of airborne particles* (ed. T. Mercer, P.E. Morrow, and W. Stober), p. 421. Charles C. Thomas, Springfield, Illinois.

Lippmann, M., D.B. Yeates, and R.E. Albert. 1980. Deposition, retention, and clearance of inhaled particles. *Br. J. Ind. Med.* **37**: 337.

McGarrity, G.J. and A.S. Dion. 1978. Detection of airborne polyoma virus. *J. Hyg.* **81**: 9.

McGarrity, G.J., L.L. Coriell, and V. Ammen. 1976. Airborne transmission of polyoma virus. *J. Natl. Cancer Inst.* **56**: 159.

McKissick, G.E., R.A. Griesemer, and R.L. Farrell. 1970. Aerosol transmission of rauscher murine leukemia virus. *J. Natl. Cancer Inst.* **45**: 625.

Moser, M.R., T.R. Bender, H.S. Margolis, G.R. Noble, A.P. Kendal, and D.G. Ritter. 1979. *Am. J. Epidemiol.* **110**: 1.

Remington, P.L., W.N. Hall, I.H. Davis, A. Herald, and R.A. Gunn. 1985. Airborne transmission of measles in a physician's office. *J. Am. Med. Assoc.* **253**: 1574.

Schlesinger, R.B. and M. Lippmann. 1972. Particle deposition in casts of human upper tracheobronchial tree. *Am. Ind. Hyg. Assoc. J.* **33**: 237.

Sellers, R.F., D.F. Barlow, A.J. Donaldson, K.A.J. Herniman, and J. Parker. 1973. Foot and mouth disease, a case study of airborne disease. In *Airborne transmission and airborne infection. VIth international symposium on aerobiology.* (ed. J.F.Ph. Hers and K.C. Winkler), p. 405. Oosthoek Publishing, Utrecht, The Netherlands.

Tinline, R. 1970. Lee wave hypothesis for the initial pattern of spread during the 1967–68 foot and mouth epizootic. *Nature* **227**: 860.

Webb, S.J., R. Bather, and R.W. Hodges. 1963. The effect of relative humidity and inositol on air-borne viruses. *Can. J. Microbiol.* **9**: 87.

Weibel, E.R. 1963. *Morphometry of the human lung.* p. 69. Academic Press, New York, NY.

COMMENTS

FIELDS: I'd like to ask the same question that I asked last time. If you take genetically altered viruses, whether it be vaccine viruses or viruses altered for different purposes, and look at the capacity of virus to survive in the droplet or be transmitted, is there any database on which to draw. For example, take the polio vaccine virus or some of the flu vaccine viruses or any other examples—Are the vaccine viruses less capable of survival in the environment?

KNIGHT: That is a very valid question. Most of the studies I cited are several years old when this was not an issue. One could probably construct viruses more resistant to loss of activity during exposure as small airborne particles.

FIELDS: Since one of the issues that emerges as you examine the field of environmental virology is the issue of what can be predicted when you genetically manipulate a virus vis-a-vis its survival in the environment, can I suggest that the system you have described of environmental production of aerosols and of testing the capacity of viruses to transmit is a nice one to begin with to think of testing specific genetic manipulations of viruses on their survival and transmissibility.

KNIGHT: One answer to your question is that when you generate viruses into aerosol, they are not replicating; so it's the chemical resistance of the virus to destruction by these forces that we must study. In contrast to genetic alteration, chemical effects would probably not discriminate missense or small base pair changes in nucleotides. Certainly we need to know the chemical means of degradation. Since viruses are going to be airborne for eons, we ought to know a little more about effects of airborne transmission on virus nucleic acid and virus survival.

KILBOURNE: I think what's worth emphasizing is that there is a nice parallel between the survival in humidity chambers of the enteroviruses and influenza viruses reciprocally related to the lower or higher relative humidity and the epidemiology, that is, the survival of the enteroviruses is greater in the summertime under conditions of high relative humidity, and the perfect environment for transmitting flu is probably a New York City apartment house in winter with a relative humidity of 10. So, I think there is that parallel.

KNIGHT: This is widely written about. Maybe one could experimentally construct viruses to resist or not resist some of these forces.

KILBOURNE: The only point I was making is that there is reason to use the experimental model that already exists of the aerosol chamber because it does have a direct parallel with the epidemiology.

CHANOCK: Except in the tropics. For example, in Hong Kong and Singapore, big outbreaks often occur.

SUMMERS: You were talking about genetically engineering viruses to improve survivability and persistence. What aspect of the virus would you engineer?

KNIGHT: Empirically, you would manufacture some viruses and put them in an aerosol chamber and see which one did the best until you found some kind of a biochemical lead that determined this property. Other methods might do it more efficiently.

MILLER: This is a question that might better be asked of some of the others here perhaps, but clearly, for anybody who is doing epidemiologic testing or other field trials of these organisms, there is going to be the possibility that the preparations are contaminated with other adventitious agents. One example that comes to mind is that some of the early polio virus vaccines were contaminated with SV-40, which wasn't detected because the virus wasn't known at the time.

KNIGHT: I'm very familiar with that.

MILLER: How likely do you think these are from some common applications that you have had experience with? What are some of the ways that in field trials for commercial purposes, for example, you might want to look for contamination with adventitious agents.

KNIGHT: I don't know. Somebody else might like to speak to that. I think you can establish whether or not a culture is pure, and that should solve the problem.

SHOPE: The question is how sure would you want to be that, for example, your virus was homogeneous, and that there weren't subpopulations of various organisms.

KNIGHT: You might want to know what subpopulations died first or had more or less virulence. There may be some gradient of virulence associated with die-off in the aerosol chamber, or something of that sort. But it could be measured.

Distribution of Viruses in the Water Environment

JOSEPH L. MELNICK AND THEODORE G. METCALF
Department of Virology and Epidemiology
Baylor College of Medicine
Houston, Texas 77030

Viruses are abundantly present in human wastes. Well over 100 different viruses are excreted in human feces and urine, find their way into sewage, and become common water pollutants. They are listed in Table 1 together with the diseases that they cause. Enteric viruses are divided into several groups, based on morphologic, physical, chemical, and antigenic differences. The most commonly studied group in natural waters has been the enteroviruses, which include polioviruses, coxsackieviruses, and echoviruses. They have been the most studied because of the ease with which they can be isolated from sewage and assayed in the laboratory.

Hepatitis type A virus has been proven beyond a doubt to be responsible for waterborne epidemics traced to sewage contamination. The virus has been shown to be a small RNA-containing virus whose size and polypeptide composition place it in the enterovirus genus, where it has been classified as type 72.

Adenoviruses are also found in the feces and have been isolated from domestic sewage. These are large viruses whose genetic material is double-stranded DNA. They cause respiratory and eye infections and have been responsible for epidemics of eye infections among bathers in nonchlorinated swimming pools. Recently, three new noncultivable types of adenovirus have been recognized as a cause of acute gastroenteritis in children.

Reoviruses also have been isolated from sewage, but their role in human disease is not understood. They are large viruses (60–70 nm in diameter) with a double-stranded RNA genome.

Improved detection methods have made it possible to detect a number of gastroenteritis viruses. An important group, the rotaviruses, have been regularly found in sewage and, in one study in a developing country, also in tap water (Hung 1984). Rotaviruses—which also are double-stranded RNA viruses—are excreted in large quantities in the stools of children with diarrhea, are a major cause of severe childhood diarrheas, and also are associated with gastroenteritis in adults. In many developing nations, diarrhea is a major cause of death among young children.

The Norwalk virus and related agents are small (27 nm) viral particles that have been implicated in several waterborne outbreaks of gastroenteritis. They may be related to human caliciviruses, which also have been associated with diarrhea. Neither group has yet been grown in the laboratory, and their characterization is incomplete.

Table 1
Human Viruses Excreted into Wastewater

Virus group	Number of types	Disease caused
Enteroviruses:		
Poliovirus	3	Paralysis, meningitis, fever
Echovirus	31[a]	Meningitis, respiratory disease, rash, diarrhea, fever
Coxsackievirus A	23[b]	Herpangina, respiratory disease, meningitis, fever
Coxsackievirus B	6	Myocarditis, congenital heart anomalies, rash, fever, meningitis, respiratory disease, pleurodynia
Types 68–71	4	Meningitis, encephalitis, respiratory disease, acute hemorrhagic conjunctivitis, fever
Hepatitis A: type 72	1	Hepatitis type A (infectious hepatitis)
Gastroenteritis viruses:		
Norwalk	2	Epidemic vomiting and diarrhea, fever
Rotavirus	4	Epidemic vomiting and diarrhea, chiefly of children
Candidate viruses		
Astrovirus	Unknown	Not clearly established
Calicivirus	Unknown	Not clearly established
Coronavirus	Unknown	Not clearly established
Minireovirus	Unknown	Not clearly established
Small round viruses	Unknown	Not clearly established
Reoviruses	3	Not clearly established
Adenoviruses	37	Respiratory disease, eye infections
Parvoviruses		
Adeno-associated viruses	5	Unknown
Parvovirus B19	1	Aplastic anemia, fever, rash
Parvovirus RA	1	Rheumatoid arthritis (?)

[a]Echovirus types are 1–9, 11–27, 29–33. Echovirus 10 has been reclassified as reovirus, echovirus 28 as rhinovirus type 1A, and echovirus 34 as coxsackievirus A24.

[b]Coxsackievirus types are A1–A22, A24. Type A23 turned out to be the same virus as the one previously identified as echovirus 9.

Members of other virus groups are known to be excreted in feces of humans, and are candidate gastroenteritis viruses. Astroviruses, coronaviruses, minireovirus, and small round viruses suspected to be parvoviruses are potential pathogens that, like the Norwalk viruses, have yet to be grown in the laboratory, and their characterization is incomplete. Members of these groups have been associated with diarrhea and gastrointestinal morbidity in children and adults. In addition, papovaviruses and cytomegaloviruses are excreted in the urine and may also find their way into raw sewage.

Some enteric viruses are excreted in concentrations of more than one billion virus particles per gram of feces, and more than 100,000 infectious virus particles per liter have been detected in raw sewage. The amount of virus present in sewage is highly variable, depending on factors such as the hygienic level of the population, the prevalence of infection in the community, and the time of year. In the United States peak levels of enteroviruses are found in the late summer and early fall. Enteric viruses survive customary secondary sewage treatment and chlorination as commonly practiced, so that they can usually be isolated at all times of the year. In the United States the average concentration of enteroviruses in sewage appears to be about 100 per liter, but much higher concentrations may be found. In other parts of the world far greater concentrations of viruses have been observed. It appears that the average concentration of enteric viruses in sewage in less developed countries of the world may be 100 times that observed in this country. However, these concentration differences may reflect to some extent the higher per capita water consumption in industrialized countries and consequently a greater dilution of viruses in wastewater.

In much of the world, human wastes are discharged into natural waters, with little or no treatment. Thus, there is little reduction from the initial input of viruses. However, in the developed countries, a large part of sewage is processed by biological and physicochemical methods before discharge into receiving waters. During biological treatment (most often activated sludge), a 90% reduction in the initial concentration of enteroviruses may occur. Physicochemical treatment involving lime or the use of other coagulating agents causes a further reduction. However, it must be borne in mind that most reported studies were conducted using laboratory grown enteroviruses which appear to adsorb more readily to flocs than do the rotaviruses. In addition, the efficiency of treatment in removing enteroviruses varies somewhat among members of this virus group. Although treatment processes can result in large reductions in the concentration of viruses present in raw sewage, substantial numbers usually remain. Disinfection of treated sewage by chlorine is sometimes practiced. Although effective in the reduction of bacterial pathogens, this treatment does not altogether eliminate viruses, as they are more resistant than bacteria, particularly in the presence of organic substances. In some studies the concentration of viruses in sewage effluent has been greater than in influent, apparently because viruses attached to solids are protected from inactivation, and then become solubilized during their passage through the treatment plant.

It has long been established that there is a significant risk of contracting infectious diseases from ingestion of sewage-contaminated water. In the case of enteric bacterial diseases, spread by this route has been largely controlled by widespread application of bacterial standards for monitoring procedures, and modern treatments to eliminate bacteria from sewage effluents and from drinking water. However, enteric viruses are less effectively removed than bacteria by many treatments, particularly by disinfection; and viruses survive longer than indicator bacteria in the environment. Since viruses may be detected in water free of fecal coliforms and streptococci, the adequacy of bacterial standards for evaluating the sanitary quality of water has been questioned as it relates to its potential for transmission of viral diseases.

Wastewater often finds its way into sources of drinking water, as illustrated in Figure 1. There are several routes by which viruses in wastewater may be hazardous to human beings: through shellfish that grow in contaminated estuary water; through food crops grown in land irrigated with wastewater or fertilized and conditioned with sludge; through recreational waters; and by contamination of drinking water. Many communities use rivers as their sources of drinking water, and

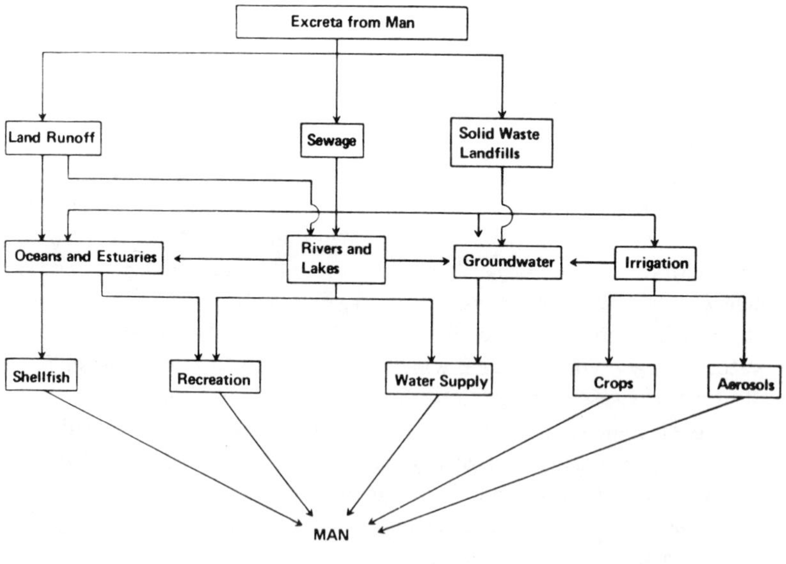

Figure 1
Routes of enteric virus transmission in the environment. Reprinted, with permission, from Melnick et al. (1978).

wastewater often is discharged into these surface waters. In most cases it is very difficult to identify the point source of viruses that are detected at any distance downstream from their discharge, but sometimes it has been done and the results show that viruses can travel long distances from the source of contamination and still be infectious. Viruses can remain viable for months in water, and longer when associated with solids found in water.

The study of the survival of enteric viruses under natural conditions in a subartic Alaskan river has contributed useful information. A site 200 miles downstream from all sources of domestic pollution was monitored. Moreover, ice totally covered the river, effectively sealing it against possible extraneous viral contamination. The mean flow time through the region was 7 days, and during this time 34% of the initial virus population survived the long trip. The ratio of enteric viruses to fecal indicator bacteria was not constant, clearly indicating that these bacteria were not a satisfactory measure of virus concentration.

Viruses also can contaminate groundwater or aquifers which serve as the water source for some communities. Over half of the waterborne disease outbreaks reported in the United States each year are caused by contaminated groundwater. Some of the factors that influence the entry, survival, and migration of viruses in groundwater are illustrated in Figure 2. For example, after spray irrigation of secondary sewage effluent onto a sandy soil, viruses can survive chlorination, sunlight, spraying, and percolation through several meters of soil.

Growing demands for available water resources by a rising world population and expanding industry are making recycling of wastewater almost inevitable, and indeed deliberate recycling is already being undertaken in some parts of the world. But in current situations in the United States, water often is being recycled inadvertently as one community pollutes the downstream water source of a second community.

Viruses, even if present in water in sufficient quantity to infect human beings, may be too few to detect readily. An important factor for detectability of a virus is whether it can be cultivated easily in the laboratory. For some viruses, cultivation has not yet been achieved, and other methods have not yet been developed to a point where they can be generally used for their detection in sewage or water. Thus, finding any type of human virus in drinking water is an indication that other viruses—including serious pathogens—may also be present.

Diseases caused by many of the enteric viruses are difficult to link positively to water sources because of the nature of the illnesses these viruses cause. Typically there are many healthy carriers. A person may acquire a viral infection by ingesting only a few virus particles in contaminated water, and then can spread the infection by other routes such as direct contacts, fecal-oral or respiratory transmission, or contamination of the environment (home, school). The infected contacts may develop serious or even fatal illness but their disease then appears to have no connection with the original contaminated water. Furthermore, when the infections

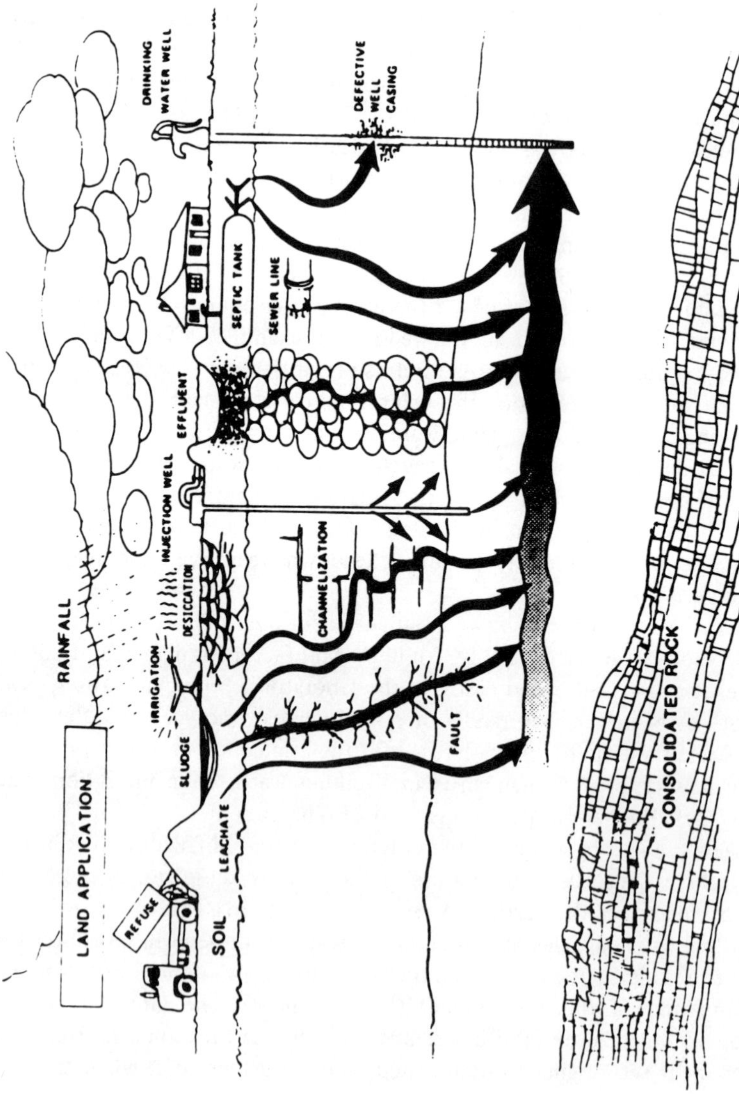

Figure 2
Factors affecting the entry, survival, and migration of viruses in groundwater. Adapted from Melnick and Gerba (1980).

do produce illness, the symptoms of infections by a single enterovirus type may vary from one patient to the next (e.g., meningitis in one patient, rash in another); therefore, the connection between cases of such illnesses is also obscure, making it even more difficult to document the original water source.

Low-level transmission of viruses by water or food may play a role in the spread of these agents between communities distant from each other, as when viruses discharged by one community contaminate the water source of another community farther downstream. Even if only a few individuals in the downstream community contract the disease directly by bathing or water consumption, they in turn can establish new foci of infection from which viruses can be spread by other routes (e.g., respiratory or fecal-oral).

Understanding environmental factors controlling enteric virus survival and transport in nature could be a key to defining the sudden occurrence of epidemics and the maintenance of infections within a population. Environmental conditions can greatly affect the survival and transport of viruses in nature, and there are many routes by which excreted virus may find its way back to the human host (Figs. 1 and 2).

Although viruses have been isolated in several countries from conventionally treated drinking water, this does not necessarily mean that these processes, when properly carried out, are inadequate for the removal of virtually all enteric viruses. In many studies, the actual water quality and water treatment conditions at the time when viruses were isolated were either not known or not stated. Additional field studies are necessary to clearly determine whether or not each step of the conventional treatment used for drinking water is adequate to guard against waterborne viral diseases. What is important to underscore is that even with our current limits on methods for their detection, viruses have been demonstrated in drinking water.

REFERENCES

Hung, T., C. Wang, Z. Fang, C. Chou, X. Chang et al. 1984. Waterborne outbreak of rotavirus diarrhea in adults in China caused by a novel rotavirus. *Lancet* i: 1139.

Melnick, J. L. and C. P. Gerba. 1980. Viruses in water and soil. *Public Health Rev.* **9**: 185.

Melnick, J. L., C. P. Gerba, and C. Wallis. 1978. Viruses in water. *Bull. W.H.O.* **56**: 499.

COMMENTS

CROWELL: You've just turned the pleasure of eating raw oysters into something very unpleasant. In the study in which you looked for survival of these different viruses, did you have any marker on the virus, such as a Sabin strain of virus?

METCALF: Yes, genetic markers were used.

CROWELL: Then these viruses really persisted for 5 months?

METCALF: Yes, no question about it?

CROWELL: What about dissecting the different tissues of the oyster or trying to identify where in the oyster you find these viruses? Has that been done?

METCALF: Yes. It is a tedious procedure. Viruses usually are found in greatest numbers in the hepatopancreatic tissue. A few may be found occasionally in adductor muscle tissue, and a few also may be found in hemolymph fluids.

CROWELL: There is presumably no virus replication going on.

METCALF: None, absolutely none. Virus carriage by shellfish is passive.

CROWELL: The only other comment I would make in this area is that a number of years ago we studied the survival of coxsackievirus B3 and human rhinovirus types 2 and 14 after drying on hard surfaces. Nobody here has mentioned that viruses may survive in the dried state in the environment for periods of time to permit transmission to humans or animals. We were surprised that a very significant amount of virus would persist for 24 hours on a dry Petri dish.

METCALF: I guess I would be too. We have not carried out experiments of that type. Generally speaking, we equate drying with inactivation of virus infectivity. I am surprised that either of the viruses you mentioned was infectious after 24 hours on a dry surface.

CROWELL: If you start with a high titer of virus and you get four or five logs of virus inactivation, that's a 99.9999% decrease, but you still have a large amount of infectious virus present.

METCALF: Enough virus may remain for infection of an individual despite inactivation of 99% or more of initial numbers. Decreases in virus numbers expressed as percent reductions should be interpreted carefully when a public health issue is involved.

CROWELL: That's right.

Epidemiology of Viruses Genetically Altered by Man—Predictive Principles

EDWIN DENNIS KILBOURNE
Department of Microbiology
Mount Sinai School of Medicine of the City University of New York
New York, New York 10029

OVERVIEW

Predictive principles for the epidemiology of viruses genetically altered by man can be derived from our knowledge of the epidemiology of contemporary viruses in nature and from the epidemiology of vaccine (genetically altered) viruses. Viruses in the laboratory and presumably in nature are intrinsically heterogeneous genetically. Viruses have evolved with their hosts and have adapted through millennia to specific ecologic virulence levels sufficient to ensure their propagation. Attenuated vaccine viruses retain partial or extremely reduced virulence and concomitantly do not survive in the environment. Genetic and ecologic constraints operate to prevent the emergence of viral "monsters" of unique virulence: Although virulence is polygenic; single point mutation can significantly alter the virulence of viruses in defined systems. Not only the epidemiology of altered viruses but of viral genes must be appreciated. Their epidemiology comprises an extended environment that includes not only cryopreserved laboratory strains but also an extended gene pool from which segmented RNA viruses can select from a repertoire of genes to produce instantly a novel virus of new genotype. Cloned viral genes must also be included in the total available gene reservoir.

INTRODUCTION

Viruses are the unwelcome but persistent evolutionary consorts of man whose visitations he would gladly forego. At the end of the 20th century only one human virus has been conquered and that one, smallpox virus, through no innovation of molecular biology, but by the tenacious and selective use of the oldest vaccine of all—the namesake of all other vaccines. If vaccinia represents a genetically altered cowpox or variola, the nature of that attenuation is not yet clear, but as a transmissible agent released into the environment, it remains an ancient prototype of the genetically altered virus. But the power of molecular biology can now alter vaccinia and other viruses nonempirically at a time when genetic determinants of viral virulence and pathogenicity are being defined at the molecular level. Is it possible, then, to derive some predictive principles for the anticipated epidemiological

Table 1
Basic Genetic Principles Relevant to the Epidemiology of Genetically Altered Viruses

A virus is intrinsically genetically heterogeneous.
Virus (RNA) mutation rates are high ($\simeq 10^{-5}$/replication) (overall genome $\simeq 10^{-1}$).
A virus is a consensus of a mixed population.
All laboratory viruses are selected mutants.
Disease severity seldom reflects viral genetic change.

behavior of viruses as yet unborn? This paper is an essay in that direction and must address an epidemiology with no past, an ambiguous present, and an unknown future.

Before considering the epidemiologic impact of deliberate genetic alterations of viruses by man, it seems appropriate to review the nature and epidemiologic significance of spontaneous viral variation (Table 1). It is not always appreciated that viruses intrinsically are genetically heterogeneous, with RNA viruses having mutation rates per gene of $\simeq 10^{-5}$ per replication. Holland and colleagues (1982) have even projected rates per genome as high as 10%. Thus, virus properties as we study them in the laboratory represent the consensus of a genetically inhomogeneous mixture of particles. Furthermore, the adaptation of a virus to laboratory cultivation inevitably selects mutants unrepresentative of that virus in nature. The fortuitous consequence of such empirical selection has been the development of all contemporary licensed live virus vaccines. Let us look at the epidemiology of these viruses, attenuated in virulence by laboratory passage if not by direct genetic alteration.

THE EPIDEMIOLOGY OF LIVE VIRUS VACCINES

Inoculation of a live virus vaccine involves not only infection of the immediate recipient, but the possibility of further transmission of the virus and its serial propagation in the community. Were a vaccine virus avirulent in all members of the population and the probability of reversion to virulence zero, then vaccine virus transmission would be desirable—the "contagious immunity" once envisioned by Albert Sabin. But frequent transmission is not usually the case. Of live virus vaccines currently licensed in the United States, only vaccinia and polio vaccine viruses are transmitted to any significant degree. It is probably significant that of the viruses listed in Table 2, only polio vaccine is administered by the natural route of infection. (The recent use of measles vaccine aerosols might well affect secondary transmission of that virus.) Experimental influenza virus vaccines are instilled in the upper respiratory tract by methods that do not simulate the small droplet nucleus deposition of virus thought to characterize natural influenza.

Table 2
Epidemiology of Vaccine Viruses (Attenuated Viral Mutants)

Vaccine virus	Spread	Transmissibility	Disease severity in secondary cases	Primary complications: incidence ($n/10^6$)	Increased viral virulence
Vaccinia	+	Low; selective (< 1%)	+ – +++	100	+[a]
Polio	+	High (8–60%)	+ – +++	1	+
Measles	0	None	–	–	–
Rubella	±[b]	Low	–	–	–
Mumps	0	None	–	–	–
Yellow fever	0	None	–	–	–
Influenza	±	Low	–	Low, variable	+

[a] With respect to laboratory markers.
[b] Horizontal and vertical

With vaccinia, transmissibility is low and selectively related to contacts with dermatoses. Vaccine-related complications in primary vaccinees are relatively high, whereas with polio vaccine transmissibility is high but complications infrequent. With vaccinia the acquisition of laboratory virulence markers has been observed in strains isolated from patients with vaccinial complications (Ehrengut et al. 1975). In the case of polio vaccines, reversion to increased human neurovirulence is undoubted. Multiple genetic changes in the vaccine viruses can occur after replication in humans (Kew et al. 1981). Even the low rate of rubella vaccine virus transmission has been worrisome because of the threat to the unborn. However, fetal infection is infrequent and teratogenic effects even in primary vaccine recipients are rare. Reversion of vaccine virus to virulence has not been documented.

The reduced level of replication and reduced viral shedding of infection with experimental influenza vaccines usually have precluded spread of this virus to contacts. Reversion of its mutants has been found in primary vaccinees.

It is notable that none of the vaccine viruses has become established in the environment. Replacement of wild-type polio virus by vaccine strains in sewage of the United Kingdom and the United States reflects the continuing administration and excretion of live virus vaccine rather than its serial propagation in the community.

CONSTRAINTS ON THE ASCENDENCY OF GENETICALLY DEVIANT VIRUSES

Natural variation appears to have explored all the potential mechanisms of genetic alteration of the viral genome including transitions, transversions, deletions, insertions, recombinations, and reassortment. Yet from this background, significantly changed viruses infrequently emerge. New diseases or variation in disease severity most often reflect altered host-virus relationships rather than intrinsic alteration of the viral genome.

As intimate and often obligate human parasites, viruses must operate under constraints that ensure their survival (Table 3). Extreme alterations of the genome will be lethal. More important is the fact that virus survival depends upon the mainte-

Table 3
Constraints on the Ascendancy of Genetically Deviant Viruses

Extreme alterations are lethal.
Virus survival requires a critical level of virulence.
Laboratory propagation in alien host systems tends to be attenuating.
Evolutionary adaptation to ecologic niches is exquisitely specific.
Pervasive human population immunity must be breached by major antigenic change.
Human infection with nonhuman (zoonotic) viruses rarely is contagious.

Table 4
Design for a Maximally Malignant "Monster" Virus (MMMV)

Attribute	Requirements	Viral archetype
Environmental stability	Non-enveloped; icosohedral nucleocapsid	Poliovirus
Antigenic mutability	RNA genome, preferably segmented	Influenza virus
Transmissibility	Infect lower respiratory tract → effective aerosol	Influenza virus
Intermediate virulence	Aerosol-generating infection without bronchiolar obstruction	Influenza virus
Wide host range	Pantropicity	Rabies virus
Persistence or latency	Integrating DNA genome; immunosuppressive	Herpes virus; HTLV

nance of a critical level of virulence. The introduction of myxomatosis virus into Australian rabbits after an initial lethal phase was followed by the evolution not only of resistant rabbits but a virus of intermediate virulence, able to sustain viremia sufficiently long to ensure its propagation by its arthropod vector (Fenner and Woodroofe 1965).

In experimental influenza in mice, transmissibility and virulence proved to be dichotomous (Schulman 1967). Aerosol transmission can be blocked by the production of inflammatory reaction capable of obstructing bronchioles (Table 4).

Adaptation to one species may be de-adapting to another, and species barriers are not readily hurdled by any but the more pantropic of zoonotic viruses such as arboviruses and rabies. Even in those cases, further transmission by contagion is rare. Especially in the case of those viruses that must replicate in both vertebrate and invertebrate hosts, viral genome capacity must be exquisitely tuned and probably incapable of tolerating large genetic alterations. Plant reoviruses adapted to continuous cultivation in plants experience deletion mutations and lose the ability to replicate in their insect vector (Reddy and Black 1977).

WORST CASE SCENARIO: DESIGN FOR A MAXIMALLY MALIGNANT (MONSTER) VIRUS (MMMV)

Because we know the primary structure of many viral genes and have the technical capacity to synthesize both genes and gene products, it would appear that the planned design of either friendly or unfriendly viruses is not too remote a possibility. If the latter should be the perverse goal of our paranoid society, can we construct a virus worse than rabies virus with its 100% case fatality rate or influenza virus with its pandemic potential for 20 million deaths worldwide? In hypothetical phenotype, yes. Our putative maximally malignant (monster) virus (MMMV) should have the environmental stability of poliovirus, the antigenic mutability of influenza virus, the unrestricted host range of rabies virus, and the latency or reactivation potential of a herpesvirus (Table 4). For maximal transmissibility MMMV, like influenza virus, should replicate in the lower respiratory tract and be disseminated by small particle aerosols, yet not so severely damage the bronchioles that aerosol generation is blocked. Survivors of the viruses' initial onslaught would be immunosuppressed and subject to persistent or reactivated infection that could renew the community burden of morbidity and mortality, and serve as foci of further transmission. Infection of other species would add to the economic losses posed by MMMV and establish additional reservoirs for its propagation and perpetuation.

If the phenotype is awesome, the creation of the MMMV genotype defies the macabre skills of Dr. Frankenstein, himself. One must somehow program a segmented RNA genome into a nonenveloped icosahedral nucleocapsid, providing it with a reverse transcriptase to make host-integrating DNA copies. Like influenza virus, MMMV must be equipped to capture novel external antigens elsewhere or (unlike any known agent) must create multiple new antigen epitopes while conserving protein structure, to circumvent population immunity (Fig. 1). But if the creation of MMMV is unlikely to the point of absurdity, we must also appreciate

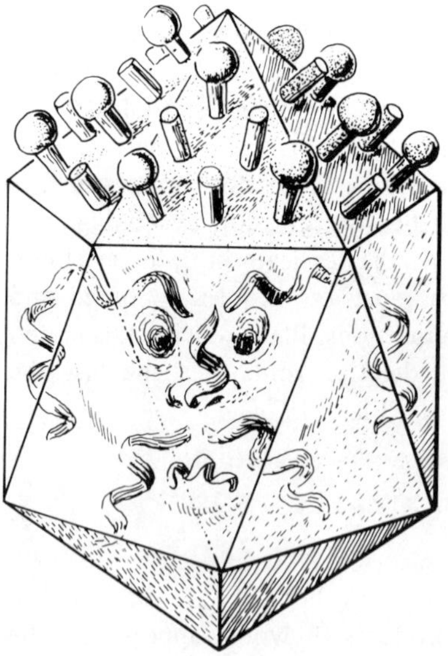

Figure 1
Genetic and structural constraints operate to contain the emergence of an extremely virulent virus.

Table 5
Single Point Mutations Can Be Critical in Alteration of Viral Properties

Virus	Gene	Mutation	Altered phenotype	Reference
Swine influenza	HA	Gly → Glu aa 155	↓ Infectivity for swine Antigenic change	Kilbourne 1978 Both et al. 1983
Influenza virus reassortant (X-31)	HA	Leu → Gln aa 226	Altered receptor Specificity	Rogers et al. 1983
Rabies	GP	Arg → X aa 333	↓ Virulence Antigenic change	Dietzschold et al. 1983
Reovirus	S1	σ1 polypeptide	↓ Neurovirulence Antigenic change	Spriggs et al. 1983

that in the proper setting single base changes may be sufficient to significantly influence viral properties, including virulence.

SINGLE POINT MUTATIONS CAN BE CRITICAL IN THE ALTERATION OF VIRAL PROPERTIES

The polygenic nature of virulence notwithstanding, single genes may be critical in altering a specific phenotype associated with attributes of virulence. A few representative examples of such mutations are shown in Table 5. These include single amino acid changes in glycoproteins of influenza and rabies viruses that pleiotropically alter virulence, receptor specificity and antigenic phenotype (Kilbourne 1978; Both et al. 1983; Dietzschold et al. 1983), and point mutations in the hemagglutinin of reovirus that determine neurovirulence (Fields and Spriggs 1982).

These studies suggest that engineered alterations of antigenic sites may have consequences related to pleiotropic effects on viral replication functions.

THE EPIDEMIOLOGY OF VIRAL GENES

The more imaginative vistas of genetic engineering are concerned less with the modification of the existing viruses than with the segregation and manipulation of single virus genes. Perforce, the epidemiology of viral genes in new environmental or even in different temporal context must be considered (Table 6).

The Extended Environment

The unexplained reappearance of a genetically unaltered 1951 influenza virus in 1977 (Nakajima et al. 1978) and present debate about the total elimination of laboratory housed variola have reminded us that our freezers are potential if unlikely sources of "new" pathogens. Such cryopreservation of viral dinosaurs constitutes

Table 6
Epidemiology of Viral Genes

The extended environment
Laboratory cryopreservation of extinct genes
The extended genotype
Special case of the segmented genome viruses
Parazoonotic gene transmission
Epidemiology of cloned viral genes
Genes out of context or in new combinations
Gene segregation and aberrant immunologic response
"Avirulent" genes in combination may → virulent phenotype

a fourth epidemiologic dimension that will assume increasing importance as efforts are directed toward the extinction of contemporary viruses. Thus, our laboratories are part of the epidemiological environment.

The Extended Genotype

Five groups of animal viruses bear their genes in separate RNA segments that are readily reassorted during dual infection with homologous viruses (Kilbourne 1981). The ecological significance of this genetic reassortment can be profound. The genes of influenza A viruses are distributed throughout a reservoir comprising domestic animals and feral migratory birds. Within this extended gene pool, gene combinations optimal for transmissibility, virulence, and antigenic dissimilarity can be continuously sampled to the extent that occasional interspecific transfer of virus can occur. There is little doubt that the pandemic influenza virus of 1968 involved the capture of a new hemagglutinin gene from an animal virus. The 1968 virus has since mated with the resurrected dinosaur of 1977 (Young and Palese 1979) to reach temporary ascendancy in the latter part of 1978.

Thus, although humans are rarely infected with animal influenza viruses, once interspecific transfer of virus occurs (in either direction), the reassortment mechanism provides for the creation of a virus in either host genetically equipped to replicate in both hosts and to transfer genes across the species barrier by what I shall call "parazoonotic" transmission.

There appears to be some precedent in nature for gene flow even among unrelated nonreassorting RNA viruses which could provide a "combination of abilities evolved in other circumstances" (Baltimore 1980). Recent evidence suggests that the extended gene pool for animal virus recombination may include plant viruses as well (Haseloff et al. 1984).

THE EPIDEMIOLOGY OF CLONED GENES

No consideration of the extended viral genotype can ignore the potential deployment of viral genes—especially those of reassorting viruses—in various vectors, cell lines, and viruses. Demonstration that a cDNA copy of poliovirus can infect mice (Racaniello 1984) sets a precedent for gene or genome retrieval even of RNA viruses. Again, the concern here is with the preservation of epidemiologically important viral genes beyond their natural tenure and out of context.

THE EPIDEMIOLOGICAL SIGNIFICANCE OF GENES OUT OF CONTEXT OR IN NEW COMBINATION

When antigenically hybrid influenza viruses bearing a hemagglutinin irrelevant to human experience are employed to induce neuraminidase-specific immunity, an anomalous immunologic response has been observed (Kilbourne 1976). That this unusual response may in fact be beneficial in avoiding the customary skewed response in favor of hemagglutinin antibody is not the point. Rather, it cautions about differences in immunization patterns that could be expected with segregated genes or gene products. The hypersensitizing but nonimmunizing effects of inactivated measles vaccine still linger in our memory.

Finally, the potential of unaltered genes in new combinations is unpredictable because such reassortment creates essentially new viruses. Although statistical probability favors the construction of viruses of reduced or intermediate virulence (Kilbourne 1969), laboratory reassortment of influenza viruses has produced viruses more virulent (by an arbitrary criterion) than either parental virus (Rott et al. 1984).

CONCLUSIONS

Viruses are intrinsically highly mutable, but in nature are for the most part constrained by the specificity of their replication and survival requirements from the attainment of extreme virulence. Any factitous combination of attributes for the generation of a maximally malignant virus could be confounded by problems of genetic nonhomology and structural incompatibility. However, single point mutations can profoundly alter viral properties and the results of new combinations of viral genes are difficult to predict. Engineered antigenic sites might pleiotropically influence viral biologic function.

Viral epidemiology in the future must take account of an extended environment that includes laboratory stocks of viruses and extinct viral genes preserved from past years. Manipulation of the genes of viruses capable of genetic reassortment must not be undertaken without awareness that these viruses comprise an extended genotype or gene reservoir available for interspecific viral gene transmission. Cloned viral genes must also be considered as a part of the extended viral genotype and the possible aberrant immunologic effects of isolated genes and their products should be recognized.

REFERENCES

Baltimore, D. 1980. Evolution of RNA viruses. In *Genetic variation of viruses. Annals N.Y. Acad. Sciences* (ed. P. Palese and B. Roizman), vol. 354, p. 492. New York Academy of Sciences, New York.

Both, G.W., C.H. Shi, and E.D. Kilbourne. 1983. Hemagglutinin of swine influenza virus: A single amino acid change pleiotropically affects viral antigenicity and replication. *Proc. Natl. Acad. Sci. U.S.A.* **80**: 6996.

Dietzschold, B., W.H. Wunner, T.J. Wiktor, A.D. Lopes, M. Lafon, C.L. Smith, and H. Koprowski. 1983. Characterization of an antigenic determinant of the glycoprotein that correlates with pathogenicity of rabies virus. *Proc. Natl. Acad. Sci. U.S.A.* **80**: 70.

Ehrengut, W., D.E. Sarateanu, U. Alswede, A. Habib, and G. Tetzlaff. 1975. Vaccinia virus variants as presumable cause of vaccinial complications. *Arch. Virol.* **48**: 229.

Fenner, F. and G.M. Woodroofe. 1965. Changes in the virulence and antigenic structure of strains of myxoma virus recovered from Australian wild rabbits between 1950 and 1964. *Aust. J. Exp. Biol. Med. Sci.* **43**: 359.

Fields, B.N. and D.R. Spriggs. 1982. Attenuated reovirus type 3 strains generated by selection of haemagglutinin antigenic variants. *Nature* **297**: 68.

Haseloff, J., P. Goelet, D. Zimmern, P. Ahlquist, R. Dasgupta, and P. Kaesberg. 1984. Striking similarities in amino acid sequence among nonstructural proteins encoded by RNA viruses that have dissimilar genomic organization. *Proc. Natl. Acad. Sci. U.S.A.* **81**: 4358.

Holland, J., K. Spindler, F. Horodyski, E. Grabau, S. Nichol, and S. VandePol. 1982. Rapid evolution of RNA genomes. *Science* **215**: 1577.

Kew, O.M., B.K. Nottay, M.H. Hatch, J.H. Nakano, and J.F. Obijeski. 1981. Multiple genetic changes can occur in the oral poliovaccines upon replication in humans. *J. Gen. Virol.* **56**: 337.

Kilbourne, E.D. 1969. Future influenza vaccines and the use of genetic recombinants. *Bull. W.H.O.* **41**: 643.

———. 1976. Comparative efficiency of neuraminidase-specific and conventional influenza virus vaccines in the induction of anti-neuraminidase antibody in man. *J. Infect. Dis.* **134**: 384.

———. 1978. Genetic dimorphism in influenza viruses: Characterization of stably associated hemagglutinin mutants differing in antigenicity and biological properties. *Proc. Natl. Acad. Sci. U.S.A.* **75**: 6258.

———. 1981. Segmented genome viruses and the evolutionary potential of asymmetrical sex. *Perspect. Biol. Med.* **25**: 66.

Nakajima, K., U. Desselberger, and P. Palese. 1978. Recent human influenza A

(H1N1) viruses are closely related genetically to strains isolated in 1950. *Nature* **274**: 334.

Racaniello, V.R. 1984. Poliovirus type II produced from cloned cDNA is infectious in mice. *Virus Res.* **1**: 669.

Reddy, D.V.R. and L.M. Black. 1977. Isolation and replication of mutant populations of wound tumor virions lacking certain genome segments. *Virology* **80**: 336.

Rogers, G.N., J.C. Paulson, R.S. Daniels, J.J. Skehel, I.A. Wilson, and D.C. Wiley. 1983. Single amino acid substitutions in influenza haemagglutinin change receptor binding specificity. *Nature* **304**: 76.

Rott, R., C. Scholtissek, and H.-D. Klenk. 1984. Alterations in pathogenicity of influenza virus through reassortment. In *Modern approaches to vaccines. Molecular and chemical basis of virus virulence and immunogenicity* (ed. R.M. Chanock and R.A. Lerner), p. 345. Cold Spring Harbor Laboratory, Cold Spring Harbor, New York.

Scholtissek, C., W. Rohde, V. von Hoyningen et al. 1978. On the origin of the human influenza virus subtypes H2N2 and H3N2. *Virology* **87**: 13.

Schulman, J.L. 1967. Experimental transmission of influenza virus infection in mice. IV. Relationship of transmissibility of different strains of virus and recovery of airborne virus in the environment of infector mice. *J. Exp. Med.* **125**: 479.

Young, J.F. and P. Palese. 1979. Evolution of human influenza A viruses in nature: Recombination contributes to genetic variation of H1N1 strains. *Proc. Natl. Acad. Sci. U.S.A.* **76**: 6547.

Spriggs, D.R. and B.N. Fields. 1982. Attenuated reovirus type 3 strains generated by selection of haemagglutinin antigenic variants. *Nature* **297**: 68.

Spriggs, D.R., R.T. Bronson, and B.W. Fields. 1983. Hemagglutinin variants of reovirus type 3 have altered central nervous system tropism. *Science* **220**: 505.

COMMENTS

SHOPE: Are you speaking only of segmented viruses when you talk about the hazard of having a cloned virus gene circulating in nature?

KILBOURNE: In my paper, I have referred to some of the recent work which illustrate the preservation and sharing of nonstructural protein genes by animal and plant viruses.

MARTIN: Besides the change in the hemagglutinin, were there changes in any other viral genes?

KILBOURNE: That really can't ever be said with certainty unless you do a complete sequencing on everything you get out and recognize the inhomogeneity of what you're getting out. I think in this case, on the basis that revertants at that position 155 switched all the properties, that it is 99.9%

certain. Then there is other evidence; that is, you can repetitively isolate these variants from different swine virus strains and they always have the lesion at that point. If you look at enough of them, you are going to find some changes at other points, but I think you prove the point by looking at revertants.

MARTIN: It is my impression from what I have seen with other viral systems that when you look for genetic determinants of virulence or tropism, several genes are involved. A particular phenotype is the product of multiple changes, and it is only the rare situation where the single change is the culprit.

FIELDS: First of all, the change in the Reo $\sigma 1$ protein associated with loss of virulence is lysine to glutamic acid; it is a single amino acid change in that protein. Second, the proof that that is the only change is a combination of the sequencing and genetics, i.e., you can take the $\sigma 1$ gene and put it into another background and show that in the crosses into other genetic backgrounds, the altered tropism phenotype, the attenuating phenotype, is conferred by the $\sigma 1$ protein.

KILBOURNE: I've forgotten my own work. That's what we did, as well. We first demonstrated the change in a reassortant, which was used for vaccine production, and then went back to wild-type virus and found the same thing. I still think that the point is valid; one always has to be concerned that something may be going on in the background.

AHMED: I was a little confused by your statement that if you used a cloned gene, you would get an altered or aberrant immune response. Would you elaborate on that?

KILBOURNE: I'm afraid that was about three jumps ahead of things, including some of the evidence. But what I'm thinking of here specifically is that we have some evidence with a neuraminidase-specific vaccine that we are studying, which shows that if we couple the neuraminidase with an irrelevant hemagglutinin, we induce a better immune response in the neuraminidase than if it is paired with a hemagglutinin which the population has seen. We apparently get antigenic competition and a skewed immune response. Now, this is relevant to the cloning of *HA* genes for vaccines that would include only a single viral protein. I think that single antigen might perpetuate the skewed response.

CHANOCK: You made the point that a cloned gene might be rescued subsequently. I think you would have to limit such an occurrence to genes of positive strand RNA viruses, because with negative strand RNA viral genes, a number of heroic attempts have been made to rescue cloned DNA in an RNA form and transfer it back into a negative strand virus. All such attempts have failed. Many scientists have tried but none have succeeded.

KILBOURNE: It remains a possibility.

CHANOCK: It is a possibility, but I think it is an extremely low-risk experiment.

KILBOURNE: I agree.

MARTIN: I don't want to ask you a loaded question, but if John Jones were playing with a molecular biological erector set in the influenza system, are there particular experiments involving the hemagglutinin gene that you would be concerned about?

KILBOURNE: I think it would be irresponsible, for example, to take hemagglutinin from an animal virus that has not yet produced disease in the human population and to use that in a live virus vaccine—the hemagglutinin, not any of the other genes—because I think we know that that particular gene and gene product are what chiefly induce specific immunity. It's only a change in that gene that is required to produce a pandemic—although what I suspect is that much more is required to produce a pandemic. I think it is terribly difficult to create a transmissible virulent virus. One of the problems that people working with live virus vaccines have is keeping the level of virulence up sufficiently to retain immunogenicity.

CHANOCK: I think you're absolutely right; but I would challenge you on two points.

KILBOURNE: Only two?

CHANOCK: The problem is I'm on your home court. However, I will cite two observations. First, I agree with you that it is very rare that a virus catastrophe develops from a single-point mutation—but it has happened within the last 2 years in the United States on the east coast.

KILBOURNE: I think Alan [Kendal] is going to discuss that.

CHANOCK: It is clear that an avian influenza A virus underwent an abrupt change in virulence from a mortality of less than 1% to a mortality of greater than 80%. The virus was initially highly transmissible in chickens and the increase in mortality occurred within a very short period of time. The change can be attributed to a single mutation which resulted in the loss of a glycosylation site on the hemagglutinin molecule. The change apparently resulted in greater exposure of the cleavage site on the hemagglutinin, thereby allowing this glycoprotein to be cleaved more readily. For example, cleavage could occur in vivo in tissues in which it ordinarily would not occur. A reasonably compelling case has been made for the explanation that increased cleavability of the hemagglutinin led to a broadening of tissue tropism with resultant spread of infection throughout the host.

KILBOURNE: But nobody did that.

CHANOCK: No, but it happened in nature. In any case, this abrupt change in virulence defines the outer limit of adverse effects that can result from a single-point mutation.

KILBOURNE: We wouldn't have the sense to make a manipulation at that site.

CHANOCK: The importance of the abrupt increase in virulence of the avian influenza A virus is that it happened and it might happen with other related viruses.

SHOPE: Are you saying by that it is irresponsible to pressure a virus with monoclonal antibodies to make a mutation because that might get loose and do something?

CHANOCK: Many different neutralization-resistant virus mutants have been selected by exposure of influenza A virus to a wide variety of neutralizing monoclonal antibodies, but the change that occurred with the avian influenza A virus has not been observed under these conditions. Fortunately, the cleavage site itself does not appear to be antigenic and hence neutralizing monoclonal antibodies directed against it have not been isolated. However, one must consider the possibility that a change in another part of the hemagglutinin molecule could bring about greater exposure of the cleavage site. The message of this discussion is that hemagglutinin mutants that are more readily cleaved should be placed on the danger list and appropriate measures should be taken to prevent their release from the laboratory.

KENDAL: I have to respond to that by stating that where we were trying to develop in vitro mutants of influenza hemagglutinin by selection of monoclonal antibodies, we insisted on using class III containment procedures. So, in answer to your question, I think that is a standard of responsibility which should be addressed.

FIELDS: You made a point that I think is a very important one—that virulence and transmissibility are separable and should be thought of as separate issues. Is there any information about the genetic basis for the transmissibility, or any mechanistic insight into how virulence and transmissibility differ? Does it involve a stability property or spread? What about the genetic makeup of the virus? Is it a property that maps to a gene or genes?

KILBOURNE: I think it is time for us to return to that particular problem. Jerry Schulman and I worked on this in terms of reassortment, using reassortants of the H2N2 virus, which appeared to be an unusually transmissible virus in mice as compared with the older strains. We related it to the hemagglutinin, but we didn't get farther than that.

FIELDS: So, from the point of view of the theme of this volume, you are saying that there are markers in flu that should now make that type of experiment potentially feasible and provide important insights into transmissibility?

KILBOURNE: It may be still more complicated than that. The only thing I can say that we have firm evidence on is that we had this inverse correlation between the degree of pathologic reaction and transmissibility. It was almost paradoxical, in the sense that the more inflammatory the reaction, the less transmissible the virus. Maximal transmissibility occurred at the peak of viral replication, preceding the closure of bronchioles with exudate. Maximally virulent strains with respect to cytopathology were less transmissible. Transmissibility is probably multigenic, involving viral adaptation.

FIELDS: Interesting.

The Effect of Influenza Virus Genetic Alteration on Disease in Man and Animals

ALAN P. KENDAL, KUNG JONG LUI, NANCY J. COX, AND KARL D. KAPPUS
Influenza Branch
Center for Infectious Disease
Centers for Disease Control
Atlanta, Georgia 30333

OVERVIEW

Quantitative data about the incidence of influenza disease in man and the host and viral factors affecting disease are discussed to establish the reasons underlying variation in virus epidemic activity. Examples of knowledge gained from molecular studies with animal influenza viruses and attenuated human live influenza vaccine strains are given to illustrate the possibility for small numbers (in some instances only single) amino acid substitutions to affect virulence or host range. Based on knowledge that influenza viruses in nature are not genetically homogeneous, but that even single isolates contain subpopulations of virions with minor genetic differences, it is proposed that this situation provides a mechanism for self-regulation by influenza viruses which maintains disease incidence within a range that is optimal for survival of the virus.

INTRODUCTION

Correlating the incidence of disease with the genetic composition of its causative virus is most realistically an objective, not an existing accomplishment. There are few instances where the disease potential of a virus is fully explainable on the basis of protein structure and function. Rather than discuss available knowledge for a variety of virus families, influenza has been selected as the topic of this paper because of the large amount of information that has become available about its major surface glycoprotein, the hemagglutinin (HA), following determination of its three-dimensional structure, as well as the existence of considerable quantitative information about the epidemiology of the virus in man, and level of disease produced in well-controlled animal models.

Because influenza viruses are renowned for their genetic variation, it is perhaps paradoxical to refer to influenza as a single disease. Kilbourne has on occasion stated that influenza is a "constant disease caused by a variable virus" (Kilbourne 1975). We shall see, however, that the constancy of the disease is arguable. It may, of course, be the case that, at least for man, the disease caused by epidemic strains of influenza is constant in a population having constant susceptibility to infection, and that different disease patterns observed in different epidemics result exclusively from host-dependent factors. Unfortunately, the logic rapidly becomes circular, as the host immunity is itself affected by the prior variation of the virus. Given these limitations by way of background, what can be learned from available knowledge about this one intensively studied virus family?

RESULTS

Annual Incidence of Human Disease

In the United States, there are presently only three data bases that have been applied in a reasonably consistent fashion from year to year over a several year period at the national level:

1. Mortality statistics generated by analysis of official death certificate records.
2. Semi-quantitative morbidity estimates by state epidemiologists. The highest category of activity is designated as "widespread" influenza. Although clearly subjective, this system permits similar criteria to be used within each state from year to year.
3. Reports of the numbers of influenza viruses isolated each week by major reference laboratories, primarily in state health departments.

The seasonal trends in mortality associated with influenza often reveal clear increases in the numbers of deaths where "influenza or pneumonia" is listed as the cause of death (Fig. 1). The difference between values in nonepidemic years, and the values in epidemic years at the same time period, represents the "excess influenza/pneumonia deaths." Excesses are also seen in total deaths compared to nonepidemic years. Although the proportionate increase in total deaths is much less than the proportionate increase in P&I deaths, the absolute numbers are higher.

Comparisons of the three data sets studied from 1978 to 1983 show that for several of the years, there is a strong similarity in morbidity and mortality reports (Fig. 2). For example, the 1980-81 season was noteworthy for an influenza A(H3N2) epidemic which caused widespread morbidity in more than half of the states and had the highest excess mortality in the period. Morbidity was also high during the influenza B/Singapore/222/79 epidemic of 1979-80, and during the epidemic of 1983-84 when viruses related to A/Victoria/7/83 (H1N1) and B/USSR/100/83 circulated to almost equal extents. Only one nonepidemic year was observed (in 1981-82) during the 6-year period. Most relevant to this discussion, however, is the occurrence in 1978 of a large type A(H1N1) epidemic in 1978-79 as judged by virus isolation numbers, but with only modest morbidity and no excess mortality. This is attributed to the fact that the A/Brazil/11/78-like viruses were responsible for many school outbreaks in affected states, but adults were in the main not involved. Those persons born before about 1955 still

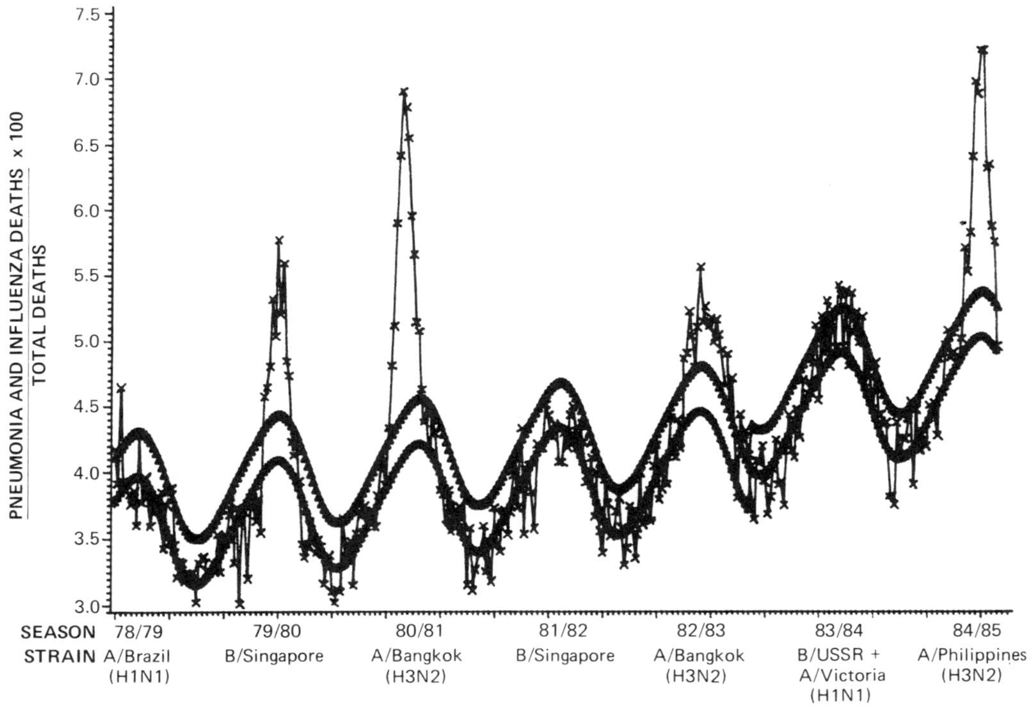

Figure 1
Pneumonia and influenza mortality surveillance, 121 U.S. cities 1978–1985

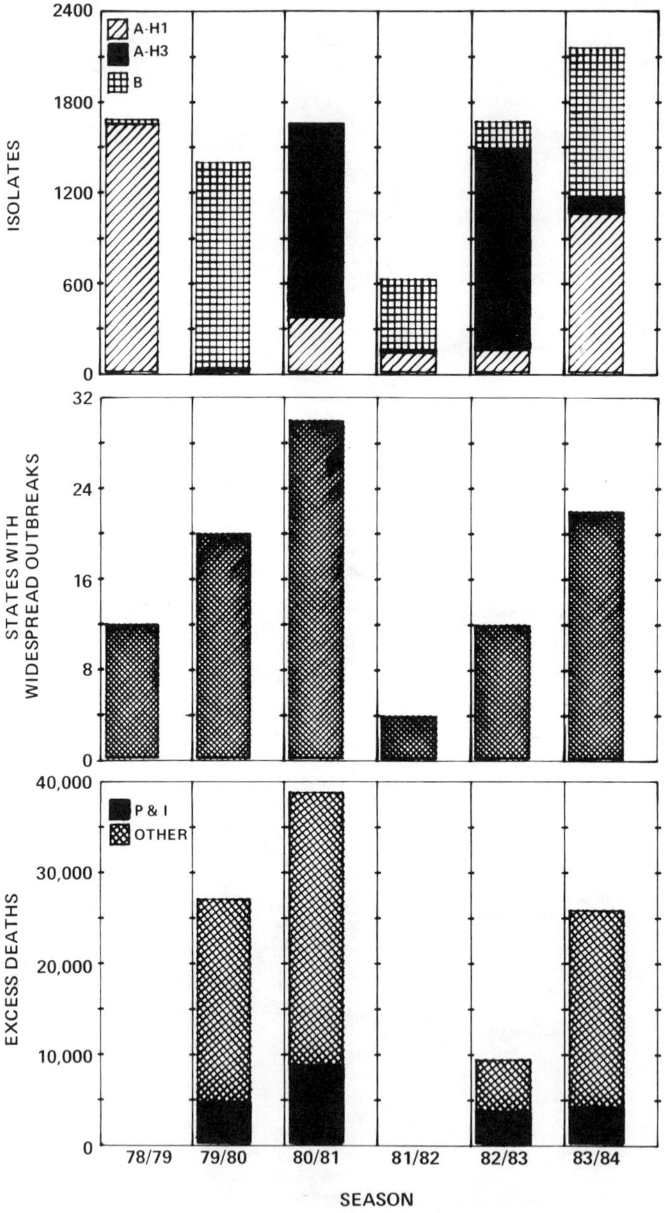

Figure 2
Indices of influenza activity in the United States, 1978–79 to 1983–84 seasons. Excess deaths are preliminary.

possessed adequate immunity to influenza A(H1N1) viruses as a result of previous infections during the pre-1957 period when such viruses were last predominant. Here then is one of the best examples that lifelong immunity to influenza can be generated, which is at least capable of preventing severe illness against virus that has not evolved antigenically to any significant extent, but which is capable of causing standard disease in fully susceptible youngsters.

Population Susceptibility

Available evidence implicates local, humoral, and cellular immune factors, as well as presumed nonspecific factors involved in interference, as relevant to disease susceptibility. Human volunteer challenge studies, in particular, suggest that the most important antibody for prevention of infection is IgA, present in respiratory secretions (Clements et al. 1986). It is also likely, based on numerous studies of the efficacy of inactivated vaccines, as well as observations during outbreaks of influenza, that increasing levels of serum antibody are associated with decreasing attack rates (Potter 1982). Studies in animal models, as well as on a limited scale in man, indicate a likely role for cytotoxic T cells in assisting in the recovery from influenza, thereby shortening the duration, and presumably the potential severity, of illness (McMichael et al. 1983a).

The role of nonspecific interference as a factor in disease incidence is suggested by analysis of the outcome in Japanese schoolchildren when influenza A(H1N1) virus appeared in the midst of an influenza A(H3N2) epidemic during the 1977-78 winter (Sonoguchi et al. 1985). Dual infections with both virus subtypes were considerably more common in schools where the H1N1 virus appeared at a different time to H3N2 virus than when both subtypes of virus were present together. Because there is no firm evidence that either heterologous antibody, or type-specific cytotoxic T cells can produce such an effect, the observations suggest that interference between influenza A viruses occurred. Such a phenomenon may play an important role in bringing about the displacement of one virus subtype by another as occurred in 1957 and 1968. The failure for the same sequence of events to be followed in 1977-78 could be due primarily to the maintenance of the H3N2 virus in adults who were not susceptible to H1N1 virus.

The example of an age-dependent occurrence of influenza outbreaks in the H1N1 epidemic of 1978-79 illustrates an extreme case of an almost "all or none" susceptibility to infection and illness in different population subgroups dependent on prior exposure to influenza virus related to a new epidemic strain. More typically, however, the population will exhibit a range of immunity to influenza varying from very low to very high, with the incidence of infection and illness decreasing as immunity increases, even though no absolute index of protection can normally be found. As new antigenic variants of influenza appear by evolution of the hemagglutinin gene, so the level of immunity is effectively lowered. An example of this is

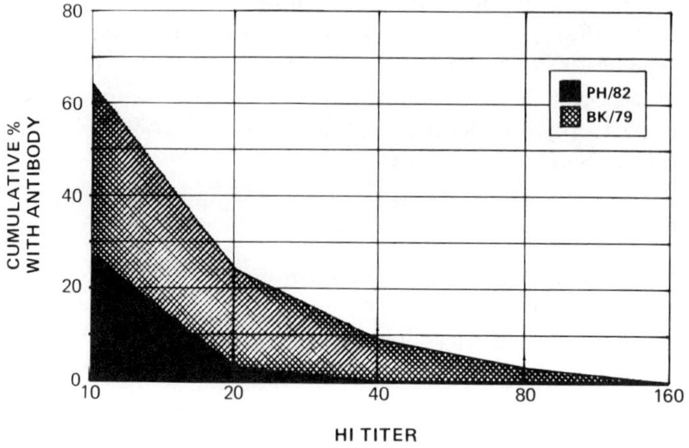

Figure 3
Effect of antigenic variation on prevalence of antibody, 1982

given in Figure 3, which compares the incidence of antibody to A/Bangkok/79 (H3N2) virus in 1982 (3 years after its introduction) with that to A/Philippines/82 (H3N2) virus at the time when it first appeared and had not yet caused many illnesses in the United States.

In addition to reduction of effective antibody levels by this phenomenon, longitudinal analysis of the level of type A influenza-specific cytotoxic T cell activity in the United Kingdom showed that the occurrence of a major epidemic in 1977–78 was followed by a rise in the level of this cellular immune response (Fig. 4). Over the next several years, when no epidemics of type A influenza were recognized in the U.K. (in contrast to the U.S. experience), the level of type A influenza-specific cytotoxic T cell response declined substantially (McMichael et al. 1983b). Because it is likely that the antigen recognized by the assays used is the nucleoprotein, or some other stable antigen of the virus, the findings indicate that population immunity changes may occur that are independent of genome evolution by the virus, confounding interpretations of the effect of virus evolution on disease patterns.

The overall outcome of the evolution of new variants and changes in population immunity are summarized in Figure 5, which illustrates the occurrence of excess mortality in relation to antigenic drift and shift of type A (H2N2) and (H3N2) viruses from 1957 to 1983. In the almost total absence of immunity in 1957, the highest levels of excess mortality in recent years were observed when Asian flu appeared. As immunity increased over the course of several years, the first of a series of variants began to appear, and major epidemics occurred in about 2 of every 3 years. In 1968, the Hong Kong flu was introduced into a population where nearly everyone was likely to have some immunity to N2 neuraminidase, a fact which

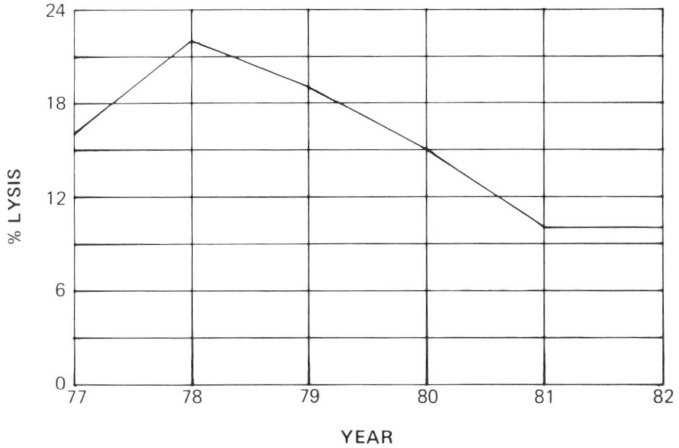

Figure 4
Change in human cytotoxic T cell activity against influenza A (U.K.). Data from McMichael et al. (1983b).

appears to have caused partial protection (Monto and Kendal 1973); and the elderly had considerable antibody to the HA, which was related to that in viruses from earlier in the century (Marine and Workman 1969). Thus, the impact of the epidemic was considerably less than in 1957. Once established, however, H3N2 viruses have similarly to H2N2 viruses produced significant epidemics in about 2 of every 3 years, in association with the periodic evolution of a new variant.

Viral Genetic Alteration

Customarily, the survival of influenza virus as an agent which can (with the present exception of type A[H1N1] virus) cause epidemics in persons of all ages has been attributed to variation in the epitopes on hemagglutinin and neuraminidase that are recognized by antibodies. Newer techniques of molecular biology have identified other types of genetic change that can be associated with the virus' ability to cause disease in man or animals. Among these are the receptor specificity of the hemagglutinin, cleavability of the hemagglutinin (a prerequisite for the membrane fusion step which permits entry of viral genomes into cells), and interactions between different gene products that affect the virus phenotype, including possibly host range.

Receptor Specificity

The receptor binding site in influenza HA contains an amino acid which determines whether 2-3 linked or 2-6 linked sialic acid substrates will be recognized (Rogers et

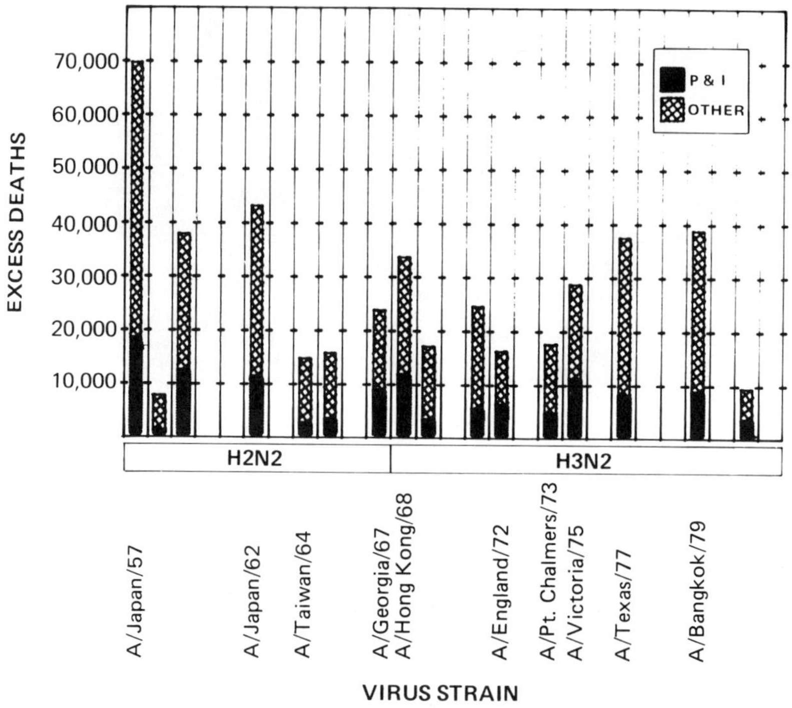

Figure 5
Excess mortality associated with influenza A epidemics in the United States, 1957-58 to 1983-84 seasons. Mortality data for the last H3N2 epidemics are preliminary.

al. 1983a,b). Human influenza A viruses, at least of the H2 and H3 subtypes, can contain subpopulations that are specific for one or the other receptor linkage. The H3 subtype viruses predominantly have the sequence recognizing the 2-6 linkage. Growth of human H3 viruses in the presence of horse serum inhibitor will select for subpopulations recognizing the 2-3 linkage. This biological selection of inhibitor-resistant viruses has previously been associated with attenuation of virulence for man (Beare 1982). Epidemic forms of influenza A H2N2 and H3N2 viruses may therefore contain subpopulations having minor amino acid differences affecting their receptor binding properties and virulence.

Host-range Selection—HA Cleavage Independent

The above was but one example of the evidence that influenza viruses causing human epidemics at any one point in time are not genetically homogenous. There are now many examples of this, and it must be accepted as fact that a single infected host normally sheds viruses with a range of genotypes exhibiting small numbers of amino acid differences from each other.

In one case, a human type A(H3N2) influenza virus being used to generate reassortants produced a variant clone with an altered HA. Unlike the parental virus, the variant was able to replicate in the duck intestine, as do naturally occurring duck influenza viruses, and then be shed in feces. Sequence analysis of the variant HA indicated that the most likely cause for the changed host range of the virus was a modification of sequence affecting amino acid residue 218 (Naeve 1984). Thus, it appears possible for a single genetic change in a human influenza virus to result in a strain that has an altered host range, and which could now be distributed in the environment through bird migration and shedding into waters occupied by ducks.

With swine influenza viruses, recent isolates, including those involved in occasional infections of man such as the limited outbreak at Fort Dix, New Jersey, in 1976, have been found to contain two antigenically distinguishable forms of HA (Kendal et al. 1977). The antigenic difference appears to involve a single amino acid residue (Kilbourne et al. 1984) and correlates with biological differences in the replication of the variants in laboratory hosts, and possibly in swine, although not apparently in man (Kilbourne et al. 1979; Beare et al. 1980). In neither of the above two cases has an explanation been found for the biological outcome of the molecular change.

Recent type B influenza viruses grown in MDCK cells have an amino acid sequence that establishes a glycosylation site on the tip of the hemagglutinin, but which is lacking on egg-grown viruses (Schild et al. 1983; Robertson et al. 1985). One can hypothesize that the patterns of disease caused by influenza B viruses are indeed affected by the virus' maintenance of a gene pool, even within a single host, that permits the virus HA to exist in differently glycosylated forms. It is possible that virions containing HA with the extra carbohydrate residue covering the antigenic site on the molecule's tip will escape neutralization by antibody directed to that site, providing a way to evade the host's immunity. This type of chameleon-like behavior could be a factor contributing to the ability of influenza B viruses to exist as successfully as influenza A viruses, but without evolving into distinct antigenic subtypes.

Host-range Selection—HA Cleavage Dependent

The H7 subtype of HA in avian influenza viruses can contain, in some strains, one or more extra basic amino acids in the HA cleavage site. This permits cleavage of HA to occur in a greater range of cell types (Klenk et al. 1984) than otherwise. Strains that possess the extra basic residue in their HA cleavage site are highly pathogenic in hosts such as the chicken, whereas strains lacking this residue are avirulent (Rott and Kleak 1986).

In 1983 a devastating series of outbreaks of avian influenza began in Pennsylvania, caused by H5N2 viruses. Comparisons of isolates from outbreaks revealed that two biologically distinct viruses existed, one highly virulent in chickens and the other of much lower virulence. Both virus types were antigenically similar and contained the same general gene composition. Sequence analysis of a number of isolates has shown, however, that the two biotypes differed consistently in the sequence of their HA at amino acid residue 23, and the virulent viruses had undergone a change which removed a potential glycosylation site (Kawaoka et al. 1984). From the known structure of the HA, it is probable that the carbohydrate side chain, when present, would be in close proximity to the HA cleavage site. Thus, its absence in the virulent variant might expose the HA to cleavage by an enzymic mechanism not otherwise possible.

Mutation of the HA at a site outside of, but physically close in the intact molecule to, the cleavage site has now been shown in a human influenza isolate to affect the host range of the virus through altering susceptibility of its HA to cleavage (Rott and Kleak 1986).

Gene Reassortment

The segmented genome of influenza virus can theoretically yield 256 different gene combinations when two completely different viruses coinfect a single cell. In the case of human influenza viruses, comparisons of the base sequence homologies of H1N1 and H2N2 viruses have shown that possibly the M, NS, NP, and one P gene (coded by RNA segment 1) were transferred from an H1N1 virus into another virus which donated the H2 and N2 genes as well as two of the P genes (Scholtissek et al. 1978). The resultant Asian influenza viruses caused the most severe epidemic of recent times.

This, however, may not have been the first time when genetic reassortment occurred between a human and presumed nonhuman influenza virus to produce a new pandemic strain. Studies of antibody prevalence in man have shown that viruses containing H3 HA but N8 NA ("equine 2 NA") circulated in man around the turn of the century. (Kendal et al. 1973). Thus, it is probable that over a period of about 60 or more years the H3 gene has, by a series of reassortment events, cycled between H3N8 virus in man, H3N? viruses in animals, and H3N2 viruses in man again.

The most recent example of reassortment involving human influenza viruses occurred in about 1978. As mentioned above, in the winter of 1977–78, type A (H1N1) viruses (the so-called Russian flu), appeared during the midst of an ongoing type A (H3N2) epidemic and some mixed influenza infections occurred (Kendal et al. 1979). The subsequent winter's epidemic in the United States was caused almost exclusively by type A(H1N1) viruses; however analysis of their genomes showed that they contained four of the genes of the 1977 H1N1 virus but that the other four genes were derived from the H3N2 virus which had circulated the previous year (Young and Palese 1979).

Analysis of the genome of influenza A(H1N1) viruses isolated around the world for the next several years showed that both "reassortant H1N1" and "nonreassortant H1N1" viruses circulated until about 1980 when the "reassortant H1N1" viruses disappeared (Cox et al. 1983). This remarkable series of events with H1N1 viruses provides the best available data to demonstrate that although genetic reassortment can lead to strains with an epidemiological advantage, for unknown reasons this is not always the case.

Gene reassortment has also been indicated as the most likely explanation for the origin of influenza A(H7N7) virus that caused a severe outbreak of influenza in harbor seals off the New England coast a few years ago. Genome analysis indicates that the seal virus was generated by reassortment between two or more avian strains (Webster et al. 1981a). The outcome was the production of a virus containing the H7 HA related to that of virulent avian influenza strains, but in a virus that could replicate well in many tissues of a marine mammal, including brain, thus causing an infection with a high mortality rate. This virus also could apparently infect the human conjunctiva (Webster et al. 1981b).

Mutational Bursts

In 1983 major epidemics in children and young adults were caused by an influenza A(H1N1) variant that exhibited a remarkably large number of oligonucleotide changes in its genome (Kendal and Cox 1985). Although it was initially suspected that another reassortment with cocirculating type A(H3N2) viruses had occurred, by partially sequencing the individual genes, all have been shown to resemble those in the 1977 "true H1N1" virus (N.J. Cox, unpubl.). The findings with type A(H1N1) viruses from 1977 to 1983 are summarized in Figure 6. Thus, it appears that in 1983 the evolution of a new variant which spread worldwide in epidemic fashion was associated with a burst of genetic mutations at a rate some three or four times greater than normally seen by oligonucleotide mapping procedures.

Attenuation

Genome comparisons have now been undertaken for two pairs of wild-type and attenuated viruses, where the attenuated virus was derived by passage at lower than normally optimal temperatures. Only four or five oligonucleotide spot differences exist between wild-type and cold-adapted mutants of A/Ann Arbor/6/60 (H2N2) virus and between wild-type and cold-adapted mutants of A/Leningrad/134/57 (H2N2) virus (Cox and Kendal 1984; Cox et al. 1985). Most of the sequence of the A/Ann Arbor/6/60 viruses has now been examined and a total of about seven amino acid substitutions detected out of about 2400 determined (Cox 1986). With this particular live vaccine strain, there is good evidence that the HA and NA genes do not contain the mutations responsible for attenuation (Kendal et al. 1981). Thus it appears that a remarkably few number of changes are needed to reduce the efficiency with which influenza virus replicates in man to the point where it loses virulence in its natural host. This type of molecular analysis supports earlier data

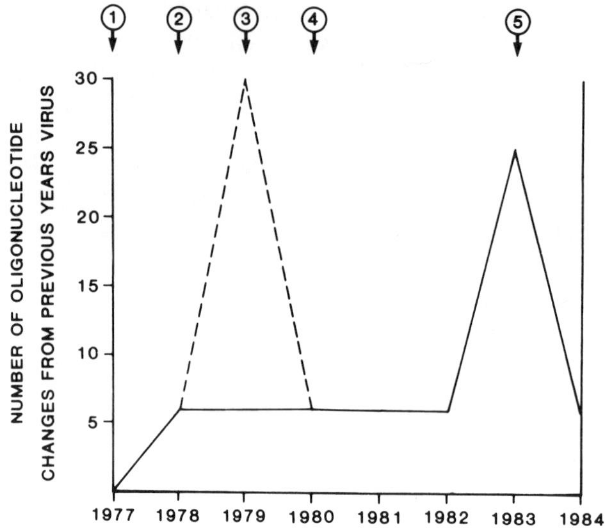

Figure 6
Schematic illustration of genome evolution in influenza A (H1N1) viruses since 1977 detected by RNA oligonucleotide mapping. Strains introduced (①) A/USSR/90/70-like;(②) A/Brazil/11/78-like; (③) A/California/10/78-like; (④) A/England/333/80-like; (⑤) A/Chile/1/83-like; (----) contains H3N2 genes; (——) lacks H3N2 genes.

that probable point mutations associated with ts phenotype could attenuate influenza viruses (Massicot et al. 1980).

Attenuation of influenza viruses has also been accomplished by reassortment of genes between two wild-type strains where at least one of the parental viruses is normally highly virulent. The outcome in terms of virulence of progeny virus appears to be quite dependent on the combination of genes derived from each parent, with different parents needed to provide different gene combinations to produce the same phenotype (Rott and Kleak 1986).

DISCUSSION

So far as is known, human influenza virus isolates do not exist in man in an exclusively avirulent form as is found for some animal influenza strains. In those cases where human influenza viruses are attenuated by selection of mutants that have altered replication or receptor binding properties, they have reduced infectiousness and transmissibility. The unpredictability of the disease patterns that will be caused from year to year in human epidemics therefore appears to result primarily from changes in the equilibrium between virus antigens and host immunity. This equilibrium can be upset by the selection of new strains exhibiting altered epitopes in their HA and NA proteins and probably also by the normal waning of population immunity, (local, humoral, or cellular) over time after epidemics.

The propensity for variation is such that viruses within the same subtype can most likely continue to be transmitted for several decades. It is in fact possible that evolution of a new antigenic subtype is unnecessary for survival of the influenza species. There may, however, be random reassortment events occurring rarely that may yield a virus having the appropriate antigenic and replicative characteristics to spread rapidly from a single focus of infection and, by a combination of immunologic advantage and interference with the previously prevalent strain, achieve the predominant status.

The situation with animal influenza viruses may be more fluid in terms of the ability for genetic alterations to affect, at least locally, disease incidence. However, based on observations of those animal influenza virus interactions which have been studied at the molecular level, it appears that single base changes are capable of modulating virulence. Therefore, survival and spread of highly virulent strains would appear to be subject to rapid natural self-control as virulent genotypes can be displaced by avirulent genotypes which should have greater long-term transmissibility.

The picture which emerges from these observations is one where the virus constantly evolves new genotypes, some of which have the potential for quite major differences in disease potential. In nature, several of these genotypes may be maintained within the virus populations being transmitted from host to host. The relative proportion of each subpopulation could fluctuate according to the characteristics of each new host encountered. In hosts such as man, a certain level of respiratory disease is probably necessary to generate infectious aerosols and maintain transmission, yet too severe a disease may damage the host to the point where transmission is adversely affected. If these various assumptions are indeed correct, then the level of disease may be maintained within a range which is optimal for survival of the virus by regulation of the relative abundance of each genetic subpopulation.

SUMMARY

Reviewing observations on the occurrence of influenza variation and disease in man and animals indicates that increased incidence of disease occurs from a variety of mechanisms, including declining background levels of immunity, antigenic variation in the viruses, probable interference between related viruses, and genetic variation which is independent of antigenic drift or shift but which affects the functioning of the major surface glycoprotein, or other viral functions. Because only small numbers of amino acid substitutions can greatly affect the virus phenotype, and because influenza viruses are almost certainly genetically heterogenous, it is possible that any one host can be infected with, or transmit to others, a mixture of virus subpopulations differing in their biological properties. This situation would appear to be advantageous for virus survival; and it is hypothesized that such a situation may enable the virus to maintain disease levels within a range which is optimum for the influenza species.

REFERENCES

Beare, A.S. 1982. Research into the immunization of humans against influenza by means of living viruses. In *Basic and applied influenza research* (ed. A.S. Beare), p. 211. CRC Press, Boca Raton, Florida.

Beare, A.S., A.P. Kendal, and J.W. Craig. 1980. Further studies in man of HSW1N1 influenza viruses. *J. Med. Virol.* **5**: 33.

Clements, M.L., R.F. Betts, E.L. Tierney, and B.R. Murphy. 1986. Comparison of inactivated and live virus vaccines. In *Options for the control of influenza* (ed. A.P. Kendal and P.A. Patriarca). Alan Liss, New York. (In press).

Cox, N.J. 1986. Progress and limitations in understanding the genetic basis for attenuation. In *Options for the control of influenza* (ed. A.P. Kendal and P.A. Patriarca). Alan Liss, New York. (In press).

Cox, N.J. and A.P. Kendal. 1984. Genetic stability of A/Ann Arbor/6/60 cold-mutant (temperature-sensitive) live influenza virus genes: Analysis by oligonucleotide mapping of recombinant vaccine strains before and after replication in volunteers. *J. Infect. Dis.* **149**: 194.

Cox, N.J., Z.S. Bai, and A.P. Kendal. 1983. Laboratory-based surveillance of influenza A(H1N1) and A(H3N2) viruses in 1980–81: Antigenic and genomic analyses. *Bull. W.H.O.* **61**: 143.

Cox, N.J., A.P. Kendal, A.A. Shilov, G.I. Alexandrova, Y.Z. Ghendon, and A.I. Klimov. 1985. Comparative studies of A/Leningrad/134/57 wild-type and 47-times passaged cold-adapted mutant influenza virus: Oligonucleotide mapping and RNA-RNA hybridisation studies. *J. Gen. Virol.* **66**: 1697.

Kawaoka, Y., C.W. Naeve, and R.G. Webster. 1984. Is virulence of H5N2 influenza virus in chickens associated with loss of carbohydrate from the hemagglutinin? *Virology* **139**: 303.

Kendal, A.P. and N.J. Cox. 1985. Forecasting the epidemic potential of influenza virus variants based on their molecular properties. *Vaccines* (in press).

Kendal, A.P., G.R. Noble, and W.R. Dowdle. 1977. Swine influenza viruses isolated in 1976 from man and pig contain two coexisting subpopulations with antigenically distinguishable hemagglutinins. *Virology* **82**: 111.

Kendal, A.P., H.F. Maassab, G.I. Alexandrova, and Y.Z. Ghendon. 1981. Development of cold-adapted recombinant live, attenuated influenza A vaccines in the U.S.A. and the U.S.S.R. *Antiviral Res.* **1**: 339.

Kendal, A.P., E. Minuse, H.F. Maassab, A.V. Hennessy, and F.M. Davenport. 1973. Influenza neuraminidase antibody patterns of man. *Am. J. Epidemiol.* **98**: 96.

Kendal, A.P., D.T. Lee, H.S. Parish, D. Raines, G.R. Noble, and W.R. Dowdle. 1979. Laboratory-based surveillance of influenza virus in the United States during the winter of 1977–1978. II. Isolation of a mixture of A/Victoria and

A/USSR-like viruses from a single person during an epidemic in Wyoming, U.S.A., January, 1978. *Am. J. Epidemiol.* **111**: 462.

Kilbourne, E.D. 1975. The Influenza viruses and influenza–An introduction. In *The influenza viruses and influenza* (ed. E.D. Kilbourne), p. 1. Academic Press, New York.

Kilbourne, E.D., G.W. Both, and W. Gerhard. 1984. Pleiotropic effects of a single amino acid change on antigenicity and biologic function of swine influenza virus hemagglutinin mutants. In *Segmented negative strand viruses* (ed. R.W. Compans and D.H.K. Bishop), p. 233. Academic Press, New York.

Kilbourne, E.D., S. McGregor, and B.C. Easterday. 1979. Hemagglutinin mutants of swine influenza virus differing in replication characteristics in their natural host. *Infect. Immun.* **26**: 197.

Klenk, H.D., W. Garten, F.X. Bosch, and R. Rott. 1984. The role of the haemagglutinin as a determinant for the pathogenicity of avian influenza viruses. In *The molecular virology and epidemiology of influenza* (ed. C.H. Stuart-Harris and C.W. Potter), p. 195. Academic Press, London.

Marine, W.M. and W.M. Workman. 1969. Hong Kong influenza immunologic recapitalization. *Am. J. Epidemiol.* **90**: 406.

Massicot, J.G., B.R. Murphy, F. Thiery, L. Markoff, K.Y. Huang, and R.M. Chanock. 1980. Temperature-sensitive mutants of influenza virus: Identification of the loci of the two ts lesions in the Udorn-ts-1A2 donor virus and the correlation of the presence of these two ts lesions with a predictable level of attenuation. *Virology* **101**: 242.

McMichael, A.J., F.M. Gotch, G.R. Noble, and A.S. Beare. 1983a. Cytotoxic T-cell immunity to influenza. *N. Engl. J. Med.* **309**: 13.

McMichael, A.J., F.M. Gotch, D.W. Dongworth, A. Clark, and C.W. Potter. 1983b. Declining T-cell immunity to influenza. 1977–82. *Lancet* ii: 762.

Monto, A.S. and A.P. Kendal. 1973. Effect of neuraminidase antibody on Hong Kong influenza. *Lancet* i: 623.

Naeve, C.W., V.S. Hinshaw, and R.G. Webster. 1984. Sequence changes in the hemagglutinin of an enterotropic H3 influenza virus. In *Segmented negative strand viruses* (ed. R.W. Compans and D.H.L. Bishop), p. 267. Academic Press, New York.

Potter, C.W. 1982. Inactivated influenza virus vaccine. In *Basic and applied influenza research* (ed. A.A. Beare), p. 119. CRC Press, Boca Raton, Florida.

Robertson, J.S., C.W. Naeve, R.G. Webster, J.S. Bootman, R. Newman, and G.C. Schild. 1985. Alteration in the hemagglutinin associated with adaptation of influenza B virus to growth in eggs. *Virology* **143**: 166.

Rogers, G.N., T.J. Pritchett, J.L. Lane, and J.C. Paulson. 1983a. Differential sensitivity of human, avian and equine influenza viruses to a glycoprotein inhibitor of infection: Selection of receptor specific variants. *Virology* **131**: 394.

Rogers, G.N., J.C. Paulson, R.S. Daniels, J.J. Skehel, I.A. Wilson, and D.C. Wiley. 1983b. Single amino acid substitutions in influenza hemagglutinin changes receptor binding specificity. *Nature* **394**: 76.

Rott, R. and H.D. Kleak. 1986. Pathogenetics of influenza virus in model systems.

In *Options for the control of influenza* (ed. A.P. Kendal and P.A. Patriarca). Alan Liss, New York. (In press).

Schild, G.C., J.S. Oxford, J.C. de Jong, and R.G. Webster. 1983. Evidence for host-cell selection of Influenza virus antigenic variants. *Nature* **303**: 706.

Scholtissek, C., W. Rohde, V. von Hoyningen, and R. Rott. 1978. On the origin of the human influenza virus subtypes H2N2 and H3N2. *Virology* **87**: 13.

Sonoguchi, T., H. Naito, M. Haro, Y. Takeuchi, and H. Fukumi. 1985. Cross-subtype protection in humans during sequential, overlapping and/or concurrent epidemics caused by H3N2 and H1N1 influenza viruses. *J. Infect. Dis.* **151**: 81.

Webster, R.G., V.S. Hinshaw, W.J. Bean, K.L. Van Wyke, J.R. Geraci, D.J. St. Aubin, and G. Petursson. 1981a. Characterization of an influenza A virus from seals. *Virology* **113**: 712.

Webster, R.G., J.R. Geraci, G. Petursson, and K. Skirnison. 1981b. Conjunctivitis in human beings caused by influenza A virus of seals. *N. Engl. J. Med.* **304**: 911.

Young, J.F. and P. Palese. 1979. Evolution of human influenza A viruses in nature: Recombination contributes to genetic variation of H1N1 strain. *Proc. Natl. Acad. Sci. U.S.A.* **767**: 6547.

COMMENTS:

FOWLE: Throughout most of this conference we will be hearing mainly about medically important viruses because they have been studied best and the most is known about them. I think that there are some parameters which should be examined from the EPA's viewpoint about the release of genetically altered viruses for environmental purposes. I wonder if you have given any thought to what one should look for, from the CPC perspective, concerning an organism that is not medically important and has not been shown to cause disease and death in human beings. Are there factors that one should be concerned about, nonetheless, before one releases such a virus for pesticide control purposes?

KENDAL: That is a very complicated topic. I think that there are certainly many people who are interested in the role of vaccinia virus to vector genes that have potential in veterinary medicine, as well as in human medicine. The whole question of a virus' ability to spread into the human host in an unanticipated way is something which is certainly of concern with that particular system. I came to this meeting without any clear guidelines as to what the EPA really wants to learn, except perhaps some basic virology, that can then be used to frame questions for risk assessment in terms of environmentally used viral vectors. There are cases, obviously, where you have human hosts, insect vectors, and animal reservoirs; and you would be very cautious about doing anything that would interfere with the balance that is achieved in those circumstances. However, I don't think that I am qualified to know what is being proposed in the field of environmental issues, using viruses that are outside the domain of either human health or animal disease. One of the discussions that I was very interested to hear was the question: Does the EPA have a role in the regulation of viral vectors that are going to be used in human hosts or in an agricultural system? The answer apparently is no. I don't know what's left, other than the baculoviruses. What are the specific examples as to how the viral vectors are planned to be used in a way that the EPA has a strong concern about? I think, given specific examples, that we can perhaps put our minds to work on subjects that require our concern. I mean, why should we be concerned about baculoviruses?

FOWLE: So then, you would argue that it should be done primarily on a case-by-case basis, depending on the application in mind, taking into account the purpose of the application and factors such as whether you want the virus to spread between hosts, what populations would be impacted (including economic stratas and age), purity of the culture, and whether it will be in the air or water.

KENDAL: After this session what seems to me to be lacking are guidelines as to the boundaries within which the environment is permitted to be altered. Maybe that has to be examined on a case-by-case basis. But, one would certainly anticipate that in the case of a human drug or a veterinary product, you would have no alternative but to look on a case-by-case basis, and there are no guidelines. You start out in exactly the same way—and this would reflect my CDC background—that any microbiologist working in a laboratory initially has a primary responsibility for considering the impact of his or her work on the safety of humans. I think the same thing applies in this situation; the person who starts out knowing most about what the important concerns are is the person who is actually taking up that particular research program or project. That person should have the responsibility to seek guidance in the areas where he or she feels that they are lacking. You end up with a kind of peer review system, which presently is very informal, that clearly operates on a case-by-case basis. I don't know how else you could do it.

DIXON: But even on a case-by-case basis, I think what I'm waiting to hear is whether the tools are generally available to characterize the virus. Can they be done with proper containment? In other words, as Jack [Fowle] said, if you take a virus that you really know very little about, are the laboratory tools well known or well described for looking at survivability, virulence, and these types of things.

FOWLE: Or, maybe even more important, with the tools available, how do you bring the various disciplines to bear, from the epidemiologists to clinicians to the molecular biologists?

KENDAL: The field is so broad–you heard a good example from Dr. Shope–that there are cases where you want the virus to spread to some extent and there are others where you don't. How are you going to set certain general guidelines? That's why I would come back to my initial point: If you're working with a human biological or drug, hopefully you start out with some idea of the benefits that you are going to achieve by the action that you are taking. You certainly know that you want the risk to be minimal in anything except the case where persons who will receive the drug or biological substance have a severe need that cannot otherwise be met. Other than that, you are looking for the most risk-free situation you can possibly have. Take the case of polio vaccine. Who are we to say whether live polio vaccine would be accepted today? One can say, on the basis of what we know presently, that we should try and have a better live type 3 polio vaccine, which is where the vast majority of the paralytic cases occurred. If you remove that problem, you actually have something which is an extremely safe program and the benefits that are obtained are absolutely immense on a worldwide or on an individual basis. When you talk about the environment, to me you are speaking about the world as a whole. I find it difficult to remove man and animals from the environment. Apparently, you can do that legally, so we're told. I am concerned about whether that actually is going to be the case. I just find it difficult to see how you can tackle this in a generic fashion.

FIELDS: It's interesting that if we think about everything that has been done or learned in the 10 years or so since the advent of recombinant DNA methodology, even though it is case-by-case, there has been a need to build a framework and construct some way of thinking of the problem in a broader sense. Maybe Mal [Martin] would like to comment on this point; I think it has been handled case-by-case. It seems that one of the things which is strikingly clear is that environmental virology has been a field that has largely been separated from molecular biology for many reasons. One of the things which is also clear is that we are talking about different things. We are talking about how viruses work in the environment, that is, what parts of viruses are playing a role in allowing a virus to spread or to alter host range (broadening it, narrowing it), and which components are involved in the mechanisms of survivability of a virus? We know so little in terms of these mechanisms. Environmental virology is one of the areas that has not been studied very much in terms of mechanisms. Therefore, when you ask about what happens to the properties of a virus when you genetically manipulate it (either by making mutants or by inserting DNA via recombinant DNA methods), it is difficult to give precise answers. I believe that it is time to begin to try to build the framework for asking what these mechanisms are, to look toward actual experiments putting together such an aerosol model and a humidifier with genetically engineered or genetically manipulated microorganisms as part of the first step—and then examining how manipulation of a particular part of a virus will affect a particular property. As of now, however, it is still case-by-case.

KAMELY: How close are we really to engineering a virus that we can deliberately release into the environment and then inactivate after it has fulfilled its function? Is it technically possible to engineer a virus so that it will not survive or will become inactive? How difficult would such a task be? Can we ultimately require that a virus engineered for environmental use have safety features and certain biological containment built into it?

KILBOURNE: I think that essentially happened with polio vaccine. Dr. Chanock brought this point up. It has not continued to be communicated in the environment. There are secondary cases from immunization, but apparently the level of replication is insufficient for environmental persistence. You can say that it "does its thing" and then is inactivated. It's "inactivated" simply because it is not transmitted further. So, I think there is a precedent for that.

KAMELY: But this happens naturally, right?

KILBOURNE: Yes.

KAMELY: How close are we to engineering such a virus?

MOSS: Maybe Max [Summers] should answer that. Perhaps a baculovirus could be made which would not form a polyhedron coat, which is necessary for it to infect insects. Then an artificial coat could be put on it for delivery to insects. It would be able to infect the insect to which it was applied, but the progeny would not be able to spread to another insect. Is that feasible?

SUMMERS: It's not that simple. The baculovirus has already anticipated that. The form that is engineered not to make a polyhedron protein produces a virus that has a different envelope and therefore has a different biological phenotype. That form of the virus is nearly 2000-fold less infectious by natural infection processes in nature than the occluded viral form. So, to encapsulate that particular form of the virus wouldn't do any good because it is not that infectious by the natural route of injection. We don't know enough about those factors that convey virulence on most strains of baculoviruses. It seems to me that it is not going to be a single gene or event, such as a single amino acid change in a surface protein of baculoviruses. There are other factors involved, such as uncoating once the virus enters a cell and the events involved in the initiation of viral replication. These are all parameters or aspects of "virulence and infectivity" that are not well defined for baculoviruses.

CROWELL: The delivery system for use of the adenovirus vaccine in the military is an enteric-coated live virus vaccine, which replicates in the intestine and does not replicate in the respiratory tract; so it's not respiratory transmitted. Measles, mumps, and rubella vaccine viruses are inoculated under the skin. This is an unnatural point of entry and therefore these viruses don't spread in the population as a respiratory disease. Thus, one could use different mechanisms of delivery of a virus to help contain it.

KAMELY: Can we engineer different mechanisms of delivery of a virus? You are quoting examples that occur naturally. However, is it possible to engineer such safeguards prior to release?

CROWELL: Obviously, only if you know enough about it to stop the transmission.

KAMELY: The answer is that we know some, but not enough.

MARTIN: Is the viral vector that you're thinking about going to have any selective advantage for survival in the environment? This is an issue that has been discussed from day one in the recombinant DNA field. No one has ever generated a bacteria, a virus, or any vector using recombinant technology that has a selective advantage in the ecosystem. It doesn't mean that you can't serendipitously come up with such a recombinant.

KAMELY: Well, I am thinking of a fertilizer that has a limited use in the environment for one season, but can then become inactive and be destroyed.

MARTIN: My point is that I think it would die out naturally unless we were smart enough to build in some selective feature guaranteeing its survival—which I seriously doubt.

KAMELY: Yes. But your suspicion is that it will die out. We have to have a greater degree of certainty or else it cannot be released into the environment. That is the main problem.

CHANOCK: It is unlikely that we have sufficient knowledge to engineer a virus mutant so specifically. What you can do is follow the course that has been adopted for the human live virus vaccines, which, without exception, either do not transmit or are unable to sustain transmission. You can test for that phenotype in the material that is being proposed for use. If the organism produces a dead-end infection or is not capable of sustained transmission, you say, "Approved." If it doesn't, you say, "Try again."

MILLER: I think the easy answer is that this lies further out on the decision tree. The crucial question is, "What is the likelihood that there is going to be a deleterious effect of some kind?" If the likelihood approaches zero, then obviously there is no need to require that the organism not be survivable. There are many examples of that in bacterial applications. Lactobacilli is added to milk; nobody cares that it replicates in the stomach, as long as it effectively, asymptomatically provides lactase for lactase-deficient people who need it. Who cares? Lactobacillus is harmless. Living organisms are also found and replicate in yogurt preparations. We all know empirically that they become stronger if you leave them at room temperature. So what? Thiobacilli are used in vast quantities to leach minerals from ore, grow, and are disseminated into the ground and into water; they appear not to cause any problem. So what? We don't require that they self-destruct.

KAMELY: I agree with you. However, often we cannot predict the transformation, persistence, and genetic alterations that a virus may undergo in the environment. For example, we learned that some of the influenza strains are totally unpredictable in their cycle of infection.

MILLER: If you're proposing to use influenza as a fertilizer, I would probably agree with that.

KNIGHT: I certainly think that the evidence is very good that most of these constructed agents will not do as well as the naturally occurring agents. Much effort has been expended to make an attenuated vaccine for a number of different organisms and they are usually less vigorous than naturally occurring viruses. Ed [Kilbourne] pointed out the probable difficulties of artificially producing increased virulence in a virus; and yet, nature has been done a time or two. The increased virulence of the virus in the chicken outbreak seems to be one of those circumstances; and I suppose in the 1918 influenza pandemic there was an event that followed by increased virulence of the virus. Experimental efforts to increase virulence of viruses for humans seem not to have been successful. Can you construct a more virulent organism? Are we going to do experiments like that? We might learn a great deal from such an enterprise if it could be done safely. I don't quite know how to do it without risk. But that, nevertheless, would be worthy of contemplation because if you know how to do that, then you'd know how not to do it.

CHANOCK: Actually, it has been done. Similar to the change that occurred in the avian influenza A virus, the change affected the cleavability of a surface glycoprotein of Newcastle disease virus (NDV), designated the fusion glycoprotein. There are variants of Newcastle disease virus in nature that express a wide spectrum of virulence. There are naturally occurring vaccine-like viruses that do not produce disease. The hemagglutinin of these viruses is not cleaved readily by host cell proteases. There are also highly virulent strains of NDV which have a fusion glycoprotein that is cleaved readily. The virulent NDV strains are able to disseminate throughout the host and produce fatal encephalitis. Dr. Rott, who did this work, produced more virulent NDV mutants in the laboratory by selecting for virus that possessed a more readily cleavable fusion glycoprotein. He exerted strong pressure on an avirulent virus by passaging it in culture, and selecting for a mutant which was cleaved more readily than its parent. Unlike its parental virus, the resulting mutant was virulent in birds.

KNIGHT: Was this fowl plague virus?

CHANOCK: No, it was Newcastle disease virus. Dr. Rott's observation provided additional evidence for an association between cleavability of a viral surface glycoprotein and virulence. It was a very difficult mutation to select experimentally.

KILBOURNE: This was X-31 influenza virus?

CHANOCK: It was Newcastle disease virus.

KILBOURNE: He's done it with X-31 too.

CHANOCK: The point is that more virulent viral surface glycoprotein mutants can be selected in the laboratory. Thus, the prescription for making a more virulent parainfluenza or influenza virus is known. One selects for a mutant whose viral surface glycoprotein is cleaved readily by host cell protease under conditions in which the parental virus is not readily cleaved. This should be considered a high-risk or forbidden experiment. This sort of mutant virus should not be constructed or selected casually. If it is constructed or selected for a valid purpose, it should be contained and not let out of the laboratory.

KILBOURNE: But that doesn't obtain with all hemagglutinins.

CHANOCK: No. This experiment was performed with a parainfluenza virus. But remember the same sort of thing happened with the avian influenza A virus under natural conditions. I would like to discuss something that was mentioned earlier, namely the problem of reassorting genes of two influenza viruses, which are attenuated for a given host, and deriving a reassortment virus that is more virulent than either parental virus. In an experiment described by Rott, neither parent was neurovirulent in mice, but a reassortment virus was neurovirulent. Cleavability of hemagglutinin again played a major role. One of the parental viruses was fowl plague influenza A virus which has a readily cleavable hemagglutinin. In its natural host, the chicken, it disseminates thoughout the body and causes fatal encephalitis. When certain genes from a mammalian influenza A virus were transferred to a virus bearing the fowl plaque hemagglutinin, the resulting reassortant virus caused fatal encephalitis in mice. I would classify this as a dangerous experiment. We now know enough about the surface glycoproteins of enveloped viruses and their effect on virulence that we can draw up a list of constructs or mutants which should not be allowed out of the laboratory.

FIELDS. I think we should keep a few things in perspective. Influenza is one of a group of viruses where virulence is genetically determined. In nature, influenza viruses alter their genetic make-up by reassortment and by mutation. When you artificially construct viruses by taking parts of one virus and putting it into regions where they don't ordinarily exist, the types of concerns are quite different and should not be confused with the concerns of either increased virulence or increased environmental impact that you are dealing with when you use influenza. In other words, by manipulating influenza genes you are simulating events that occur in nature but by taking SV-40 or retrovirus, or even intact vaccinia virus, and putting in a piece of a foreign viral gene, the likelihood of the new construct's being of increased virulence is quite low. I think it's very important to keep these experiments separate. Many of the experiments using natural strains are trying to understand the functions of the different viral proteins in virulence and involve taking strains with known differences in virulence and altering them to try to understand virulence. Thus the outcome is increased or decreased virulence. That is not the same as inserting a foreign gene into a vector, placing it into the environment, and saying, "My God, we're going to create this monster!"

KILBOURNE: On the other hand, there is a continual problem of covariation. You may alter a virus for, say, an environmental property, such as stability, and then find that it has become more virulent. So it is just really not predictable.

KENDAL: I tend to think that one of the issues that the EPA needs to face up to is, as with USDA, that you need a safe proving ground. You need to find some island somewhere or other where the wind direction is such that whatever goes in is not going to end up landing somewhere that it shouldn't. The only way you're really going to answer many of the questions is by having the equivalent of an IND and doing a control study. In the human case, you can lock people up in a motel room or a volunteer facility. With USFDA you can put them on Plum Island. I don't know where EPA is going to do the experiments, but that is what should be done.

FOWLE: I was not an organizer of this conference, but I would hope that its purpose isn't only to address the Environmental Protection Agency's needs. Rather we should be looking at a range of issues from a scientific standpoint. I think very interesting questions have come up, such as as trying to identify those potential risks that might be trivial from those that might be more important and should be focused on, from a research perspective, to help address whether there really is a concern.

KENDAL: We were told earlier in the conference that EPA would like a framework for research guidelines to enable some kind of quantitative risk assessment. I have been involved in at least one recent risk assessment procedure where the Public Health Service has contracted to assess the risks in a number of different decisionmaking processes that the PHS has been involved in, one of which is the swine influenza vaccination program. What you come down to in that area is going to be what has already been mentioned—that scientific risk assessment and public risk assessment are two different things. As scientists, we probably recognize that in many cases we don't have the adequate data to make a quantitative risk assessment in these areas, that the risk assessments are going to be made by public and political processes anyway. That would be a cycle which it would be nice to break. Your answer, "Yes, we should really be talking about science," is equally true; but there are issues at hand which impact on us all one way or another, in terms of using modern technology products in order to do the jobs we perceive need to be done. How are we going to educate the public as to risks when we're unable to quantitate them accurately ourselves? I don't think it's flippant to say that the EPA needs somewhere where it can prove whether or not a genetically engineered product is likely to have a hazard or not. The best evidence that something is safe is to do the experiment and show it. I don't know what is magic about 10 acres of land. Don't ask me on what basis anybody can statistically say that if you have a "controlled release" or an organism over 10 acres of land that is safe; but if it is on 20 acres it isn't. It beats me.

MARTIN: I would like to comment on something that Bernie [Fields] said before, and I also have a question that you, Bob [Chanock], raised about ad hoc decisions and ad hoc risk assessments. Historically, the NIH Guidelines were written as an omnibus document to deal with animal, plant, and insect viruses. But we found very early that, because of the large number of different classes of viruses, this was going to be an impossible job. It also turned out that of all the eukaryotic viruses, only a few are actually used as viral vectors. For example, nobody uses unfluenza virus as a vector. A system gradually evolved to deal with different classes on an ad hoc basis. This was also useful because as we collected more information we developed "case law" and could begin to make some generalizations. Later, in 1981, I think, as novel animal vectors were being developed and used, the concept of nondefective versus defective viral recombinants virus was generated. Today we assess viral recombinants on the basis of whether or not they are replication-competent. For example, some vaccinia viral vectors that Bernie Moss works with are clearly replication competent and investigators using them must consider the impact of the vector and its insert if they are somehow released into the environment. There is no simple way to evaluate every single vaccinia recombinant. If you really wanted to be a stickler about it, every recombinant vaccinia made might have to be tested. On the other hand, SV-40 vector-based systems, with the possible exception of situations in which only the enhancer or other regulatory elements are altered or exchanged, are, by definition, defective. The concern regarding this class of viral vector system is much lower because nothing occurs beyond that first replication

The bottom line is that there is no single risk assessment experiment that will address all the concerns. But, if you first limit the evaluation to those agents and the vectors that people are using or plan to use—they're not going to use hepatitis B, or influenza, or Korean hemorrhagic fever virus vectors—you can quickly eliminate many viruses right off the bat and focus on just the ones that people are using. I think the emphasis on *E. coli* K-12 between 1975 and 1980 is a good example to follow.

MILLER: Is there any experience with environmental microcosms with viruses and viral systems?

FOWLE: Clint [Kawanishi] or Max [Summers], do you know? You've done some work with microcosms.

SUMMERS: Are you talking about genetically engineered viruses?

KAMELY: The Drinking Water Program at EPA funded and published data on the survival of viruses in drinking water. There are no reports on genetically engineered viruses in drinking water.

FIELDS: I think that is a key point because when you look through the experimental virology textbooks, there are many interesting facts cited; but there are few experiments relating environmental features to the biochemical or molecular biological properties of viruses. Yet, in many ways, many of the environmental phenotypes are going to have direct relevance to the issues of virulence, involving viruses in hosts. Environmental virology is not a field that has been connected with modern biochemical virology.

KILBOURNE: I think the questions have to be very carefully framed and very specific. It is useful at a meeting like this to have this kind of general discussion, but one simply cannot talk in generalities about virulence; it has to be highly specific in order to be predictive.

Session 3: Tropisms

Cellular Receptors as Determinants of Viral Tropism

RICHARD L. CROWELL, KEVIN J. REAGAN, MAGGIE SCHULTZ,
JOHN E. MAPOLES, JANET B. GRUN, AND BURTON J. LANDAU
Department of Microbiology and Immunology
Hahnemann University School of Medicine
Philadelphia, Pennsylvania 19102

OVERVIEW

Specific receptors on the plasma membranes of cells serve as major determinants of the host range of viruses. It is now recognized that the receptor specificity for the different species and genera of picornaviruses sort these viruses according to their original classification based on disease and histopathology in humans and animals (Table 1). Thus, cellular receptors are considered to be important determinants of virus tropism in the pathogenesis of human and animal diseases. Studies from our laboratory have described many properties of receptors from HeLa cells for prototype group B coxsackieviruses (CB) including the recent purification of a receptor protein of 49.5 kd. In addition, we have isolated host range virus variants of CB following serial passage in the human rhabdomyosarcoma (RD) cell line. Results of competition studies between variant and prototype viruses for receptors on HeLa cells suggested that the variant viruses may have acquired a second site for binding to an additional receptor. The CB-RD variants also acquired the capacity for hemagglutination. Results of comparative studies of parental and variant viruses in three inbred strains of young adult mice revealed major differences in the capacity of CB3-RD to produce pathology in the heart, pancreas, and skeletal muscle. An in vitro model of skeletal muscle infection has been studied using the rat L_8 myogenic cell line. CB3 was unique among the CB viruses for binding to receptors and infecting L_8 cells, suggesting the possibility of another receptor. Probably more than one type of receptor for binding virus, which can influence virus tropism for different cell types within a given animal or human, exists on cell surfaces. Further studies are required to characterize the different viral receptors and to determine their distribution and function on different cell types during stages of differentiation.

INTRODUCTION

Studies of the early events in human and animal cell-virus interactions have been reviewed extensively in recent years (Howe et al. 1980; Lonberg-Holm and Philipson

Table 1
Receptor Families for Prototype Picornaviruses Based on Virus Competition for Cell Receptors

Receptor family	Reference
1. Poliovirus types 1–3	Crowell (1966)
2. Coxsackievirus, group B, types 1–6	Crowell (1966)
3. Coxsackievirus, group A, types 2 and 5; 13, 15,18	Schultz and Crowell (1983); Crowell (1976)
4. Human rhinovirus	
A. Types 1A,2,44,49	Lonberg-Holm et al. (1976)
B. Types 3,5,9,12,14,15,22,32,36,39,41, 51,58,59,60,66,67,89, Cox A21	Abraham & Colonno (1984)
5. FMDV A_{12}, O_{1B}, C_{3Res}, SAT_1, SAT_2, SAT_3	Sekiguchi et al. (1982)
6. Echovirus, type 6	Crowell (1966)
7. Cardioviruses	Burness and Pardoe (1983)

1980, 1981; Bukrinskaya 1982; Dimmock 1982; Tardieu et al. 1982; Crowell and Landau 1983). In brief, the early interactions of viruses with cells begin with the attachment of the virions to specific receptors on the plasma membrane. The virions penetrate the plasma membrane by fusion or endocytosis. A decrease in the pH of the endosome generally aids disassembly of the viral capsid and uncoating of the genome for many viruses (Pastan and Willingham 1981). An understanding of the details of these events in molecular terms for each of the different groups of viruses is currently being sought (Crowell and Lonberg-Holm 1986). Of particular interest is a definition of the virion attachment protein (VAP), which is located on the surface of the virion in multiple copies, and serves to recognize a specific receptor, which occurs as repeating units on the cell surface. Comparative studies of the fine structure of the VAP on prototype and variant viruses may reveal differences that result in altered viral tropisms and altered pathologic changes in animals following infection. These differences would be mirrored in the receptor structure of genetically different strains of inbred mice or species of animals, thereby permitting selection of one variant virus over another. Thus, we have proposed that certain individuals in a population may select a variant virus because of their unique receptor structure. This variant virus may result in a severe disease, whereas another individual with a different receptor configuration may select another virus strain and have no significant disease. In essence, receptor variation may account for selection of virus variants, which can exist in large numbers (Holland 1984) and lead to diverse expression of disease among individuals in the population.

The results of experiments to be reported herein include a brief account of the purification of a receptor protein from HeLa cells which specifically binds CB viruses; the isolation of host range variants of CB viruses and their altered tropism in inbred strains of mice; and the finding that only CB3 among the CB viruses infects the differentiating myogenic L_8 rat cell line. Each of these observations represent parts of a continuing research program aimed at defining the role of cellular receptors as determinants of CB virus tropism in the pathogenesis of infection.

RESULTS AND DISCUSSION

Properties and Purification of a Receptor Protein for CB Viruses

A number of studies from our laboratory have focused attention on the characterization of the HeLa cell receptor for the group B coxsackieviruses (CB) and on the virus-receptor interactions leading to productive infection (Crowell et al. 1983). A summary of our current knowledge of the properties of this receptor is given in Table 2. The HeLa cell contains about 10^5 copies of the receptor (Crowell 1966) and it is considered to be an intrinsic glycoprotein complex of the plasma membrane with an approximate size of 275 kd (Krah and Crowell 1985). Results of very recent studies from our laboratory (Mapoles et al. 1985) have identified a HeLa cell receptor protein (Rp-a) of approximately 49.5 kd. This protein has been purified from a virus-receptor complex (VRC) and shown to combine specifically with only CB viruses. The receptor site on the cell is likely to be comprised either of multimeric units of Rp-a or of multiple proteins (Crowell and Landau 1979). We believe the latter suggestion is more likely to be correct, since a mouse IgG-2a class monoclonal antibody prepared against the HeLa cell receptor for CB3 (R. Crowell et al.,

Table 2
Properties of Receptors from HeLa Cells for Prototype Group B Coxsackieviruses

1. Chymotrypsin sensitive; trypsin resistant
2. Inactivated by periodate, α, B glucosidases, and α mannosidase
3. Regeneration requires mRNA and protein synthesis
4. Under genetic control of cell
5. Integral membrane protein of P.M. (0.2% DOC)
6. Approximately 10^5 sites/cell
7. Approximately 275 Kd-Seph. 4B
8. Density on sucrose 1.06
9. Virus species specificity
10. Stable at pH 1, 2°C, 10 min; 60°C, 30 min
11. Antibodies to receptor block virus attachment
12. Dual function: attach and eclipse virus
13. Variant receptors select virus variants

in prep.) did not immunoprecipitate Rp-a. Further studies, however, are needed to establish the nature of the receptor site.

Most receptors constitute only a small fraction of the total membrane protein of the cell, and their purification can be difficult. Our approach was to use labeled CB3 as an affinity surface, allowing the detergent solubilized receptor (Crowell and Siak 1978) to bind to the virions, and to purify the VRC with methods normally used for virus purification. A major effort was given to identify a detergent mixture which would extract plasma membrane-bound virus without disrupting the virus-

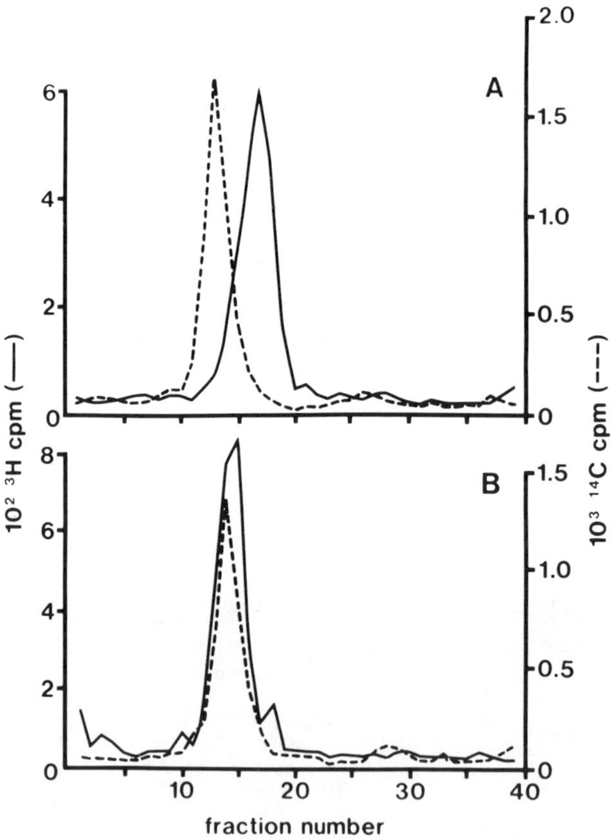

Figure 1
Specific inhibition of VRC formation by preincubation of cell lysate with unlabeled CB3. Co-sedimentation of ^{14}Caa-CB3 in a detergent extract of ^{3}H-uridine-labeled-CAV on a 5–20% sucrose gradient. (*A*) Cell lysate, ^{3}H-CB3 with marker ^{14}C-CB3; (*B*) cell lysate preincubated with saturating amounts of unlabeled CB3 and then treated as *A*. Sedimentation was from right to left. (——) Extracted ^{3}H-CAV, (----) ^{14}C-CB3 marker. Reprinted, with permission, from Mapoles et al. (1985).

receptor bond. Triton X-100 (1%) sodium deoxycholate (0.5%) and sodium dodecylsulfate (1%) were used together. The VRC sedimented more slowly (140S) than native CB3 virions (155 S) when cocentrifuged on 5-20% sucrose gradients (Fig. 1A). Preincubation of HeLa cells with receptor-saturating amounts of unlabeled CB3 prevented formation of a labeled virus VRC (Fig. 1B) to provide evidence that the virions derived from the detergent-extracted HeLa cells were in the form of a VRC.

The VRC was purified and iodinated with ^{125}I using chloramine-T. The iodinated VRC was repurified by gel chromatography and sucrose gradient centrifugation and analyzed by SDS-PAGE on slab gels. The results in Fig. 2 (lane *B*) revealed the iodinated major virus capsid proteins of CB3 as well as one additional

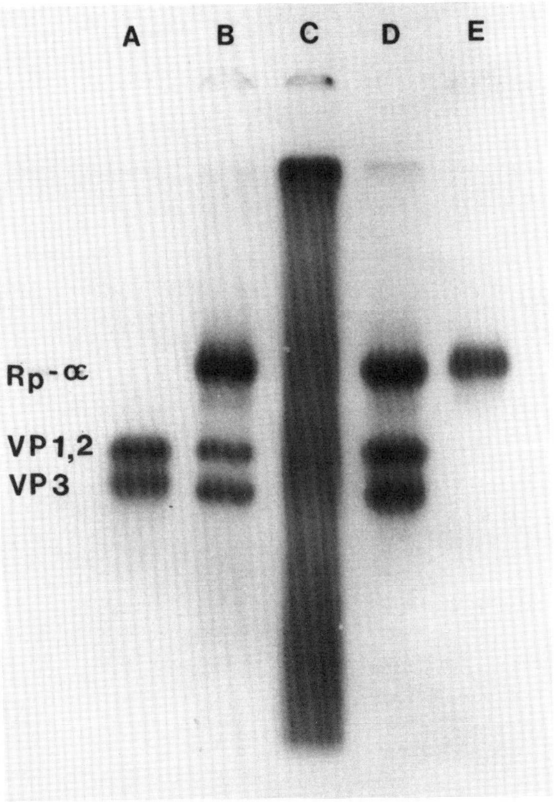

Figure 2
Autoradiogram of the iodinated coxsackievirus B3-receptor complex following SDS-PAGE. Samples were prepared for electrophoresis and added to the respective lanes as follows: (Lane *A*) 7500 cpm ^{125}I-CB3, pelleted; (lane *B*) 14,000 cpm ^{125}I-VRC; (lane *C*) 40,000 cpm of ^{125}I control sample, pelleted; (lane *D*) 13,000 cpm of ^{125}I-VRC, immunoprecipitated with anti-CB3 serum; and (lane *E*) 4500 cpm of ^{125}I-Rp-a from sucrose gradient (Fig. 3A). Reprinted, with permission, from Mapoles et al. (1985).

protein (Rp-a) of 49.5 kd. In this gel system, VP1 and VP2 of the virus comigrated (lane *A*). Lane *C* is a control sample obtained from detergent-solubilized cells (labeled with ^{3}H-leu) and iodinated in parallel with samples recovered from the VRC. It is clear that Rp-a is absent. The labeled VRC also was immunoprecipitated with rabbit anti-CB3 antiserum (lane *D*).

To show that Rp-a possessed specificity for virus attachment characteristic of the receptor, this protein was tested for capacity to bind CB1, CB3, and poliovirus T1. Our approach to this problem was to elute Rp-a from the virions, incubate the dissociated (labeled) receptor with fresh (unlabeled) virus, sediment this preparation on sucrose gradients and examine the virus for presence of associated ^{125}I label. The results shown in Figure 3 reveal that the disrupted VRC had contained an

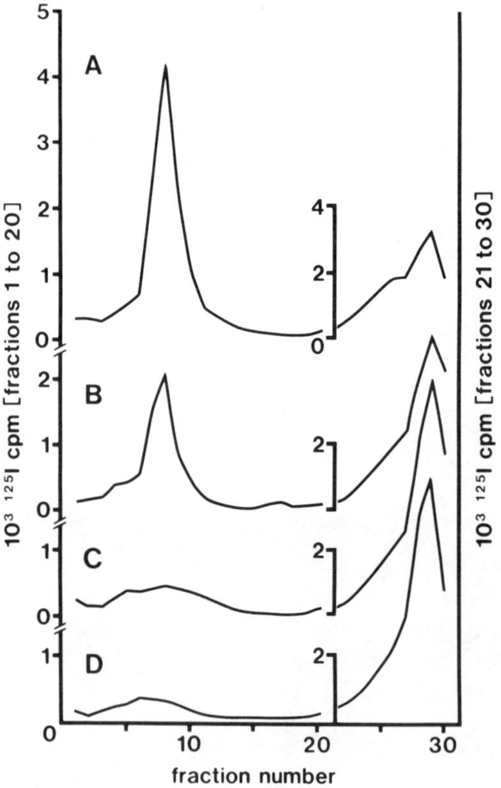

Figure 3
Sedimentation in 5–30% sucrose gradients of ^{125}I-labeled cellular receptor protein, Rp-a, which had reassociated with group B coxsackieviruses. 100 μl samples of purified ^{125}I-VRC were heat dissociated (2 hr at 45°C) and incubated for 48 hrs at 4°C with (*A*) 10^9 PFU of CB3; (*B*) 10^9 PFU of CB1; (*C*) 10^9 PFU of poliovirus; and (*D*) PBS as a control. Sedimentation was from right to left. A total of 40,000 cpm of ^{125}I was applied to each gradient, and about 10,000 cpm pelleted in each gradient. Reprinted, with permission, from Mapoles et al. (1985).

iodinated protein that bound to both CB3 and CB1 (Fig. 3A and 3B, respectively), whereas there was no apparent binding to poliovirus (Fig. 3C). The gradient shown in Figure 3D contained PBS and no added virions. When the fresh VRC (CB3) peak from the sucrose gradient (Fig. 3A) was isolated and dialyzed overnight, only one iodinated protein, Rp-a was found (Fig. 2, lane *E*). This fresh VRC also was immunoprecipitated by anti-CB3 serum and found to contain only Rp-a.

These results (Mapoles et al. 1985) provide the first identification and purification of a cellular receptor protein for picornaviruses, other than that for the encephalomyocarditis (EMC) virus receptor derived from erythrocytes (Burness and Pardoe 1983). Further studies are proposed to determine whether additional proteins are present in the receptor site on cells and to determine whether purified receptors possess the dual function for attachment and eclipse of virus infectivity (Crowell and Siak 1978).

Selection of RD Variants of CB Viruses and Comparative Pathology Produced in Inbred Strains of Mice

Serial "blind" passages in RD cells of prototype viruses from each of the six immunotypes of CB viruses resulted in the isolation of intratypic variants of CB1, CB3, CB5, and CB6 (Fig. 4) (Reagan et al. 1984). Each variant virus strain acquired the capacity for hemagglutination and produced small plaques on HeLa cells, although their serological specificity remained unchanged. The CB3-RD variant was plaque purified on RD cells and studied for receptor interactions on both HeLa and RD cells. Parental CB3 did not attach to RD cells, whereas CB3-RD attached well to both RD and HeLa cells. In virus attachment competition assays, HeLa cells saturated with CB3-RD blocked attachment of CB3, whereas saturation of cells with unlabeled CB3 failed to block the attachment of ^{35}S-met-labeled CB3-RD (Table 3). This unidirectional receptor blockade suggested that a second site was acquired by the CB3-RD variant for attachment to a second set of receptors on HeLa cells. Thus, more than one virus receptor specificity may be operative in the selection of host range virus mutants (Reagan et al. 1984).

An analogous virus-receptor relationship may occur between parental and variant reoviruses. L cells serve as host cells and may possess a group receptor for the three reovirus immunotypes (Lee et al. 1981; Epstein et al. 1984; Gentsch and Hatfield 1984); however, reovirus type 1 has a distinct histotropism from that of reovirus type 3. This difference is based on the specificity of the viral sigma-1 protein for a specific receptor on ependymal cells for reovirus type 1, which differs from that on neurons for reovirus type 3 (Sharpe and Fields 1985). Thus, in target cells in animals, it also appears that a second type of receptor may have important biological significance for reoviruses.

The rat L_8 myogenic cell line and its nonfusing clonal derivative L_8 CL3-U provide an in vitro system to study the course of infection by coxsackieviruses with the developing cell. The in vivo correlates of myotube formation through differentiation are retained by the L_8 cell line. Assay of these cells at various stages of muscle

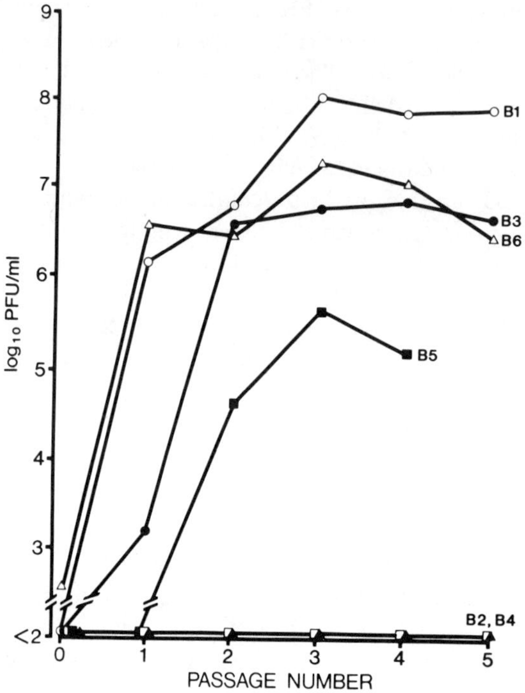

Figure 4
Serial passage of group B coxsackieviruses in RD cell monolayers. Monolayer cultures of RD cells were infected with parental CB viruses, incubated for 24–72 hr and frozen-thawed, clarified extracts used to infect fresh cultures. Aliquots of culture fluids were assayed for plaque production on RD cells. Input inocula (PFU) of parental CB viruses as determined on HeLa cells were CB1 (6×10^8), CB2 (4×10^8), CB3 (3×10^9), CB4 (3×10^8), CB5 (1×10^8), CB6 (8×10^7). Reprinted, with permission, from Reagan et al. (1984).

Table 3
Assay for Viral Attachment Interference on HeLa Cells by Coxsackievirus B3

	Amount of virus attached (%)		
^{35}S-Labeled virus	Normal cells	CB3-Treated cells	Inhibition (%)
CB3	36	5	86
CB3-RD	82	81	1

CPM added: CB3 5150; CB3-RD 5416

cell development revealed that presumptive L_8 myoblasts in prefusion cultures and L_8CL3-U cell cultures were resistant to coxsackievirus infection. Yet undifferentiated (nonfusing) cells had a specific receptor for coxsackieviruses A2 (CA2) and CA5, which differed from that for binding CB3 (Schultz and Crowell 1983). It was shown that L_8CL3-U cells had a restriction at the virus eclipse phase. Enhanced susceptibility to infection was acquired, however, as presumptive myoblasts differentiated into postmitotic myoblasts with fusion capabilities. Well-fused immature myotube L_8 cultures, expressing full eclipse activity, readily supported the replication of representative group A coxsackieviruses (A1-5,7,10-13,15-18, 22,24) and CB3, as well as CB3-W and CB3-W-RD (CB3-W strain was obtained from Jack Woodruff (Woodruff and Kibourne 1970) and passaged in RD cells for derivation of CB3-W-RD).

These results suggested that the development and expression of functional receptors on differentiating muscle cells appears to be a two-stage process characterized by viral attachment to the membrane and subsequent cellular eclipsing activity. Further studies are needed to determine whether other developing embryonic tissues can retain attached virus until the appropriate stage of differentiation occurs and the cells become susceptible to infection. Restriction events have been noted in other virus-differentiating cell systems to occur after attachment, penetration, and uncoating (Cox et al. 1977; Dutko and Oldstone 1981; Laporta and Taichman, 1981).

It is interesting to note that well-fused L_8 muscle cultures supported the replication of only CB3 and not any of the other CB viruses (Table 4). Reagan et al. (1984) found that the RD cell line failed to support the growth of CB2 and CB4. Yet these CB viruses have been found in earlier studies (Crowell 1976) to share the same receptor specificity on HeLa cells. Perhaps, differentiated or partially differentiated cell types express unique receptor sites that are not shared by all of the CB viruses.

To determine if the restriction of susceptibility of L_8 cells to CB viruses was due to the lack of cellular receptors, the rate of attachment of CB1 was measured. The results (Fig. 5) revealed that L_8CL3-U cells did not bind CB1, whereas CB3 used as a positive control attached to the cells. These findings provide further evidence for receptor variation on cell types other than HeLa cells for binding receptor-related viruses. It is postulated that there are different epitopes on the receptor proteins which comprise a multicomponent receptor (Crowell and Landau 1979). The development of monoclonal antibodies against the different cellular receptors for picornaviruses (Campbell and Cords 1983; Minor et al. 1984; R. Colonna et al., in prep.; R. Crowell et al., in prep.) will be most useful to probe the respective receptor proteins and their epitopes.

The results of preliminary studies suggested that the CB-RD virus variants were less virulent and had altered tissue tropisms in mice as compared to their respective parental viruses (Reagan et al. 1984). Thus, the disease potential of a prototype

Table 4
Comparative Recoveries of Group B Coxsackieviruses from Infected Postfusion L_8 Cell Cultures[a]

	Amount of virus (PFU/ml)							
	Cox B1	Cox B2	Cox B3	Cox B3W	Cox B3-W-RD	Cox B4	Cox B5	Cox B6
Input virus titers	3.8×10^6	1.6×10^6	2.1×10^7	1.3×10^9	2.9×10^8	2.7×10^6	5.1×10^8	3.5×10^6
Serial passage no.								
1	9.0×10^4	1.0×10^1 [b]	1.4×10^7	2.3×10^7	2.4×10^7	2.0×10^4	2.5×10^6	9.0×10^2
2	1.8×10^2	1.0×10^1	4.0×10^6	2.3×10^6	2.2×10^7	6.2×10^3	5.0×10^4	1.0×10^1
3	1.0×10^1	1.0×10^1	4.6×10^6	1.3×10^6	2.5×10^7	1.0×10^1	2.7×10^2	1.0×10^1
4	1.0×10^1	1.0×10^1	2.7×10^6	NT	NT	NT	NT	NT

[a]Postfusion L_8 cells were washed and inoculated with 0.2 ml of the appropriate group B coxsackievirus. Cultures were incubated at 37°C for 24 hr.
[b]No plaques were detected in undiluted samples.
NT = not tested

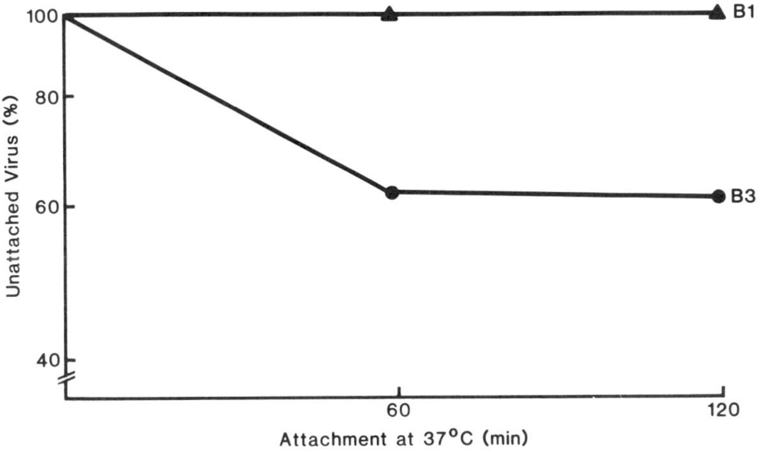

Figure 5
Comparative attachment rates of coxsackieviruses B1 and B3 to L_8-CL3-U cells. Monolayer cultures containing 3×10^6 cells/60 mm dish were inoculated with 1 PFU/cell. After the attachment period, the supernatant fluids were assayed for unattached virus.

coxsackievirus B3 (CB3-W) was tested and compared to that of a variant (CB3-W-RD) obtained by passage of the prototype strain through RD cell cultures. Studies were done by infecting each of three strains of inbred mice with these viruses and then assaying selected tissues for their capacity to support virus multiplication. A determination was made of the temporal relationship between virus titer in a given organ and occurrence of pathologic change.

The data summarized in Table 5 suggest that host-cell range variants were present in the CB3-W virus strain. Plans are made to determine whether these variants were selected by different forms of cellular receptors located on different cell types within a given organ or tissue.

The intraperitoneal inoculation of 8-week-old male C3H/HeJ and SJL/J mice with 5×10^6 PFU, respectively, and BALB/c mice with 5×10^2 PFU, of CB3-W (parental strain), resulted in development of myocarditis in each of these animals (Table 5). In addition, these animals developed frank pancreatitis characterized by acinar cell destruction although the islet cells were spared. These changes correlated well with the levels of virus found in both pancreata and hearts infected with CB3-W (data not shown).

We also observed prominent focal myositis only in infected SJL/J mice. Whether these differences relate to receptors on skeletal muscle cells, which were either more plentiful or more active than those potentially present on skeletal muscle in C3H/HeJ or BALB/c mice, remains to be determined (Schultz and Crowell 1983).

Table 5
Comparative Occurrence of Pathologic Changes in Selected Organs of Inbred Mouse Strains Infected with Parental and Variant Strains of Coxsackievirus B3

Mouse strain[a]	Virus[b,c]	Occurrence of Pathologic Changes[d] Heart	Pancreas	Skeletal muscle
C3H/HeJ	CB3-W	+	+	–
	CB3-W-RD	–	+	–
SJL/J	CB3-W	+	+	+
	CB3-W-RD	–	+/–	+/–
BALB/c	CB3-W	+	+	–
	CB3-W-RD	–	+	–

[a]8-week-old males
[b](CB3-W) Parental strain coxsackievirus B3 obtained from Dr. Jack Woodruff of Cornell Medical School; (CB3-W-RD) variant of CB3-W obtained by passage in RD cells.
[c]Groups of C3H/HeJ mice and SJL/J mice were infected with 5×10^6 PFU of CB3-W or CB3-W-RD, respectively. Groups of BALB/c mice were inoculated with 5×10^2 PFU of CB3-W while others were inoculated with 5×10^6 PFU of CB3-W-RD.
[d]14 days postinfection

Serial passages of CB3-W in RD cell cultures resulted in selection of a virus population, CB3-W-RD, lacking apparent myocardiotropic potential as evidenced by the absence of myocarditis and virus replication in the hearts of tested mice. The CB3W-RD strain, however, may have contained pancreatotropic variants that caused pancreatitis in C3H/HeJ and BALB/c mice without inducing pronounced pancreatitis in the SJL/J mouse (although virus replication occurred in this organ).

Experiments also were done to compare the virulence (lethality) of CB3-W and CB3-W-RD in each of the three inbred strains of mice. The results showed that the most striking differences occurred in the BALB/c mice, where an inoculum of 5×10^4 PFU CB3-W killed most mice within 1 week and all mice by 14 days. Even an inoculum of 5×10^2 PFU killed 60% of these mice by 14 days postinfection and the remaining mice were moribund (Table 6). In contrast, BALB/c mice infected

Table 6
Comparative Susceptibility of BALB/c and CeH/HeJ Mice to CB3W and CB3W-RD

Mouse strain[a]	Virus strain	Virus inoculum	% Surviving at 14 days P.I.
BALB/c	CB3W	5×10^2	40
	CB3W	5×10^4	0
	CB3W-RD	5×10^6	100
C3H/HeJ	CB3W	5×10^6	100
	CB3W-RD	5×10^6	100

[a]8-week-old males

with as much as 5×10^6 PFU of CB3-W-RD virus all survived for at least 14 days postinfection. Differences in virulence of CB3-W and CB3-W-RD were not observed in either C3H/HeJ or SJL/J mice since both survived infection when inoculated with as much as 5×10^6 PFU of either virus strain. It is not clear why the BALB/c mice are highly susceptible to lethal infection by CB3-W virus. It is possible that this virus has a greater range of organotropism in these mice than in the C3H/HeJ or SJL/J strains, or that another virus variant, as yet undescribed, emerges from the virus population. The data described above give rise to a number of questions. For example, what factors account for the variability in the extent and severity of CB3-W and CB3-W-RD virus-induced pathology among animals of the same inbred strain as compared to differences seen in animals of different strains? Could such differences reflect either qualitative or quantitative variations in the distribution of receptors on a specific organ from one animal to another within a mouse strain as well as between mouse strains? We believe that efforts to correlate the presence of receptors with occurrence and extent of virus-induced pathology will be based on development of a library of monoclonal antibodies against the different receptor proteins.

SUMMARY

The specificity of receptors on HeLa cells for the different picornaviruses sort these viruses according to their original classification based on disease. Thus, cellular receptors are considered to be important determinants of virus tropism. A receptor protein, Rp-a, (49.5 kd) from HeLa cells has been purified and found to bind specifically to group B coxsackieviruses. The isolation of host range variants of CB viruses following passage in RD cells, however, resulted in the acquisition of a second binding site on the virion for attachment to a second type receptor on HeLa cells. The CB-RD variants were relatively avirulent and amyocarditic in young adult mice. In general, CB-RD variants and the parental virus strain had altered tissue tropisms that were influenced by the genetic predisposition of inbred mouse strains. CB3 was unique among the CB viruses for binding to receptors and infecting the rat L_8 myogenic cell line during the fusogenic stage of differentiation. The results suggest that more than one type of receptor for binding virus may exist on cell surfaces and influence virus tropism.

ACKNOWLEDGMENTS

Studies from our laboratory were supported by U.S. Public Health Service Research Grant AI-03771 from the National Institute of Allergy and Infectious Diseases, and grants from the Muscular Dystrophy Association and the American Heart Association (Pennsylvania Affiliate).

REFERENCES

Abraham, G. and R.J. Colonno. 1984. Many rhinovirus serotypes share the same cellular receptor. *J. Virol.* **51**: 340.

Bukrinskaya, A.G. 1982. Penetration of viral genetic material into host cell. *Adv. Virus Res.* **27**: 141.

Burness, A.T.H. and I.U. Pardoe. 1983. A sialoglycopeptide from human erythrocytes with receptor-like properties for encephalomyocarditis and influenza viruses. *J. Gen. Virol.* **64**: 1137.

Campbell, B.A. and C.E. Cords. 1983. Monoclonal antibodies that inhibit attachment of group B coxsackieviruses. *J. Virol.* **48**: 561.

Cox, N.J., M.C. O'Neill, and A.P. Kendal. 1977. Replication of animal viruses in differentiating muscle cells: Influenza virus A. *J. Gen. Virol.* **37**: 161.

Crowell, R.L. 1966. Specific cell-surface alteration by enteroviruses as reflected by viral-attachment interference. *J. Bacteriol.* **91**: 198.

——. 1976. Comparative generic characteristics of picornavirus-receptor interactions. In *Cell membrane receptors for viruses, antigens and antibodies, polypeptide hormones and small molecules* (ed. R.F. Beers, Jr. and E.G. Bassett), p. 179. Raven Press, New York.

Crowell, R.L. and B.J. Landau. 1979. Receptors as determinants of cellular tropism in picornavirus infections. In *Receptors and human disease* (ed. A.G. Bearn and P.W. Choppin), p. 1. Josiah Macy Fdn. Press, New York.

——. 1983. Receptors in the initiation of picornavirus infections. *Comp. Virology.* **18**: 1.

Crowell, R.L. and K. Lonberg-Holm (ed.). 1986. ASM Conference on Virus Attachment and Entry into Cells. *Microbiology* (In press).

Crowell, R.L. and J.S. Siak. 1978. Receptors for group B coxsackieviruses: Characterization and extraction from HeLa cell membranes. In *Perspectives in virology* (ed. M. Pollard), p. 39. Raven Press, New York.

Crowell, R.L., D.L. Krah, J. Mapoles, and B.J. Landau. 1983. Methods of assay of cellular receptors for picornaviruses. *Methods Enzymol.* **96**: 443.

Dimmock, N.J. 1982. Initial stages in infection with animal viruses. *J. Gen. Virol.* **59**: 1.

Dutko, F.J. and M.B.A. Oldstone. 1981. Cytomegalovirus causes a latent infection in undifferentiated cells and is activated by induction of cell differentiation. *J. Exp. Med.* **154**: 1636.

Epstein, R.L., M.L. Powers, R.B. Rogart, and H.L. Weiner. 1984. Binding of ^{125}I-labeled reovirus to cell surface receptors. *Virology* **133**: 46.

Gentsch, J.R. and J.W. Hatfield. 1984. Saturable attachment sites for type 3 mammalian reovirus on murine L cells and human HeLa cells. *Virus Res.* **1**: 401.

Holland, J.J. 1984. Continuum of change in RNA virus genomes. In *Concepts in viral pathogenesis* (ed. A.L. Notkins and M.B.A. Oldstone), p. 137. Springer-Verlag, New York.

Howe, C., J.E. Coward, and T.W. Fenger. 1980. Viral invasion: Morphological and biophysical aspects. *Comp. Virology.* **16**: 1.

Krah, D.L. and R.L. Crowell. 1985. Properties of the deoxycholate-solubilized

HeLa cell plasma membrane receptor for binding group B coxsackieviruses. *J. Virol.* **53**: 867.
Laporta, R.F. and L.B. Taichman. 1981. Adenovirus type-2 infection of human keratinocytes: Viral expression dependent upon the state of cellular maturation. *Virology* **110**: 137.
Lee, P.W.K., E.C. Hayes, and W.K. Joklik. 1981. Protein sigma-1 is the reovirus cell attachment protein. *Virology* **108**: 156.
Lonberg-Holm, K. and L. Philipson. 1980. Molecular aspects of virus receptors and cell surfaces. In *Cell membranes and viral envelopes* (ed. H.A. Blough and J.M. Tiffany), p. 789. Academic Press, New York.
———. (ed.). 1981. *Receptors and recognition,* series B, vol. 8, *Virus receptors* part 2. Chapman and Hall, London.
Lonberg-Holm, K., R.L. Crowell, and L. Philipson. 1976. Unrelated animal viruses share receptors. *Nature (Lond.).* **259**: 679.
Mapoles, J.E., D.L. Krah, and R.L. Crowell. 1985. Purification of a HeLa cell receptor protein for the group B coxsackieviruses. *J. Virol.* **55**: 560.
Minor, P.D., P.A. Pipkin, D. Hockley, G.C. Schild, and J.W. Almond. 1984. Monoclonal antibodies which block cellular receptors for poliovirus. *Virus Res.* **1**: 203.
Pastan, I.H. and M.C. Willingham. 1981. Journey to the center of the cell: Role of the receptosome. *Science* **214**: 504.
Reagan, K.J., B. Goldberg, and R.L. Crowell. 1984. Altered receptor specificity of coxsackievirus B3 after growth in rhabdomyosarcoma cells. *J. Virol.* **49**: 635.
Schultz, M. and R.L. Crowell. 1983. Eclipse of coxsackievirus infectivity: The restrictive event for a non-fusing myogenic cell line. *J. Gen. Virol.* **64**: 1725.
Sekiguchi, K., A.J. Franke, and B. Baxt. 1982. Competition for cellular receptor sites among selected aphthoviruses. *Arch. Virol.* **74**: 53.
Sharpe, A.H. and B.N. Fields. 1985. Pathogenesis of viral infections. Basic concepts derived from the reovirus model. *N. Engl. J. Med.* **312**: 486.
Tardieu, M., R.L. Epstein, and H.L. Weiner. 1982. Interaction of viruses with cell surface receptors. *Int. Rev. Cytol.* **80**: 27.
Woodruff, J.F. and E.D. Kilbourne. 1970. The influence of quantitated post-weaning undernutrition of Coxsackievirus B3 infection of adult mice. I. Viral persistence and increased severity of lesions. *J. Infect. Dis.* **121**: 137.

COMMENTS

KHOURY: I have two related questions. First, when you showed us the tissue pathology, could that be clearly ascribed to receptor differences?

CROWELL: Perhaps, but we really haven't looked at that yet. We have solubilized mouse brains and can measure receptors in an in vitro assay. We also have looked at mouse hearts in the same way. Although they're harder to solubilize, we have not been able to demonstrate receptors. Since myocardial cells are susceptible to infection, I think the difficulty may be with the sensitivity of the assay.

KHOURY: The second part of my question is: Is it correct to assume that this receptor is really the receptor for some normal ligand and that the virus has usurped it?

CROWELL: Yes, that is a fascinating point. I think as we know more and more about receptor functions, it will be found that viruses probably have evolved to take advantage of hormone receptors or specialized antigens on the cell surface. Bernie Fields obviously will comment on the reovirus receptor as the B-andrenergic hormone receptor in his presentation. Probably the development of many monoclonal antibodies to each receptor for the picornaviruses will be needed to really get a good handle on our receptor story. I can envision that there will be multiple epitopes on receptors and that we will need monoclonal antibodies to determine the relationship between receptors on HeLa cells and on mouse cells. This relationship is still unknown. It will also be important to trace the formation and distribution of receptors with age and differentiation of different cell types. Monoclonals will also be useful to probe for the presence of variant receptors that would select variant viruses from the population.

MARTIN: As a follow-up to George's [Khoury] question, is it known what the basis of the neurotropism associated with polio is?

CROWELL: No, it is not known for certain, although some studies point to mutations in the capsid region of the genome. There is only one monoclonal, identified by Philip Minor and colleagues, which blocks receptors for both wild-type and vaccine strains of polioviruses. This group has not been able to identify a protein for that receptor, however, and it has not been ruled out whether there are specific receptor epitopes that are capable of selecting a virulent virus over an avirulent virus.

MARTIN: Could the neurotropism be due to a viral gene other than the one that interacts with the receptor? Are we talking about the cell receptor as a passive doorkeeper—once you get in, other viral genes then affect virulence?

CROWELL: It's hard to know. What has really hindered the story of receptors for many years is the observation of a certain amount of nonspecific virus binding to cells and the question of whether or not cells can actually have nonspecific binding that results in infection. We know from the work of Holland and his associates, who really started this whole area in their studies of picornaviruses years ago, that there is a virus-receptor specificity. For example, it has been established beyond any doubt that mouse L cells have no receptors for coxsackieviruses or polioviruses and they are resistant to natural infection. However, you can put viral RNA into them and bypass this resistance. Thus, if the cell binds virus nonspecifically, it is probably not going to be uncoated in a productive way, even if the virus does get inside the cell.

KILBOURNE: About a hundred years ago, I was studying the effects of Coxsackie B1 on pancreatic cells in mouse and getting the typical acinar destruction and sparing of the islets, and for reasons I have now forgotten I wrote to Gil Dalldorf for another strain of B1. I was absolutely unable to get any neurotropic effects. I then learned that this was a strain that had been passed serially only in the brain, so there was an absolute dichotomous specificity there. I couldn't tell from your presentation whether you specifically addressed the problem of variation in organ specificity within a host, specifically neurotropicity.

CROWELL: That's a very good point. Everything gets rediscovered. I remember those papers of Pappenheimer and of Dalldorf and yourself. There is no question in my mind that there is real specificity for certain cell types. I think that Craighead and then Notkins and his group have established this beautifully with EMC virus in a mouse model causing a diabetic syndrome with islet cell destruction. I believe that coxsackieviruses can do that also. Thus, it is not unlikely that there can be selection of virus variants by specific cells. We have not studied neutrotropism with our viruses. Now, the challenge remains to determine what kind of receptors are involved. I suspect it is not going to be as easy as I would have predicted. If there are two different receptors for a single virion and there are receptor variations to select variant viruses, we are going to be busy for a long time.

KILBOURNE: In the flu system there may be the same receptor chemically, in the sense of being neuraminic acid, but the linkages may be different, that is, the oligosaccharide linkages (either 2-3 or 2-6) may determine specificity, even relating to species. I think there is some precedent for very high selectivity.

CROWELL: We believe that the receptor we are studying is a glycoprotein, but there is no glycoprotein on picornaviruses. There is quite a difference between the hemagglutinin of a flu virus and a picornavirus, but I believe that similar chemical specificities exist. However, we are not that far along. The kind of x-ray crystallography work of Rueckert and Rossman is very exciting because of their observation that a cleft exists on the rhinovirus virion for binding the receptor. This cleft has dimensions that are too small for antibodies to penetrate and when the virion is disrupted, there is no protein which binds specifically to cells. Thus, we don't have a single protein that can be used to look for receptors for picornaviruses. We would have to use a virion which is 8 million M.W. It's just too big to use as a good probe. Thus, we must probably wait for the development of more monoclonal antibodies against receptors.

MARTIN: I'd like to ask a question regarding the influenza hemagglutinin. If somebody were cloning the hepatitis A virus and decided to purposely alter the receptor gene, would that bother you? What sorts of things would go through your mind?

CROWELL: Altered receptor tropism has resulted in our selection of the naturally occurring RD virus variants, which retained their identical neutralizing antigens as measured by polyclonal antiserum. What you are asking is whether you can change the virus tropism, presumably away from the liver cell so that you don't get the disease, and maybe restrict the virus to replicating in the intestine or something like that. That virus might make a good vaccine. Holland originally thought that that type of mechanism accounted for the avirulence of Sabin vaccine strains of poliovirus. However, the subsequent studies of Harter and Choppin suggested that another mechanism was operative. So, I don't know that answer, other than that Mother Nature has already presented us with many variant viruses.

MARTIN: Are you saying that these things are unpredictable?

CROWELL: Yes, it is not predictable. I would suggest that one has to use subhuman primates for a model study of altered viruses to be confident as to where that virus might go in the body and be excreted into the environment.

Genetic Alterations in Reovirus and Their Impact on Host and Environment

MARK A. KEROACK, RHONDA BASSEL-DUBY, AND BERNARD N. FIELDS
Department of Microbiology and Molecular Genetics
Harvard Medical School
Boston, Massachusetts 02115

OVERVIEW

The mammalian reoviruses have been used to study the genetic basis for the interactions of a virus with its host and environment. We have identified viral genes that play a role in determining 1° multiplication, spread in the host, tissue tropism, virulence, and persistance. Studies of variants of the cell-binding protein, $\sigma 1$, have identified a region of the genome encoding the cell-binding epitope. Intertypic differences in transmissibility between type 1 and 3 reovirus have allowed us to perform preliminary mapping experiments and we have assigned the differences in transmission to the L2 gene. Further investigations should further define the molecular basis of viral pathogenesis as well as the determinants involved in viral spread in the environment.

INTRODUCTION

Evolutionary survival of viruses depends on their successful transmission from one host to the next, a complex chain of events that begins with entry into the host, evasion of nonspecific, humoral and cellular immune defenses, multiplication at a primary site and subsequent spread and growth in the host (Fig. 1). In order to maintain the cycle, virus must then be shed from one host and transmitted to another. Transmission may occur directly (via aerosols, droplets, contact) or indirectly after being introduced for a time into the environment. In this site outside host, viruses may persist within a vector or may reside in an environmental site such as water or on the ground. Finally, the virus must enter its new host through the appropriate portal and at titers high enough to continue the chain of infection.

It is apparent from this brief outline that a virus infection includes several stages within its host and that upon release into the environment, there are a number of possible outcomes. It is thus not surprising that, while there is a great deal known about descriptive aspects of the virus-host interaction and release of a virus into the environment, much less is known concerning the precise genetic and molecular events that influence the outcome of the interactions of the virus with its host or environment. The fact that there are striking differences in the way certain

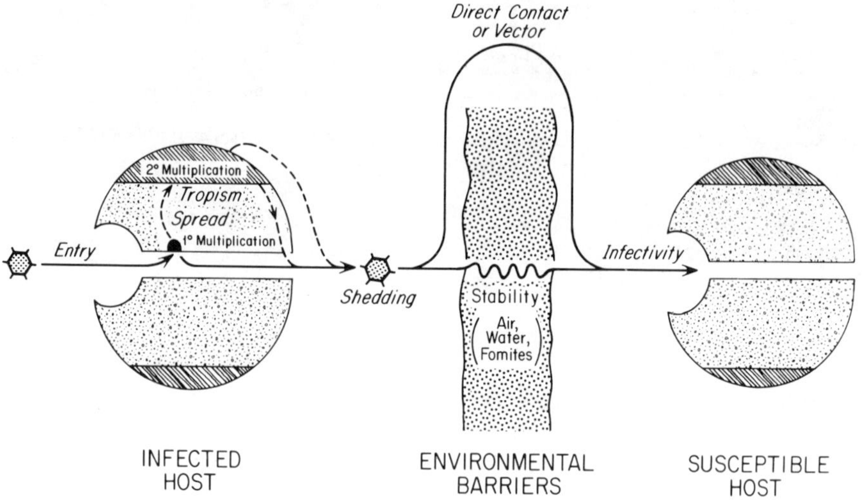

Figure 1
The life cycle of an enteric virus in the host and environment

viruses interact with the host or environment, even between viruses of the same family, suggests that important differences in viral genomes are responsible for many of these properties.

The genomic structure of reovirus makes it a useful model for the study of the functions of viral genes and the consequences of viral genetic alterations. The reovirus genome consists of ten segments of double-stranded RNA (dsRNA) grouped by size into large (L_1-L_3), medium (M_1-M_3), and small (S_1-S_4). These encode eight structural and two nonstructural proteins (large [λ], medium [μ] and small [σ]) (McCrae and Joklik 1978). The genome is enclosed in an inner protein shell and an outer icosahedral capsid. The λ2 protein forms spikes, which span from the core through the capsid and anchor the viral hemagglutinin, σ1, at the vertices of the icosahedron (Fields and Greene 1982). The remainder of the outer capsid is composed of the σ3 protein and the μlc cleavage product of the μl protein. Simultaneous infection of mouse L cells with two different serotypes of reovirus results in the reassortment of dsRNA segments, with the consequent formation of hybrid virions (Sharpe et al. 1978). The parental origin of each dsRNA segment of the hybrid can be identified by its electrophoretic mobility. By analysis of a number of reassortants, the gene, and hence the protein, responsible for any trait that distinguishes the two serotypes can be identified.

Viral Entry

Like the closely related rotavirus, reovirus is spread by the fecal-oral route (Stanley 1967). Successful infection, therefore, depends upon the ability of the virus to withstand stomach acidity, secretory IgA, and proteolytic enzymes within the small intestine. Both type 1 Lang and type 3 Dearing reovirus are human stool isolates, but in suckling mice, only type 1 grows to high titer following oral inoculation. Type 1 virus enters the gut mucosa through M cells, specialized antigen-sampling cells overlying Peyer's patches (Wolf et al. 1981; Bye et al. 1984). Primary multiplication occurs within Peyer's patches. Type 3 grows poorly at this site and gradually loses titer in the gut. Through analysis of reassortants, the ability to reach the mucosa and grow efficiently in Peyer's patches was found to be conveyed by the M2 gene of type 1, which encodes the μlc capsid protein (Rubin and Fields 1980). The μlc protein of type 1 but not type 3 is resistant to degradation by chymotrypsin, and this property may play a role in survival within the gut lumen. Variants with alterations in the M2 gene may have enhanced or diminished infectivity by the oral route, while those possessing the M2 of type 1 can be expected to be infectious. Clone 9, a murine isolate which is serotypically type 3, has a protease-resistant μlc and is highly infectious by the oral route.

Spread from the Site of 1° Multiplication

Following 1^0 replication, reovirus may fail to exit Peyer's patches, cause intestinal disease only, or spread to distant organs. At high viral inocula, reovirus infects intestinal epithelial cells, causing inflammatory changes in the submucosa and occasionally perforation (Rubin et al. 1985). At lower inocula, intestinal pathology is minimal and spread of type 1 occurs via lymphatics to mesenteric lymph nodes and spleen. Type 3 spreads only to mesenteric lymph nodes, and the intertypic variation is determined by the S1 gene, which encodes the viral hemagglutinin, σ1 (Kauffman et al. 1983). The σ1 of type 1 may mediate binding to lymphocytes, which then transport the virus to distant tissues. Restriction of spread is also age-dependent. Adult and 10-day-old mice require the type 1 σ1 for spread to the spleen, while 2-day-old mice will show disseminated infection despite a type 3 σ1 if the μlc is derived from type 1 or is protease-resistant (Rubin and Fields 1980).

Construction of reovirus variants with altered spread from the gut of adult mice would require directed mutation at the S1 gene. Conversely, mutations directed at other genes which spared the type 3 S1 should be restricted in their spread. In newborn animals, however, barriers to spread are less well developed and the behavior of altered viruses would depend on the nature of the M2 gene.

Immune Defenses

The host mounts nonspecific, humoral, and cellular immunity to combat viral infection. For reovirus, nonspecific defenses include proteases acting on the μlc protein. Type-specific immunity, however, is directed against the σ1 protein. Both neutralizing antibody and cytotoxic T lymphocytes recognize regions on the σ1 protein as the major immune determinant (Weiner and Fields 1977; Finberg et al. 1980). Studies with monoclonal antibodies to σ1 have dissected the protein still further. There are three major epitopes: one involved in immune recognition and viral tissue binding, one controlling hemagglutination, and one demonstrating neither property (Burstin et al. 1982). Since the site governing tissue tropism and that affecting immune recognition is one and the same, mutants that alter tropism (Spriggs et al. 1983) affect immune recognition (Finberg et al. 1980).

Reovirus Tissue Tropism

Key factors that determine viral-tissue tropism are the viral cell attachment protein and the host cell receptor. The cell attachment protein is generally located on the surface of the virion such that the cell binding site is accessible to its host cell. For reoviruses, the $\sigma 1$ protein (viral hemagglutinin), which is an outer capsid protein, has been identified as the reovirus cell attachment protein (Weiner et al. 1977; Lee et al. 1981).

Presently an effort is being made to identify viral receptors on host cell surfaces. A common feature of identified putative viral receptors is that they are not synthesized de novo but are existing receptors with cellular biological functions (Helenius et al. 1978; Lentz et al. 1982; Dalgleish et al. 1984; Fingeroth et al. 1984; Inada and Mims 1984). Thus, it appears that many viruses have evolved to utilize already existing receptors to gain entry into cells. Recently, the cell receptor to which reovirus 3 binds has been isolated (Co et al. 1985a). It is chemically similar to the β-adrenergic receptor (Co et al. 1985b).

Reovirus type 1 and reovirus type 3 show distinct patterns of infection. Upon injecting reovirus type 3 intracerebrally into newborn mice, viral infection is demonstrated in mouse neurons (Margolis et al. 1971). This infection produces a highly lethal encephalitis. Thus, reovirus type 3 is neurotropic. However, intracerebral injections of reovirus type 1 into newborn mice result in a viral infection in the ependymal cells that is nonlethal (Kilham and Margolis 1969). Using intertypic reassortants (serotype 1 $\times$ 3) the S1 gene which codes for the $\sigma 1$ protein was shown to be responsible for this tropism (Weiner et al. 1977). An extension of reovirus type 3 neurotropism is that after intracerebral injection, reovirus type 3 spreads to the eye where it infects retinal ganglion cells and grows exponentially. Reovirus type 1, however, grows poorly in the eye after intracerebral inoculation and does not damage ganglion cells. Intertypic reassortants identified the S1 gene product to be responsible for retinal ganglion cell injury (Tyler et al. 1985).

Reovirus also demonstrates tropism within the pituitary gland whereby reovirus type 1 infects the anterior pituitary and reovirus type 3 does not. Using intertypic reassortants, this tropism was shown to be mediated by the S1 gene (Onodera et al. 1981).

Viral tropism also exists with muscle cells. Injecting reovirus type 3 into skeletal muscle of newborn mice produced a severe necrotizing myositis, whereas reovirus type 1 injection produced only a mild inflammatory response in muscle. It is the

S1 gene, which encodes the σ1 protein, that has been identified by the use of intertypic reassortants as being responsible for muscle injury (Tyler et al. 1984).

To further study the reovirus cell attachment protein, reovirus variants were selected. Variant selection was based on the ability of the variant to resist neutralization by reovirus type 3 σ1 protein neutralizing monoclonal antibody. The isolated reovirus type 3 variants not only demonstrated a resistance to neutralization but biologically they demonstrated an altered tropism (Spriggs and Fields 1982). The variants showed an inability to cause a fatal neurological disease when injected intracerebrally in newborn mice. Studies performed to analyze the anatomic distribution of injury produced by the variants indicated that the variants were altered in their capacity to injure selected parts of the brain (Spriggs et al. 1983).

To define the alteration in the cell attachment protein, the S1 gene of one avirulent variant (variant K), was sequenced. A comparison of the sequence of the S1 gene of the K variant to the wild-type parent showed that only one nucleotide base was altered (Bassel-Duby et al. 1985). The σ1 of the K variant contains an amino acid substitution of a lysine for a glutamic acid at amino acid 419 in the cell attachment protein (which is 455 amino acid residues). Thus, the alteration of the cell attachment protein of the avirulent variant identified the carboxy-terminal region of the cell attachment protein as containing a binding site which determines neurotropism.

VIRULENCE AND PERSISTENCE

Once a virus has reached a target organ, the extent of damage inflicted by the infection depends on the number of cells involved and the degree to which normal cellular functions are disrupted. Types 2 and 3 reovirus profoundly depress normal cellular activity in cultured mouse L cells. The σ1 protein of type 3 is responsible for inhibition of host DNA synthesis, whereas the σ3 protein of type 2 inhibits host RNA and protein synthesis (Sharpe and Fields 1981, 1982). In animal experiments, reovirus infection causes an acute infection. If the animal survives, virus eventually disappears. However, in cell culture, mutations in the S4 gene, which encodes the σ3 protein, can result in persistent infection (Ahmed and Fields 1982). Persistent infection has not yet been observed in animals, but variants with alterations in the S4 gene might be expected to demonstrate changes in the extent of cytopathology in affected organs or in the capacity of the virus to persist.

Variation in the virulence of reoviruses has been observed among members of a serotype. Hrdy et al. (1982) studied the type 3 (H/Ta) variant and type 3 Dearing following intracerebral injection in suckling mice (Hrdy et al. 1982). Both viruses infected neurons; however type 3 (H/Ta) was less virulent since it showed less efficient growth in the brain and required higher doses of virus to cause lethal infection. The attenuation of neurovirulence of type 3 (H/Ta) was a property of the M2 gene of type 3 (H/Ta) which encodes the μ1c protein. Thus, the manifestations of infection could be altered not only by altering tropism ($\sigma1$ protein) but also by altering the capacity of the virus to grow efficiently and destroy cells (μlc protein).

TRANSMISSION

Prevalence studies indicate that reovirus is transmitted with high efficiency. A variety of strains have been isolated from avian and mammalian species (Hrdy et al. 1979), and in man neutralizing antibody is detected in 50-80% of adults (Stanley 1977). Seroconversion usually occurs without symptoms or in the setting of a mild diarrheal illness. Like the closely related rotavirus, transmission occurs via the fecal-oral route (Stanley 1967). The most widely studied strains, type 1 Lang and type 3 Dearing, are human stool isolates. Although the mechanism for transmission is understood for many viruses, the viral determinants which affect transmissibility are poorly understood.

We have studied the transmission of type 1 and type 3 reovirus among members of a litter of newborn mice. After inoculating two members of the litter, titers of virus were determined in their littermates. Type 1 transmitted to a high percentage of littermates, whereas type 3 transmitted only rarely. A natural murine isolate, clone 9, transmitted in nearly all cases. The frequency of transmission decreased with smaller inocula and in smaller litters. Virus appeared first in the intestines, consistent with fecal-oral transmission. Preliminary analysis of type 1-type 3 reassortants indicate that the L2 gene (encoding the $\lambda2$ spike protein) confers the high transmissibility phenotype (unpublished results).

Transmission may be divided into three stages (Fig. 1): shedding from the host, survival outside the host and infection of another susceptible animal. Each stage may be studied separately in the reovirus system to determine its relative importance in the overall process. Intestinal washings were examined at intervals following infection to determine intertypic differences in shedding. Highly transmitted strains were shed in the stool at several $\log_{10}$ pfu higher than strains showing low transmission. Further studies are needed to determine if shedding and transmission maps to the same gene.

If survival for long periods of time outside the host were required for reovirus transmission, then infection of a new host would depend upon the ability of the virus to withstand environmental stresses. These might include drying and high ionic strength, as well as resistance to inactivation by bile salts and bacterial products present in stool. Reovirus strains differ in their susceptibility to chemical agents; and by analysis of recombinant viruses, the genetic basis for a number of these differences was determined (Drayna and Fields 1982). The S1 gene ($\sigma1$ protein) is responsible for sensitivity to pH 11 and 2.5 M guanidine; the M2 gene (μlc protein) controls inactivation by chymotrypsin or phenol and ethanol; and the S4 gene ($\sigma3$ protein) determines the response to 1% SDS or heating to 55°C.

Variants with alterations in a gene encoding a particular capsid protein might differ in their stability in response to chemical and physical agents. Such changes in stability might have profound effects on survival in the environment. Furthermore, alterations that are known to modify a viral trait, such as oral infectivity or tissue tropism, might have unintended effects on environmental stability and hence transmissibility.

The final step in transmission is entry into a susceptible host by a route and at a dose that allows for successful infection. Some of the factors affecting reovirus infectivity have been discussed under Entry.

DISCUSSION

The behavior of genetically altered reoviruses in the murine infection model illustrates several important points about viral pathogenesis and transmission. First, it is possible to identify individual proteins playing important roles in various stages of the infectious process. These include oral infectivity, spread with host, tissue tropism, and virulence. Preliminary results suggest that it is also possible to assign a specific viral protein to the phenotype of transmissibility, an important trait affecting the impact of a virus on the environment.

By discovering the specific role of a given gene in the infectious process, it is possible to predict the traits in which a virus bearing a variant of that gene would differ from a wild-type strain. Variants of the S1 gene, the cell-binding protein, alter tissue tropism and immune recognition. Variants of M2 demonstrate differences in oral infectivity and in neurovirulence. Conversely, variants at a given gene would not be expected to affect properties controlled by other genes. Any reassortant or variant bearing the S1 gene of type 3 Dearing will demonstrate neuronal binding if the virus is able to reach the brain. Viruses bearing the M2 gene of type 1 Lang will spread from the gut in 2-year-old mice regardless of the presence of other outer capsid proteins of T3 Dearing. Variants that possess the L2 gene of type 1 Lang demonstrate transmission from one mouse to another under controlled environmental conditions, even though the behavior of the virus in the individual host is largely controlled by the outer capsid proteins. The known repertoire of certain reovirus variants and reassortants is presented in Table 1.

Finally, genetic alterations of a given gene may have no effect or may affect some or all of the important properties controlled by that gene. In the case of S1, different epitopes control hemagglutination and neutralization. Variants at the neutralization epitope that fail to bind type-specific antibody still demonstrate type-specific hemagglutination. On the other hand, the neutralization and the cell-binding epitope are inseparable, so that all the variants studied as of this time, which are altered in their response to type-specific antibody, demonstrate altered tissue tropism as well. Since S1 also controls sensitivity to certain chemical agents, variants of S1 may have altered stability in the environment as well.

Table 1
Impact of Specific Reovirus Genes on Host and Environment

Gene (Protein)	Function	Effect of genetic alteration: On host	Effect of genetic alteration: On environment
S1 (σ1)	Receptor interactions	Change in tissue tropism	Sensitivity to alkali
M2 (μlc)	Yield in differentiated tissues	Altered efficiency of growth and extent of cell injury	Sensitivity to proteases, ethanol
S4 (σ3)	Inhibition of host RNA and protein synthesis	Persistent vs. lytic infection	Sensitivity to heat, detergents
L2 (λ2)	Part of transcriptase, core spike	Generation of defective virus	? Titer of virus shed into environment

CONCLUSIONS

Studies using the mammalian reoviruses have helped elucidate several principles of viral pathogenesis and transmission. Specific viral traits are encoded by specific genes. Traits under genetic control include infectivity, spread, tissue tropism, virulence, shedding from the host, and stability to chemical and physical agents. Alterations in a given gene may alter the viral traits encoded by that gene. Some genes control several properties, and a single genetic alteration may affect a number of traits. Nevertheless, an understanding of the molecular basis of viral phenotypes is an important step in predicting and controlling the effects of genetic alterations.

ACKNOWLEDGMENTS

We thank our colleagues for helpful discussion. This work was supported by NIAID, grant 5RO1 A1 13178 and partial support from the Shipley Institute of Medicine.

REFERENCES

Ahmed, R. and B.N. Fields. 1982. Role of the S4 gene in the establishment of persistent reovirus infection in L-cells. *Cell* **28**: 605.

Bassel-Duby, R., A. Jayasuria, D. Chatterjee, N. Sonnenber, J. V. Maizel, and B.N. Fields. 1985. Sequence of reovirus haemagglutinin (cell attachment protein) predicts a coiled-coil structure. *Nature* **315**: 421.

Burstin, S.J., D.R. Spriggs, and B.N. Fields. 1982. Evidence for functional domains on the reovirus type 3 hemagglutinin. *Virology* **117**: 146.

Bye, W.A., C.H. Allan, and J.S. Trier. 1984. Structure distribution and origin of M cells in Peyer's patches of mouse ileum. *Gastroenterology* **86**: 789.

Co, M.S., G.N. Gaulton, B.N. Fields, and M.I. Greene. 1985a. Isolation and biochemical characterization of the mammalian reovirus type 3 cell surface receptor. *Proc. Natl. Acad. Sci. U.S.A.* **82**: 1494.

Co, M.S., G.N. Gaulton, A. Tominaga, C.J. Homcy, B.N. Fields, and M.L. Greene. 1985b. Structural similarities between the mammalian beta-adrenergic and reovirus type 3 receptors. *Proc. Natl. Acad. Sci. U.S.A.* (in press).

Dalgleish, A.G., P.C.L. Beverley, P.R. Clapham, D.H. Crawford, M.F. Greaves, and R.A. Weiss. 1984. The CD4 (T4) antigen is an essential component of the receptor for the AIDS retrovirus. *Nature* **312**: 763.

Drayna, D. and B.N. Fields. 1982. Genetic studies on the mechanism of chemical and physical inactivation of reovirus. *J. Gen. Virol.* **63**: 149.

Fields, B.N. and M.I. Greene. 1982. Genetic and molecular mechanisms of viral pathogenesis: Implications for prevention and treatment. *Nature* **300**: 19.

Finberg, R., D.R. Spriggs, and B.N. Fields. 1980. CTL recognize the major neutralization domain of the viral hemagglutinin. *J. Immunol.* **129**: 2235.

Fingeroth, J.D., J.J. Weis, T.F. Tedder, J.L. Stominger, P.A. Psiro, and D.T. Fearon. 1984. Epstein-Barr virus receptor of human B lymphocytes as the C3d receptor CR2. *Proc. Natl. Acad. Sci. U.S.A.* **81**: 4510.

Helenius, A., B. Morein, E. Fries, K. Simons, P. Robinson, V. Schirrmacher, C. Terhorst and J. Strominger. 1978. Human (HLA-A and HLA-B) and murine (H-2K and H-2D) histocompatibility antigens are cell surface receptors for Semliki Forest virus. *Proc. Natl. Acad. Sci. U.S.A.* **75**: 3846.

Hrdy, D.B., L. Rosen, and B.N. Fields. 1979. Polymorphism of the migration of double-stranded RNA genome segments of reovirus isolates from humans, cattle and mice. *J. Virol.* **31**: 104.

Hrdy, D.B., D.H. Rubin, and B.N. Fields. 1982. Molecular basis of reovirus neurovirulence: Role of the M2 gene in avirulence. *Proc. Natl. Acad. Sci. U.S.A.* **79**: 1298.

Inada, T. and C.A. Mims. 1984. Mouse Ia antigens are receptors for lactate dehydrogenase virus. *Nature* **309**: 59.

Kauffman, R.S., J.L. Wolf, R. Finberg, J.S. Trier, and B.N. Fields. 1983. The σ1 protein determines the extent of spread of reovirus from the gastrointestinal tract of mice. *Virology* **124**: 403.

Kilham, L. and G. Margolis. 1969. Hydrocephalus in hamsters, ferrets, rats and mice following inoculations with reovirus type 1. *Lab. Invest.* **21**: 183.

Lee, P.W.K., E.C. Hayes, and W.K. Joklik. 1981. Protein σ1 is the reovirus cell attachment protein. *Virology* **108**: 156.

Lentz, T.L., T.G. Burrage, A.L. Smith, J. Crick, and G.H. Tignor. 1982. Is the acetylcholine receptor a rabies virus receptor? *Science* **215**: 182.

Margolis, G., L. Kilham, and N. Gomatos. 1971. Reovirus type III encephalitis: Observations of virus-cell interactions in neural tissues. I. Light microscopy studies. *Lab. Invest.* **24**: 91.

McCrae, M.A. and W.K. Joklik. 1978. The nature of the polypeptide encoded by each of the 10 double-stranded RNA segments of reovirus type 3. *Virology* **89**: 578.

Onodera, T., A. Toniolo, U.R. Ray, A.B. Jensen, R.A. Knazek, and A.L. Notkins. 1981. Virus-induced diabetes mellitus. XX. Polyendocrinopathy and autoimmunity. *J. Exp. Med.* **153**: 1457.

Rubin, D.H. and B.N. Fields. 1980. Molecular basis of reovirus virulence. Role of the M2 gene. *J. Exp. Med.* **152**: 853.

Rubin, D.H., M.J. Kornstern, and A.O. Anderson. 1985. Reovirus serotype 1 intestinal infection: A novel replicative cycle with ideal disease. *J. Virol.* **53**: 391.

Sharpe, A.H. and B.N. Fields. 1981. Reovirus inhibition of cellular DNA synthesis: Role of the S1 gene. *J. Virol.* **38**: 389.

———. 1982. Reovirus inhibition of cellular RNA and protein synthesis. Role of the S4 gene. *Virology* **122**: 381.

Sharpe, A.H., R.F. Ramig, T.A. Mustoe, and B.N. Fields. 1978. A genetic map of reovirus. I. Correlation of genome segments between serotypes 1, 2 and 3. *Virology* **85**: 531.

Spriggs, D.R. and B.N. Fields. 1982. Attenuated reovirus type 3 strains generated by selection of hemagglutinin antigenic variants. *Nature* **297**: 68.

Spriggs, D.R., R.T. Bronson, and B.N. Fields. 1983. Hemagglutinin variants of reovirus type 3 have altered central nervous system tropism. *Science* **220**: 505.

Stanley, N.F. 1967. Reoviruses. *Br. Med. Bull.* **23**: 150.

——— . 1977. Diagnosis of reovirus infections–Comparative aspects. In *Comparative diagnosis of viral disease* (ed. E. Kurstak and K. Kurstak), p. 385. Academic Press, New York.

Tyler, K.L., W.C. Schoene, and B.N. Fields. 1984. A single viral gene determines distinct patterns of muscle injury: Genetics of reovirus myositis. *Neurology (Suppl. 1)* **34**: 191.

Tyler, K.L., R.T. Bronson, K.B. Byers, and B.N. Fields. 1985. Molecular basis of viral neurotropism. *Neurology* **35**: 88.

Weiner, H.L. and B.N. Fields. 1977. Neutralization of reovirus: The gene responsible for the neutralization antigen. *J. Exp. Med.* **146**: 1305.

Weiner, H.L., D. Drayna, D.R. Averill, Jr., B.N. Fields. 1977. Molecular basis of reovirus virulence: Role of the S1 gene. *Proc. Natl. Acad. Sci. U.S.A.* **74**: 5744.

Wolf, J.L., D.H. Rubin, R. Finberg, R.S. Kauffman, A.S. Sharpe, J.S. Trier, and B.N. Fields. 1981. Intestinal M cells: A pathway for entry of reovirus into the host. *Science* **212**: 471.

COMMENTS

KILBOURNE: Bernie [Fields], aren't there phytoreoviruses which, when serially passaged in plants, apparently lose virulence for insects?

FIELDS: Sure.

KILBOURNE: And I think they concomitantly lose one or two of their segments. Is that understood, because it seems to me a possible model not only for change in host specificity, but perhaps for environmental adaptation?

FIELDS: We have done a number of studies involving passaging of mammalian reoviruses. Rafi Ahmed did the largest number of passages of reovirus in cell culture and showed that there were multiple genetic changes. When Dan Hrdy studied an avirulent strain, he was able to show the lesion was in the M2 gene. He then passaged the virus back to high virulence and did the genetics. There were multiple changes in the virus, but the gene that conferred the increased virulence was M2. Thus, there were specific mutations in M2.

KHOURY: What is the proposed role of M2?

FIELDS: I think that it plays a critical role in virus endosomes by being cleaved. The data that have come out are by a student, Max Nibert, who has shown that as reovirus moves into endosomes, the M2 product is a dimer that is cleaved to form a delta product. Delta is a monomer. We think this cleavage allows reovirus to get close to the membrane and may play a role in cell trafficking. Thus, the M2 product is processed during early events in passage through endosome. So, I think it is at some level in the endosome that M2 is playing its role.

HOPKINS: Are the effects of M2 and S4 tissue-specific, and could they ever be the sole determinant of the tissue tropism of the virus?

FIELDS: We have asked that. We keep saying: "What is our enhancer?" I like to think maybe that M2 is the closest so far to an enhancer. It is obviously a totally different mechanism. But, there is no way we can do an experiment that shows tissue tropism where the S1 gene isn't the gene that segregates with where the virus localizes in host tissues. The S1 determines where the virus is going to go while M2 gene says how well it is going to grow.

HOPKINS: But are they better or worse in different tissues, more or less in different tissues?

FIELDS: I think that they are going to be very tissue-specific and maybe very analogous to tissue-specific protease effects, but we don't know much yet about that.

HOWLEY: Is the hemagglutinin interacting with the beta adrenergic receptor, and if so, how do you account for the different tropisms for ependymal cells and neurons?

FIELDS: That's a separate story. The beta adrenergic receptor is one of the least well understood receptors. It has never been isolated in enough quantities to get a sequence of it, to really look at its genetic basis. I don't know the answer to your question, but I think that we are dealing with a gene family that has organ-specific configurational changes. I think I would answer that question in part by giving a slightly different interpretation to what you are saying about two receptors. I think we may be dealing with gene families. The virus may be interacting with an organ-specific configuration of a receptor on the surface of a target cell that is a variable component of a gene family; but we don't know.

SUMMERS: I have a question of a general nature because I don't know your system that well. With the specific tissue tropisms that you document in vivo, are these tropisms similar in cultured nerve cells?

FIELDS: The closest answer to that is an experiment that Howard Weiner did with Mark Dichter. They took ependymal cells and neural cells in explants, and did binding studies. Type 1 bound to ependyma; type 3 bound to neurons. Most cell cultures, like the L cell, have no specificity at all. For example, neuroblastoma cells will support both type 3 and type 1–type 3 grows a little better. But again, these are cell lines that are growing out in vitro, and it is well known that you can lose differentiated properties.

SUMMERS: Are you making the assumption then that if the receptor isn't there, the viruses do not attach and enter the cells? Is it possible that the virus enters other tissues by circulating in the blood?

FIELDS: That is a different issue. Type 3 spreads through neurons; it does not circulate in the blood. Type 1 spreads in association with lymphoid tissue in the blood. It is a different path depending on the route on introduction.

SUMMERS: Are you making the assumption that the viruses would not enter cells for which there would not be a receptor?

FIELDS: I don't know yet.

SUMMERS: We've been doing some studies with John Couch with EPA at Pensacola, on the susceptibility of some cultured aquatic vertebrate cells to baculoviruses. We were quite surprised to see that a lot of baculovirus DNA enters the nucleus of those cells rapidly. We had made the assumption that if the cells did not appear susceptible to viral infection, efficient virus entry would not occur. We were wrong. Much of the viral DNA penetrates and it persists for a sufficient period of time to allow it to be expressed and perhaps to integrate.

FIELDS: I would love to find another interpretation of the S1 segregation experiment; but the tissue distribution maps so clearly to the S1 gene that we can't provide another explanation.

AHMED: Do you find the same thing with the natural isolates of type 3?

FIELDS: Yes, all of the type 3 strains.

AHMED: Yes, but many of them will attach to neurons but will not grow.

FIELDS: Yes.

AHMED: Is this due to defects in other genes?

FIELDS: Yes. The M2 is the one gene we identified in that sense.

KENDAL: Would you like to comment on suppression and reversion and the hazards of making predictions about how genetically engineered microorganisms will revert?

FIELDS: I'm not sure how to answer the question of suppression. One point is that you can get protein-protein interactions very frequently, so that you have to do the genetics to make sure that a mutational change is in the gene that you think it is, because second site mutations can lead to changes in configuration and change function. Would you repeat the question about altered microorganisms?

KENDAL: I think that part of your thesis in this paper is, if I understand the relationship between structure and function, you can make accurate predictions about the biological behavior of the virus. I question whether that is a safe approach in terms of suppression and reversion.

FIELDS: What I'm basically saying is that mutants in M2 are not going to tell type 3 to go to ependyma or not, but they will determine yield. We have looked at a spectrum of M2 mutants and they vary yield. They will not change the localization of tissue injury. In terms of the in vivo experiments, it is the S1 and the M2 that have given us insight in terms of those predictions to allow us to say that mutations in one gene might predictably do what we think. This is only one way of looking at this problem: trying to understand the function of each of the proteins and then seeing where it fits into our broader understanding of the virus as a whole. I think that you have to be very careful with the genetic experiments. You have to do backcrosses constantly to make sure the gene you are studying is in fact involved, because you can get second site revertants that will revert phenotypes to look like wild-type. But, I'm not sure what to say beyond that. I think we are at a stage of such enormous ignorance about this field that we just need to continue learning.

MARTIN: Let me ask a very general question: Are you or anybody else aware of a situation where a virus can interact with a cell receptor that is not involved in the adsorption/penetration phase of infection?

FIELDS: Other receptors?

MARTIN: Yes.

FIELDS: The Epstein-Barr virus is binding to one of the complement receptors, C3d. There is a report that has been questioned, but I think there is some interesting evidence about it, in terms of the acetylcholine receptor in rabies. The recent sequence of the binding site of the rabies' glycoprotein has striking homology to ligands and acetylcholine. On the HTLV-III, Robin Weiss' data say that the T4 receptor binds the virus.

GALLO: But that's not the question, is it?

FIELDS: Is that what you're asking?

MARTIN: I heard a lecture about a month ago involving the HTLV family of viruses. The speaker referred to the interaction of virus and lymphocyte receptors but not in the context of entry.

GALLO: I thought that's what your question was.

FIELDS: You're specifically asking about entry.

MARTIN: No, I said something other than entry.

MOSS: Vaccinia virus makes a growth factor which binds to the epidermal growth factor receptor. That probably causes a possible stimulation of cell growth, but as far as we know, that has nothing to do with entry of the vaccinia virus.

KAMELY: How realistic or how feasible is it to set up a test system in tissue culture and then extrapolate to organs or animals with respect to tissue specificity? Does it make sense to use such an extrapolation approach?

FIELDS: I think that's one area where from the little we know, it is very dangerous to extrapolate from tissue culture into in vivo situations. I think one of the reasons that I highlighted that and spoke about these mutants that grew perfectly well in L cells and not in the brain was to illustrate the difference of an L cell from a neuron.

GALLO: I don't think anybody would debate that, but there are opposite examples. In the case of HTLV-I, I cannot think of a single exception to the similarities of the in vitro transformed cells to what we see in vivo in HTLV-I leukemias. The characteristics of the cells in vitro are just like what we see in vivo in many ways.

FIELDS: Yes. Well, there is your answer. In other words, if you can show cell culture growth mimicking in vivo situations, then you're okay; but you had better be careful.

KAMELY: You were talking about established cell lines such as L cells and 3T3 cells. What if you use differentiated cells? There are some cells that you can grow in culture–like liver cells or some organ explants–then proceed to examine those.

FIELDS: I can tell you about reo. For reo you can take explants of brain, as Mark Dichter has done. Type 3 grows in neurons and type 1 grows mainly in background cells. Those are primary explants, and they come much closer to what happens in the brain. But then you take neuroblastoma, which you can differentiate, and once you differentiate it, it doesn't support type 3. However, undifferentiated neuroblastoma supports type 3 and type 1. So I just don't think we know enough yet for this particular group of viruses to be able to say how closely we could trust cell cultures. I know that I wouldn't trust in vitro systems without also testing them in an animal model.

KILBOURNE: I would look at your work another way. I would say that it is teaching us that organ specificity has to be segmented into compartments. This was observed earlier, again with some of the neurotropic flu strains, in that they would be ependymal. Those strains which had the incomplete replicative cycle were multiplying in parenchymal cells. However, I think your work elegantly shows that to say something is neurotropic is meaningless; it has to be broken down.

The Role of Enhancer Elements in Viral Host Range and Pathogenicity

LIONEL FEIGENBAUM AND GEORGE KHOURY
Laboratory of Molecular Virology
Division of Cancer Etiology
National Cancer Institute
National Institutes of Health
Bethesda, Maryland 20205

Studies with the small DNA viruses have provided insights into a number of regulatory elements used by eukaryotic genes for the synthesis and processing of mRNA. About 4 years ago, the elucidation of enhancer elements (Gluzman and Shenk 1983; Khoury and Gruss 1983) represented a breakthrough which has subsequently had important implications for our understanding of eukaryotic gene regulation. Enhancers are transcriptional control elements, working in concert with other promoter elements with the following characteristics (see Table 1):

1. Enhancers are short sets of nucleotides (50–100 bp) often tandemly repeated.
2. They are *cis*-acting regulatory elements that must be physically linked to the set of promoter elements which they influence.
3. By a mechanism that is not yet clear, enhancers significantly raise the overall transcriptional efficiency of the nearest promoters.
4. Enhancer function is relatively independent of position and orientation.
5. Although enhancers may contain short stretches of recognizable "core" sequences, in general, they are substantially different from one another in nucleotide sequence.
6. Whereas some enhancers work in a variety of cell types, many manifest a tissue- or species-specificity.

This last property of host-cell or species-specificity of an enhancer may play a role in determining the host range of certain viruses in which enhancers have been identified. In addition, enhancer specificity appears to be a critical factor in determining the disease potential of some of these viruses. This characteristic of viral enhancers will be considered in detail in this paper.

Table 1
Viral Enhancer Elements

Virus[a]	Number of repeats	Repeat length (in bp)	Sequence[d]
SV40*	2	72	GGTGTGGAAAGTCCCCAGGCTCCCCAGCAGGCAGAAGTA
JCV*	2	98	CTGTATATATAAAAAAAAGGGAAGGGATGGCTGCCAGC
BKV*	3	68	CACAGGGAGGAGCTGCTTACCCATGGAATGCAGCCAAAC
LPV*	2	60	CCAAATGGCGGGCTAATTTAAAAAAGGCGGGCTTCTTGG
Mo-MSV*	2	73/72	ACAGCTGAATATGGGCCAAACAGGATATCTGTGGTAAG
Mo-MuLV*	2	75	AACAGATGGAACAGCTGAATATGGGCCAAACAGGATAT
T^+ MuLV[b] (B tropic)	2	53	GGAAAGTACCGGGACTAGGGCCAAACAGGATATCTGTG
Fr-MuLV	2	65/74	ACGTTGGGCCAAACAGGATATCTGTGGTAAGCAGTTTCG
FeSV	1	63	GTTAGAGGCTAAAACAGGATATCTGTGGTTAAGCACCTG
SL3-3*	2.5	72	CCCGGCCCAGGGCCAAGAACAGATGGTCCCCAGACCGCT
Akv	2	99	ACAGAGAGGCTGGAAAGTACCGGGACTAGGGCCAAACA
HTLV-1	2	50	GGCCCAGACTAAGGCTCTGACGTCTCCCCCCAGAGGGAC
PrRSV* [A][c]	1	62	CCAAATAAGCAGGCAAGACAGCTATTGTAACTGCGAAA
PrRSV* [B][c]	1	101	AATGTAGTCTTATGCAATACTCCTGTAGTCTTGCAACAT
PrRSV* [C][c]	1	87	CATGCCGATTGGTGGTAGTAAGGTGGTACGATCGTGCCT

[a]Sequence of putative and confirmed (designated with an asterisk) enhancer elements
[b]T^-MuLV (N-tropic) contains only one 53-bp repeat.
[c]A, B, and C represent contiguous domains of the PrRSV enhancer. In combination, [A+B], [B+C], dimer B and trimer B were shown to function as enhancer elements.
[d]References for the sequences and/or demonstration of enhancer elements: SV40 (Benoist and Chambon 1981; de Villiers and Schaffner 1981; Gruss et al. 1981; Fromm and Berg 1982); JCV (Kenney et al. 1984); BKV (Rosenthal et al. 1983); LPV (M. Pawlita, pers. comm.); Mo-MSV (Laimins et al. 1982); Mo-MuLV (Linney et al. 1984); T^+MuLV (DesGroseillers et al. 1983a); Fr-MuLV (Chatis et al.1984): FeSV (Hampe et al. 1983); SL3-3 (Celander and Haseltine 1984); Akv (Lenz et al. 1984); HTLV-1 (Josephs et al. 1984); and PrRSV (Laimins et al. 1984)

Sequence[d]
TGCAAAGCATGCATCTCAATTAGTCAGCAACCA
CAAGCATGAGCTCATACCTAGGGAGCCAACCAGCTAACAGCCAGTAAACAAAGCACAAGG
CATGACCTCAGGAAGGAAAGTGCATGACT
CGGCGCTGATGTAAATGAGTA
CAGTTCCTGCCCCGCTCAGGGCCAAGAACAGATGG
CTGTGGTAAGCAGTTCCTGCCCCGGCTCAGGGCCAAG
GTCAAGCACTAGGGC
GCCCCGGCCCGGGGCCAAGAACAGAT
GGCCCCGGCTTGAGGCCAAGAACA
AACGACAGGATATCTGTGGTTAAGCACTAGGGC
GGATATCTGTGGTCAAGCACTAGGGCCCCGGCCCAGGGCCAAGAACAGATGGTCCCCAGAA
AGCTCAGCACC
TACGCTTTTGCATAGGGAGGGGGA
GCAACATGCTTATGTAACGATGAGTTAGCAATATGCCTTACAAGGAAAGAAAAGGCACCGTG
TATTAGGAAGGTATCAGACGGGTCTAACATGGATTGGACGAACCACTG

ENHANCER ELEMENTS AND RETROVIRUS-INDUCED DISEASE

The presence of enhancer elements in the long terminal repeats (LTRs) of retroviruses (see Fig. 1) was first suggested by the observation of Dhar et al. (1980) that Moloney murine sarcoma virus (MSV) contained a 72-bp repeated sequence in the U3 region of the LTR which was known to harbor transcriptional control elements. Although the MSV repeat sequence was fundamentally different from the SV40 72-bp tandem repeat (Fig. 2), Levinson et al. (1982) demonstrated that it could substitute for a deleted simian virus 40 (SV40) enhancer in activating early gene expression and allowing for SV40 replication in monkey kidney cells. Studies in several laboratories also demonstrated that the retrovirus enhancer elements function with host-cell specificity (Laimins et al. 1982; Kriegler and Botchan 1983). For example, the MSV enhancer, in contrast to the SV40 enhancer, functions more efficiently in rodent cells than in primate cells. Even et al. (1983) compared the transformation potential of the feline sarcoma virus, FeSV, (which contains the *fes* oncogene within the LTRs of feline sarcoma virus) with an isologous chimeric construct containing the U3 region from MSV in place of the FeSV U3 region. Whereas both viruses grew equally well in mink cells, the exchange of the transcriptional regulatory sequences from FeSV for those of MSV provided for a 50–100-fold increase in the transformation frequency of the recombinant virus on mouse NIH 3T3 cells. The increased transformation frequency correlated directly with an enhanced transcriptional efficiency of the recombinant virus in mouse cells.

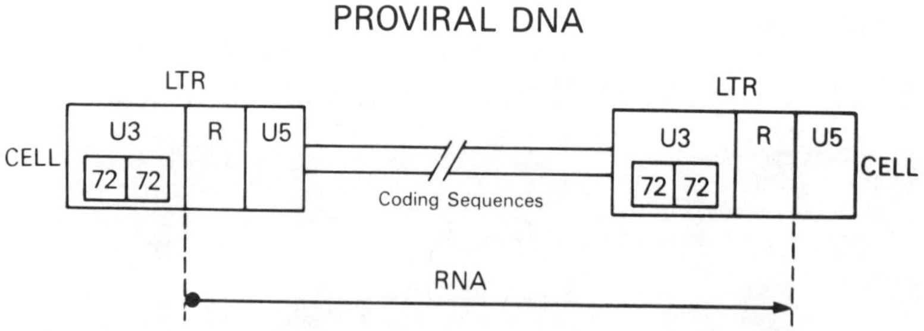

Figure 1
Schematic diagram of retrovirus proviral DNA. An integrated copy of proviral DNA is shown to contain two flanking long terminal repeats (LTRs) and a central set of coding sequences. The LTR is divided into three regions. Region R is so named because it is repeated at both ends of the viral RNA transcript; it begins at the cap site for transcription and ends at the poly(A) addition site. The U3 sequences are unique to the 3′ end of the viral transcript whereas the U5 sequences are unique to the 5′ end of the viral transcript. In proviral DNA, the U3 sequences represent nontranscribed upstream elements adjacent to the initiation site for transcription. They have been shown to contain the regulatory sequences, including enhancers (e.g., 72-bp repeats), which govern proviral transcription events.

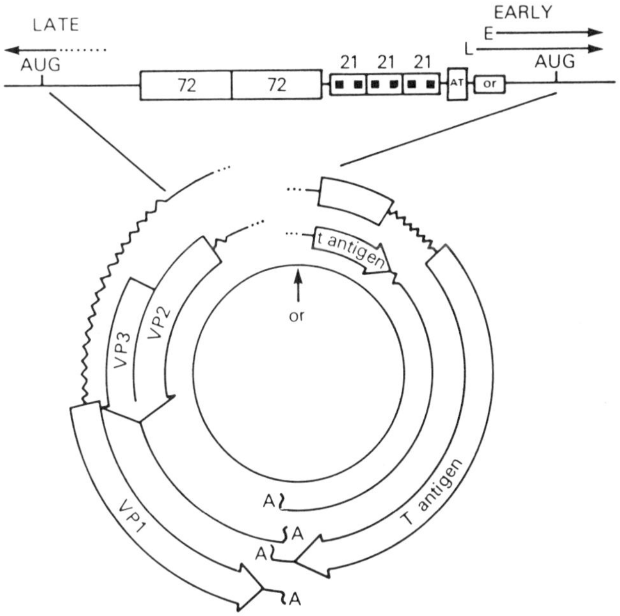

Figure 2
Transcriptional control region of papovavirus SV40. Above the circular genome of SV40 is presented a schematic diagram of the control region for SV40 transcription. Positioned between the early transcription unit (encoding large T- and small T-antigens) and the late transcription unit (encoding viral capsid proteins VP-1, 2, and 3) is situated a set of control sequences which includes the origin for viral DNA replication or the Goldberg-Hogness or TATA box (AT), three 21-bp repeats containing two G-C rich hexanucleotides in each copy, and the enhancer element, a tandem 72-bp repeat. Similar sets of regulatory sequences have been demonstrated in the control regions of the other papovaviruses. The enhancer element has been shown to be as an essential transcriptional regulatory sequence for early SV40 gene transcription.

In an analogous study, Anderson and Scolnick (1983) inserted the transforming gene of Rous sarcoma virus (*src*) into a construct containing the control sequences of a murine retrovirus. The recombinant virus, introduced as a pseudotype intravenously into NIH Swiss mice, resulted in the appearance of splenic foci, spenomegaly, and anemia. Infectious transforming viruses recovered from the spleens of the diseased animals encoded an active pp60*src* gene product. Since the avian sarcoma virus, Rous sarcoma virus (RSV), never produces this disease in rodents, the likely interpretation of these experiments is that the host range of the recombinant, as well as the tissue tropism was determined by the murine retroviral control elements, and the disease was induced by the *src* gene product. More recently, Feuerman et al. (1985) have constructed a recombinant retrovirus in which the Rous v-*src* gene is expressed as a *gag-src* fusion protein in Moloney murine leukemia virus

(Mo-MuLV). Although the infectivity of the retrovirus for mice was closely related to the Mo-MuLV, the production of fibrosarcomas at the site of inoculation indicated a contribution of the *src* oncogene as well.

Analogous experiments have been conducted with competent retroviruses, which generally do not contain an oncogene, to localize the sequences responsible for their host range. Several groups have carried out elegant experiments which demonstrate convincingly that the LTRs of Friend murine leukemia virus and Mo-MuLV are responsible for the disease spectra of these two retroviruses. In one study, the 3′ LTR of the Friend viral genome containing presumed regulatory elements was replaced by the analogous sequence from Mo-MuLV (Chatis et al. 1983). When introduced into mice, the recombinant virus produced thymic lymphomas almost exclusively. Conversely, the recombinant in which the Friend viral regulatory elements replaced the corresponding region from Mo-MuLV, induced principally erythroleukemias. Thus, in both cases the disease spectrum corresponded to the viral LTR retained in the recombinant genome. A subsequent study by these investigators (Chatis et al. 1984) definitively localized the sequences responsible for the disease specificities to the U3 region of the LTR, presumably the transcriptional enhancer elements.

In parallel experiments, which were designed to investigate the disease specificity of two murine leukemia virus variants, DNA recombinants were constructed between the two viral strains; one strain replicated efficiently in thymocytes (T^+) whereas the other strain (T^-) did not (DesGroseillers et al. 1983a). Since it seems likely that the ability of the murine leukemia viruses to replicate efficiently in thymocytes parallels their leukemogenicity, it was predicted that the efficiency of replication of the T^+/T^- recombinants might identify segments of the genome responsible for pathogenicity. Infection of mice with a number of T^+/T^- reciprocal recombinants indicated that sequences from the LTR of the T^+ strain were required for efficient replication in thymocytes. Recently, investigators have more precisely localized the critical elements to the direct repeat sequence present in the LTR of the T^+ strain (DesGroseillers and Jolicoeur 1984). It seems likely that this sequence corresponds to an enhancer element with a specificity for the thymocytes. In a similar study, these same investigators have acquired evidence from recombinant viruses to indicate that the high leukemogenic potential of the Gross passage A murine leukemia virus maps to sequences in the LTR (DesGroseillers et al. 1983b).

A detailed analysis of a comparable pair of murine leukemia viruses was carried out by Haseltine and his colleagues (Lenz et al. 1984). The Akv virus is not leukemogenic whereas a similar virus SL3-3 is highly leukemogenic. Recombinants between the two viruses showed clearly that the leukemogenic capacity could be localized to sequences in the LTR of the SL3-3 strain. More recently, it has been demonstrated that these LTR sequences represent transcriptional control elements which greatly enhance the expression of SL3-3 in thymocytes. Sequence comparison of the LTRs of these two viruses indicates that they differ in a number of single-point mutations in the enhancer region.

Whereas Mo-MuLV replicates efficiently in most mouse cells, infection of undifferentiated embryonal carcinoma cells leads to the integration of proviral DNA but not to virus production. Although a number of models might explain this phenomenon, it seems most likely that the Mo-MuLV enhancer is nonfunctional in undifferentiated mouse cells. In support of this supposition, Linney and his colleagues inserted a mutant of the polyoma virus enhancer into the retroviral LTR (Linney et al. 1984). This mutant polyoma virus enhancer functions efficiently in undifferentiated embryonal carcinoma cells. Using a transient assay, these investigators demonstrated that substitution of the polyoma sequences into the retroviral LTR led to activation of the Mo-MuLV promoter in the undifferentiated cells. Davis et al. (1985) have demonstrated, in addition, that insertion of the polyoma viral enhancer into a Moloney virus LTR abolishes the leukemogenicity of Mo-MuLV without substantially decreasing the potential for the virus to replicate. Perhaps, the best explanation for this result is that the polyoma virus enhancer alters the tissue-specificity of the recombinant virus preventing it from replicating efficiently in the target cell for leukemic transformation.

Two RSV transformation-defective variants (i.e., viruses which have lost the *src* oncogene) differ substantially in their disease spectra. The transformation-defective (td) Schmidt-Ruppin strain induces thymic lymphomas while the td Prague strain causes a number of other diseases (e.g., osteopetrosis) but induces lymphomas only with very low frequency (Robinson et al. 1982). Differences in the enhancer regions of these closely related RSV variants may contribute to the diseases they cause in chickens. The 3′ terminal region of the Prague strain contains three distinct enhancer domains that comprise two functional enhancer elements. Two of these domains, located in the U3 region of the 3′ long terminal repeat, are almost identical to those found in the Schmidt-Ruppin strain (Luciw et al. 1983). The third enhancer region (XSR), located upstream from the LTR may be important in the activation of downstream genes. This third domain, located outside the LTR, is quite distinct in the two strains of virus, suggesting a possible model for differential activation of cellular genes by either Schmidt-Ruppin or Prague RSV (Laimins et al. 1984).

Although many avian retroviruses lacking an oncogene such as Rous-associated virus (RAV-1) and the td mutants of RSV are able to cause diseases by integrating near a proto-oncogene, the Rous-associated virus, RAV-0, is not. Recently, Weber and Schaffner (1985) have shown that RAV-0 lacks an enhancer. This further supports the model of viral pathogenicity through enhancer insertion.

The human T cell leukemia/lymphoma (HTLV) retroviruses appears to cause its pathogenic effect in human T lymphocytes. This group of viruses includes HTLV-I, which induces human T cell leukemia/lymphoma; HTLV-II, which has been associated with hairy cell leukemia; bovine leukemia virus (BLV); and perhaps HTLV-III (LAV or AIDS retrovirus), associated with the acquired immune deficiency syndrome or AIDS. It is not yet entirely clear whether the enhancer elements associated with these retroviruses are the principal determinants of their tissue tropism (Sodroski et al. 1984; Derse et al. 1985; Rosen et al. 1985; Sodroski et al. 1985). It has been suggested that virus encoded *trans*-acting proteins, the pX or LOR gene products, are responsible for activation of transcription from the viral LTRs. Thus, it is possible that the pX gene product is expressed in a tissue-specific way and that this protein in turn may further stimulate viral RNA expression in the same cell.

ENHANCER ELEMENTS IN THE HOST RANGE OF PAPOVAVIRUSES

Several studies have shown that the enhancer elements of SV40 (Fig. 2) and polyoma virus contribute to the host range of these viruses (de Villiers et al. 1982; Laimins et al. 1982). Nevertheless, both enhancer elements are somewhat promiscuous and function at reasonable levels in most cells except undifferentiated cells (see below). Thus, while the level of early gene expression controlled by these enhancer elements may be maximal in the cell which serves as the lytic host for the virus, the enhancer does not appear to be the principal determinant of host range in these cases. It seems more likely that replication-specific proteins interact with the polyoma virus replication origin in mouse cells and with the SV40 replication origin in monkey kidney cells, thus permitting full expression of the lytic cycle.

In contrast, there are situations in which a papovavirus enhancer element specifically determines the ability of a virus to be expressed in a particular host cell. One of these mentioned above is the expression of polyoma virus mutants in embryonal carcinoma cells. Although the wild-type virus is both unable to express its early gene product efficiently or to replicate in undifferentiated cells, mutants in the enhancer region (Py-EC mutants) overcome both of these restrictions (Katinka et al. 1980; Fujimura et al. 1981; Sekikawa and Levine 1981). The polyoma virus enhancer element contributes both to the expression of early viral RNA (which encodes the replication-required protein, T-antigen) and to a separate but essential requirement for viral DNA replication (Fujimura et al. 1981; de Villiers et al. 1984; Veldman et al. 1985). Thus, wild-type polyoma virus containing the normal enhancer element will not replicate in undifferentiated EC cells, even in the presence of functional levels of Py T-antigen generated by a Py-EC mutant. Whether the specific role of the enhancer in DNA replication is linked to its ability to stimulate transcription and/or its function in chromatin organization is unclear.

Human papovavirus JCV is closely associated with the degenerative neurologic disease progressive multifocal leukoencephalophathy (PML). The virus replicates efficiently only in human glial cells in tissue culture and has been repeatedly isolated from demyelinated plagues found in the brains of patients with this disease. Recent studies have shown that the JCV enhancer element functions efficiently only in glial cells, but not in other primate cells such as HeLa or AGMK (Kenney et al. 1984). Thus, it would appear that the expression of JCV T-antigen, under the control of the early viral promoter, is restricted to glial cells. Whether or not there are additional restrictions (such as replication-associated proteins) that limit the growth of JCV to human glial cells is a question presently under investigation. Major and his colleagues (1985) have recently obtained an SV40-transformed human glial cell line which constitutively produces SV40 T-antigen, and continues to express a number of cell markers specific for glial cells. Although JCV replicates efficiently in these cells, its replication appears to be dependent upon the production of JCV T-antigen from the transfected genome and not on the endogenous SV40 T-antigen. This observation appears to conflict with recent studies of Li and Kelly (1985) who have shown that in vitro replication of JCV DNA can be supported by SV40 T-antigen with 20% of the efficiency of SV40 DNA replication in vitro. It is clear, however, that the present in vitro replication assay scores for only a subset of the determinants required for efficient in vivo replication. For example, there is no additional benefit seen in vitro after addition of transcriptional control sequences (such as the 21-bp repeats of SV40) to the minimal SV40 origin of replication (J. Li and T. Kelly, pers. comm.). Yet in vivo, addition of these sequences stimulates replication at least fivefold (Bergsma et al. 1982).

CONCLUSIONS

Clearly, the host range and tissue-specificity of the papovaviruses and retroviruses is complex. The life cycle includes a number of stages such as penetration and uncoating, early and late gene expression, replication, packaging, and release. Any of these stages in the life cycle can affect the relative efficiency of expression of the virus in a particular tissue. Nevertheless, it is clear that one of the most important steps involves expression of "early" viral genes, which for these groups of viruses is generally under the control of a viral enhancer element (Fig. 2).

It is sometimes difficult to predict from the tissue-specificity of an enhancer element in tissue culture how the virus will behave in animal systems. For example, recent experiments with transgenic mice carrying the SV40 genome in every cell suggested that the expression of the SV40 early genes (as evidenced by T-antigen production and tumor formation) occurs only in the chorid plexus of the cerebral ventricles (Brinster et al. 1984). More recent studies in which the SV40 T-antigen has been linked to regulatory elements from the insulin gene showed that gene expression was restricted to the predicted tissue, namely, the β-cells of the pancreas (Hanahan 1985). A more complex picture results from the linkage of the mouse mammary tumor virus (MMTV) LTR to the *myc* gene (Stewart et al. 1984). The MMTV transcriptional regulatory elements require induction by steroid hormones; in fact, Stewart et al. (1984) observed tumor production in the mammary tissue of transgenic mice carrying this chimeric construct but only during pregnancy (presumably in response to high levels of hormone). These few examples serve to underscore the potential ability to target genes in animal systems for tissue-specific expression. When such regulatory elements are coupled to viral genomes, it is clear that they can affect the tissue in which the virus expresses its gene products. As such, these transcriptional regulatory elements (e.g., enhancers and promoters), which show tissue-specific expression, represent both a powerful tool and conceivably a potential hazard. The linkage of particular transcriptional control sequences to specific genes and the subsequent introduction of these recombinant molecules into animals, either through viral vectors or transgenic manipulations, deserves judicious planning and appropriate precautions.

ACKNOWLEDGMENT

We thank P. Howley and L. Laimins for helpful advice and M. Priest for preparation of the manuscript.

REFERENCES

Anderson, S.M. and E.M. Scolnick. 1983. Construction and isolation of a transforming murine retrovirus containing the SRC gene of Rous sarcoma virus. *J. Virol.* **46**: 594.

Benoist, C. and P. Chambon. 1981. *In vivo* sequence requirements of the SV40 early promoter region. *Nature* **290**: 304.

Bergsma, D.J., D.M. Olive, S.W. Hartzell, and K.N. Subramanian. 1982. Territorial limits and functional anatomy of the simian virus 40 replication origin. *Proc. Natl. Acad. Sci. U.S.A.* **79**: 381.

Brinster, R., H. Chen, A. Messing, T. Van Dyke, A. Levine, and R. Palmiter. 1984. Transgenic mice harboring SV40 T-antigen genes develop characteristic tumors. *Cell* **37**: 367.

Celander, D. and W.A. Haseltine. 1984. Tissue-specific transcription preference as a determinant of cell trophism and leukaemogenic potential of murine retroviruses. *Nature* **312**: 159.

Chatis, P., C. Holland, J. Hartley, W. Rowe, and N. Hopkins. 1983. Role of the 3′ end of the genome in determining the disease specificity of Friend and Moloney murine leukemia viruses. *Proc. Natl. Acad. Sci. U.S.A.* **80**: 4408.

Chatis, P., C.A. Holland, J.E. Silver, T.N. Frederickson, N. Hopkins, and J.W. Hartley. 1984. A 3′ end fragment encompassing the transcriptional enhancers of nondefective Friend virus confers erythroleukemogenicity on Moloney leukemia virus. *J. Virol.* **52**: 248.

Davis, B., E. Linney, and H. Fan. 1985. Suppression of leukaemia virus pathogenicity by polyoma virus enhancers. *Nature* **314**: 550.

Derse, D., S. Caradonna, and J. Casey. 1985. Bovine leukemia virus long terminal repeat: A cell-type specific promoter. *Science* **227**: 317.

DesGroseillers, L. and P. Jolicoeur. 1984. Mapping the viral sequences conferring leukemogenicity and disease specificity in Moloney and amphotropic murine leukemia viruses. *J. Virol.* **52**: 448.

DesGroseillers, L., E. Rassart, and P. Jolicoeur. 1983a. Thymotropism of murine leukemia virus is conferred by its long terminal repeat. *Proc. Natl. Acad. Sci. U.S.A.* **80**: 4203.

DesGroseillers, L., R. Villemur, and P. Jolicoeur. 1983b. The high leukemogenic potential of Gross passage A leukemia virus maps in the region of the genome corresponding to the long terminal repeat and to the 3′ end of *env*. *J. Virol.* **47**: 24.

de Villiers, J. and W. Schaffner. 1981. A small segment of polyoma virus DNA enhances the expression of a cloned β-globin gene over a distance of 1400 base pairs. *Nucl. Acids Res.* **9**: 6251.

de Villiers, J., L. Olson, C. Tyndall, and W. Schaffner. 1982. Transcriptional "enhancers" from SV40 and polyoma virus show a cell type preference. *Nucl. Acids Res.* **10**: 7965.

de Villiers, J., W. Schaffner, C. Tyndall, S. Lupton, and R. Kamen. 1984. Polyoma virus DNA replication requires an enhancer. *Nature* **312**: 242.

Dhar, R., W.L. McClements, L.W. Enquist, and G.F. Vande Woude. 1980. Nucleotide sequences of integrated Moloney sarcoma provirus long terminal repeats and their host and viral junctions. *Proc. Natl. Acad. Sci. U.S.A.* **77**: 3937.

Even, J., S.J. Anderson, A. Hampe, F. Galibert, D. Lowy, G. Khoury, and C.J. Sherr. 1983. Mutant feline sarcoma proviruses containing the viral oncogene (v-*fes*) and either feline or murine control elements. *J. Virol.* **45**: 1004.

Feuerman, M.H., B.R. Davis, P.K. Pattengale, and H. Fan. 1985. Generation of a recombinant Moloney murine leukemia virus carrying the v-*src* gene of avian sarcoma virus: Transformation *in vitro* and pathogenesis *in vivo*. *J. Virol.* **54**: 804.

Fromm, M. and P. Berg. 1982. Deletion mapping of DNA regions required for SV40 early region promoter function *in vivo*. *J. Mol. Appl. Genet.* **1**: 457.

Fujimura, F.K., P.L. Deininger, T. Friedmann, and E. Linney. 1981. Mutation near the polyoma DNA replication origin permits productive infection of F9 embryonal carcinoma cells. *Cell* **23**: 809.

Gluzman, Y. and T. Shenk (ed.). 1983. *Enhancers and eukaryotic gene expression—Current communications in molecular biology*. Cold Spring Harbor Laboratory, Cold Spring Harbor, New York.

Gruss, P., R. Dhar, and G. Khoury. 1981. Simian virus 40 tandem repeated sequences as an element of the early promoter. *Proc. Natl. Acad. Sci. U.S.A.* **78**: 943.

Hampe, A., M. Gobet, J. Even, C.J. Sherr, and F. Galibert. 1983. Nucleotide sequences of feline sarcoma virus long terminal repeats and 5′ leaders show extensive homology to those of other mammalian retroviruses. *J. Virol.* **45**: 466.

Hanahan, D. 1985. Heritable formation of pancreatic β-cell tumours in transgenic mice expressing recombinant insulin/simian virus 40 oncogenes. *Nature* **315**: 115.

Josephs, S.F., F. Wong-Staal, V. Manzari, R.C. Gallo, J.G. Sodroski, M.D. Trus, D. Perkins, R. Patarca, and W.A. Haseltine. 1984. Long terminal repeat structure of an American isolate of type I human T-cell leukemia virus. *Virology* **139**: 340.

Katinka, M., M. Yaniv, M. Vasseur, and D. Blangy. 1980. Expression of polyoma early functions in mouse embryonal carcinoma cells depends on sequence rearrangements in the beginning of the late region. *Cell* **20**: 393.

Kenney, S., V. Natarajan, D. Strike, G. Khoury, and N.P. Salzman. 1984. JC virus enhancer-promoter active in the brain cells. *Science* **226**: 1337.

Khoury, G. and P. Gruss. 1983. Enhancer elements. *Cell* **33**: 313.

Kriegler, M. and M. Botchan. 1983. Enhanced transformation by a simian virus 40 recombinant virus containing a Harvey murine sarcoma virus long terminal repeat. *Mol. Cell. Biol.* **3**: 325.

Laimins, L.A., P. Tsichlis, and G. Khoury. 1984. Multiple enhancer domains in the 3′ terminus of the Prague strain of Rous sarcoma virus. *Nucl. Acids Res.* **12**: 6427.

Laimins, L.A., G. Khoury, C. Gorman, B. Howard, and P. Gruss. 1982. Host-specific activation of gene expression by 72 base pair repeats of simian virus 40 and Moloney murine leukemia virus. *Proc. Natl. Acad. Sci. U.S.A.* **79**: 6453.

Lenz, J., D. Celander, R. Crowther, R. Patarca, D. Perkins, and W. Haseltine. 1984. Determination of the leukaemogenicity of a murine retrovirus by sequences within the long terminal repeat. *Nature* **308**: 467.

Levinson, B., G. Khoury, G. Vande Woude, and P. Gruss. 1982. Activation of SV40 genome by 72-base pair tandem repeats of Moloney sarcoma virus. *Nature* **295**: 568.

Li, J.J. and T.J. Kelly. 1985. Simian virus 40 DNA replication *in vitro*: Specificity of initiation and evidence for bidirectional replication. *Mol. Cell. Biol.* **5**: 1238.

Linney, E., B. Davis, J. Overhauser, E. Chao, and H. Fan. 1984. Non-function of a Moloney murine leukaemia virus regulatory sequence in F9 embryonal carcinoma cells. *Nature* **308**: 470.

Luciw, P.A., J.M. Bishop, H.E. Varmus, and M.R. Capecchi. 1983. Location and function of retroviral and SV40 sequences that enhance biochemical transformation after microinjection of DNA. *Cell* **33**: 705.

Major, E.O., A.E. Miller, P. Mourrain, R.G. Traub, E. De Widt, and J. Sever. 1985. Establishment of a line of human fetal glial cells that supports JC virus multiplication. *Proc. Natl. Acad. Sci. U.S.A.* **82**: 1257.

Robinson, H.L., B.M. Blais, P.N. Tsichlis, and J.M. Coffin. 1982. At least two regions of the viral genome determine the oncogenic potential of avian leukosis viruses. *Proc. Natl. Acad. Sci. U.S.A.* **79**: 1225.

Rosen, C., J. Sodorski, R. Kettman, A. Burny, and W. Haseltine. 1985. Transactivation of the bovine leukemia virus long terminal repeat in BLV-infected cells. *Science* **227**: 320.

Rosenthal, N., M. Kress, P. Gruss, and G. Khoury. 1983. BK viral enhancer element and a human cellular homolog. *Science* **222**: 749.

Sekikawa, K. and A. Levine. 1981. Isolation and characterization of polyoma host range mutants that replicate in nullipotential embryonal carcinoma cells. *Proc. Natl. Acad. Sci. U.S.A.* **78**: 1100.

Sodroski, J.G., C.A. Rosen, and W.A. Haseltine. 1984. Trans-acting transcriptional activation of the long terminal repeat of human T lymphotrophic viruses in infected cells. *Science* **225**: 381.

Sodroski, J., C. Rosen, F. Wong-Staal, S. Salahuddin, M. Popovic, S. Arya, R. Gallo, and W. Haseltine. 1985. Trans-acting transcriptional regulation of human T-cell leukemia virus type III long terminal repeat. *Science* **227**: 171.

Stewart, T.A., P.K. Pattengale, and P. Leder. 1984. Spontaneous mammary adenocarcinomas in transgenic mice that carry and express MTV/*myc* fusion genes. *Cell* **36**: 627.

Veldman, G.M., S. Lupton, and R. Kamen. 1985. Polyomavirus enhancer contains multiple redundant sequence elements that activate both DNA replication and gene expression. *Mol. Cell. Biol.* **5**: 649.

Weber, F. and W. Schaffner. 1985. Enhancer activity correlates with the oncogenic potential of avian retroviruses. *Eur. Mol. Biol. Organ. J.* **4**: 949.

COMMENTS

MOSS: Is it possible to replace the SV-40 enhancer with another enhancer and have a replication competent virus?

KHOURY: Yes. In fact, we did that experiment several years ago. Barbara Levinson put the MSV enhancer into an SV-40 plasmid from which the SV-40 enhancer had been removed. The recombinant was a replication competent virus. It was much less efficient than wild-type SV-40 and produced much less T-antigen in monkey kidney cells. We thought that perhaps we had created a recombinant virus which could now extend its host range to mouse cells. It did not replicate in murine cells presumably because there are replication-essential proteins in monkey cells but not in mouse cells which recognize the SV-40 origin. Yet, there was production of as much, if not more, T-antigen in mouse cells from the recombinant virus. That was our first insight into tissue specificity of regulatory elements like enhancers.

AHMED: Does the T-antigen bind to the enhancer?

KHOURY: That's a good question. Using standard in vitro experimental techniques, T-antigen does not bind to the enhancer. There are experiments conducted in Peter Tegtmeier's laboratory, however, which suggest that under certain conditions T-antigen may bind to the enhancer.

AHMED: How do you explain its working and turning on the late genes?

KHOURY: I have suggested that there might be a site in the late region where either an attenuator or a repressor-type protein binds, and the interaction of T-antigen with the control region might somehow alter the template to overcome this effect. In addition, as I have suggested, there may be indirect means by which T-antigen activates late transcription.

The Genetic Basis of Leukemogenicity and Disease Specificity in Nondefective Mouse Retroviruses

NANCY HOPKINS
Center for Cancer Research
Department of Biology
Massachusetts Institute of Technology
Cambridge, Massachusetts 02139

OVERVIEW

Naturally occurring mouse C-type retroviruses differ in their ability to induce leukemia and in the type of leukemias they induce. Multiple viral and host genes contribute to these two phenotypes. By using recombinant DNA technology, it is possible to construct new viruses with novel combinations of genes derived from leukemogenic and nonleukemogenic viruses, and so to study the role of each component in leukemogenesis. Most important has been the finding that viral transcriptional signals can determine what type of cancer the virus will cause. This is probably because transcriptional signals encode host range and hence, organ tropism. Novel recombinant viruses that one constructs can have unexpected biological properties; however most manipulations yield viruses that are less leukemogenic than those that are selected by the process of tumor induction in the animal.

INTRODUCTION

Mice inherit many DNA copies of C-type RNA tumor viruses. Although the majority of these genes probably remain silent in the animal, some can be expressed. Sometimes virus expression culminates in the synthesis of complete infectious virus, and sometimes the end result of this is to cause cancer (Gross 1951, 1970; Rowe and Pincus 1972; Rowe 1973). Given their inheritance in all mice, their ability to be expressed, and their high recombination frequencies, it isn't surprising that a great many different C-type retroviruses have been isolated from mice. These viruses are closely related genetically but they differ dramatically in their ability to induce disease when injected back into animals. Many are nonpathogenic; many induce leukemia; a few induce other forms of cancer; and rare isolates induce a neurological disease. Among isolates that induce leukemia, the majority induce lymphoid neoplasms of T cells; however, some isolates induce tumors of B cells and others induce erythroleukemias.

My laboratory, in collaboration with Dr. Janet W. Hartley and the late Dr. Wallace P. Rowe of the National Institutes of Health, has been interested in the genetic basis for the difference in disease-inducing potential of murine leukemia viruses and in the mechanisms by which they induce disease. In particular, we have been interested in three questions: What viral genes make a mouse retrovirus leukemogenic? What viral genes determine whether a virus will induce T cell, B cell, or erythroleukemia? What is the mechanism of leukemogenesis? These studies in the context of major advances in the field of retrovirus-induced cancers have given us a reasonably clear picture of how nondefective mouse retroviruses induce various forms of leukemia. Several important conclusions have emerged:

1. Multiple viral genes contribute to the leukemogenicity of nondefective mouse retroviruses. (Lung et al. 1980, 1983; Kelly et al. 1983; Holland et al. 1985a).
2. As originally proposed by Hartley and Rowe, leukemogenic viruses of high leukemic inbred mice arise by recombination between different C-type viruses (Hartley et al. 1977; Cloyd et al. 1980). Multiple recombination events are involved. Thus, endogenous retroviruses serve as a reservoir of genes that can be used to construct novel viruses, some of which prove to be highly oncogenic (Thomas and Coffin 1982; Lung et al. 1983).
3. Murine leukemia viruses integrate adjacent to oncogenes in the T cells they transform, a result predicted from analogous, seminal findings with avian leukosis viruses (Hayward et al. 1981; Neel et al. 1981; Payne et al. 1981; Tsichlis et al. 1983; Corcoran et al. 1984; Cuypers et al. 1984; Li et al. 1984; and Steffen 1984).
4. Perhaps most novel and surprising, the type of leukemia induced can be determined by sequences that encode the viral transcriptional signals, and this is probably because these sequences confer host range including tissue tropism (Chatis et al. 1983, 1984; DesGroseillers et al. 1983a; Laimins et al. 1982; Holland et al. 1985b).
5. Finally, much of the work we have done has gone to confirm and provide molecular explanations for biological observations made many years ago, namely, the ability of nondefective retroviruses to induce a high incidence of T cell leukemia in mice is correlated with their ability to replicate extensively, within the first few weeks of life, within the thymus where tumor cells will arise (Lieberman and Kaplan 1959).

In terms of this volume, a relevant aspect of these studies is the question of how genetically altered retroviruses behave. When we construct variant viruses in a system that we now understand quite well, can we predict their disease-inducing profile? The answer is that although we are certainly getting better at it, quite often we are surprised. Faced with a new retrovirus with novel disease-inducing properties, one should have an open mind when trying to deduce the genetic basis for its phenotype.

RESULTS

The Contribution of Multiple Viral Genes to the Leukemogenicity of Nondefective Mouse Retroviruses

High leukemic inbred mice, such as AKR, inherit DNA copies of retroviruses that are expressed to yield infectious viruses soon after birth. As a result, the animals develop viremia and ultimately T cell lymphomas and leukemias (Gross 1951; Rowe 1972; Rowe and Hartley 1972). Some years ago it was shown that although the inherited virus that replicates readily in the mouse is responsible for the ultimate development of leukemia, nonetheless, this virus itself is nonleukemogenic. Rather, it undergoes recombination with other inherited viral sequences to generate a novel recombinant, called an MCF virus, that actually induces the leukemia (Hartley et al. 1977). MCF viruses were discovered because they differ from their nonleukemogenic progenitors in their envelope glycoproteins and thus have a new host range on tissue culture cells (Fischinger et al. 1975; Hartley et al. 1977).

Lung has analyzed the genomes of numerous MCF viruses isolated from individual mice by Hartley and Rowe (Lung et al. 1980, 1983). As might have been expected, the genome of each MCF virus was unique, reflecting the particular recombination events involved in its generation. Furthermore, as predicted, the glycoprotein genes of the MCF viruses always differed from their nonleukemogenic progenitors. Unexpectedly, however, there were other consistent genetic changes. Figure 1 is a diagram that shows the genetic structure of leukemogenic MCF viruses relative to their nonleukemogenic progenitors. The two types of viruses differ in the amino terminal two-thirds of their glycoprotein coding genes, in the sequences that encode the last few amino acids of Prp15E (the transmembrane protein that anchors the glycoprotein to the surface of the virion) and in U3, the portion of the LTR that encodes the viral transcriptional signals. In addition, many (possibly all) MCF viruses have alterations in the 5′ half of their genomes, but the methods of genome analysis used did not reveal if identical sequences are involved in these changes. Analysis of the genomes of additional MCF viruses isolated by O'Donnell revealed that the midportion of the Prp15E gene must be derived from the nonleukemogenic virus to make a leukemogenic MCF virus (M. Lung, N. Hopkins, and P. O'Donnell, unpubl.).

The studies described above strongly implied that at least three genetic elements were involved in determining the leukemogenic phenotype of MCF viruses: gp70, Prp15E, and the U3 region of the LTR (Lung et al. 1980, 1983; Kelly et al. 1983). To confirm this observation and to analyze the role of each segment in the leukemogenic process, we wished to construct retroviruses in which just one, two, or all three of these elements from a leukemogenic virus were introduced into the genome of a nonleukemogenic virus. This was done by Holland using molecular clones of an MCF virus and of its nonleukemogenic progenitor, called Akv virus

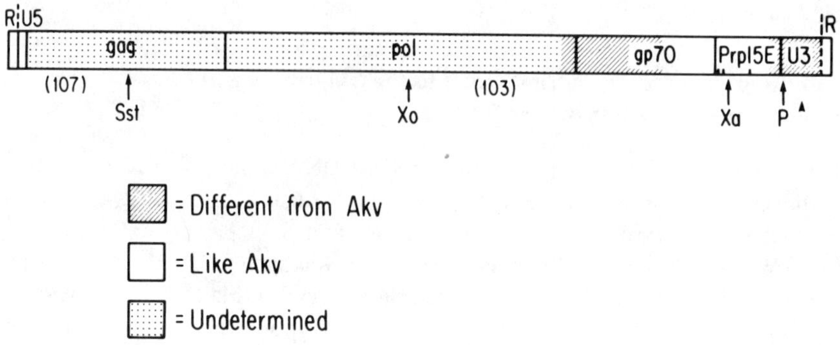

Figure 1
Diagrammatic representation of the RNA genome of the MCF 247 mouse retrovirus. This highly leukemogenic virus was isolated from an AKR mouse (Hartley et al. 1977). It arose by recombination between a nonleukemogenic but rapidly replicating virus, called Akv, and other endogenous mouse retrovirus sequences. The portions of the MCF genome that are identical to Akv are indicated in white and include the C terminal portion of the envelope glycoprotein gp70, the amino terminal portion of the transmembrane protein Prp15E, and the R and U5 portions of the genome that become part of the LTR in proviral DNA. Portions of the genome that are derived from endogenous viruses other than Akv are shown in black and include the amino terminal portion of gp70, the extreme C terminus of Prp15E, and the U3 portion that encodes the transcriptional signals used to transcribe viral RNA from integrated proviral DNA. The stipled area contains many sequences in common with Akv but these are interrupted by sequences known to be derived from other endogenous viruses (as indicated by the numbers 107 and 103). The precise structure of this region has not been determined by nucleotide sequencing. *Sst, Xo, Xa*, and *P* indicate positions of restriction endonuclease cleavage in proviral DNA that were useful in constructing recombinants between molecular clones of Akv and MCF 247 viruses. Reproduced, with permission, from Holland et al. 1985a.

(Lowy et al. 1980; Holland et al. 1985a,b). Holland was able to use restriction enzyme sites common to these two viruses to separate what we believed to be the critical portions of the genome (Lung et al. 1980, 1983; Chattopadhyay et al. 1981). The viruses she constructed were then tested for their leukemogenicity by Hartley. The results of these tests are shown in Figure 2. They clearly show that multiple viral genes contribute to the leukemogenicity of an MCF virus. As we had suspected, genome segments encoding gp70, Prp15E, and U3 are important, but to our surprise, even genes in the 5′ half can make a contribution to the phenotype (compare recombinants called 16 and 1 that differ in having their 5′ halves derived from either Akv or MCF).

Additional studies indicate that when viruses with only a few of their genes derived from the leukemogenic MCF virus succeed in causing leukemia, they do so by recombining with endogenous viruses to regenerate fully leukemogenic MCF viruses (Holland et al. 1985a).

A question of great interest is how the viral genes that have been identified in these studies contribute to the leukemogenicity of the virus. Although we have no

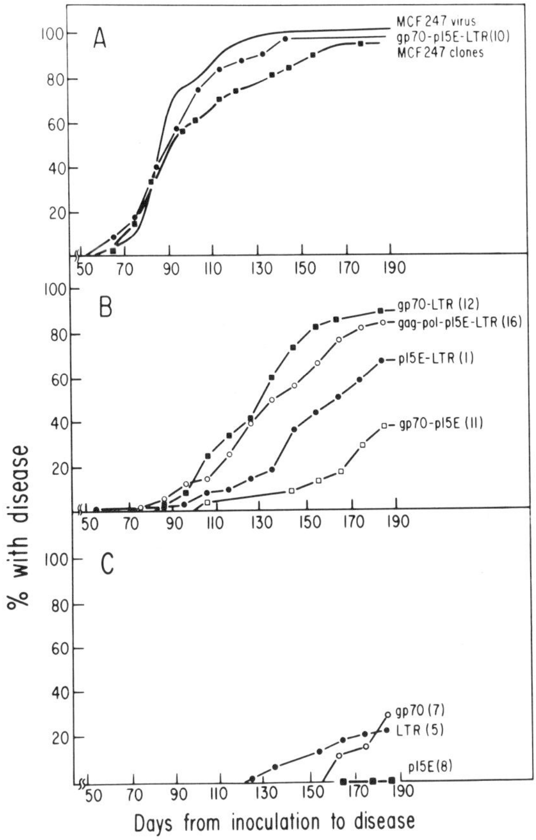

Figure 2
Induction of T cell lymphomas by MCF 247 virus and by recombinants between this virus and the nonleukemogenic Akv virus. AKR mice less than 2 days old were injected with virus and followed for 6 months for signs of disease. Studies are terminated at 6 months since AKR mice begin to develop spontaneous lymphomas after this time. The graph shows the percent of injected mice developing disease as a function of time. Each curve is the result for a single virus. To the right of the curve, viral genes derived from the leukemogenic MCF virus are indicated, with the remainder of the genome being derived from Akv. The number in parentheses is the name that was assigned to each recombinant virus. Curves represent between 22 and 91 mice. (*A*) The induction of disease with MCF virus itself is shown: 90–95% of the animals develop thymic lymphoma within about 4–5 months. If the entire 3′ half of MCF 247 is joined to the 5′ half of Akv, the resulting virus is also highly leukemogenic. In contrast, when just one or two genes from the MCF parent are inserted in place of Akv genes, the resulting viruses are less leukemogenic. (*B*) Viruses with two MCF genes are shown to be moderately to quite highly leukemogenic. (*C*) Viruses with just one gene from MCF are shown to be nonleukemogenic (C terminal portion of Prp15E from MCF) or weakly leukemogenic (gp70 or LTR from MCF). Reprinted, with permission, from Holland et al. 1985a.

evidence to indicate the role of the envelope glycoprotein in the process, we have evidence that the MCF type p15E and LTR increase the thymotropism of the virus (see below).

The LTR as a Determinant of the Type of Leukemia A Mouse Retrovirus Will Induce

In the course of the studies described above, we happened upon the observation that the MCF LTR encoded a host range property involving preferential plating on two different cell types in vitro. This result was surprising since previously determinants of retrovirus host ranges had been mapped to protein coding genes. Because this in vitro host range phenotype of MCF viruses was correlated with thymotropism, our results led us to ask whether LTRs might be able to determine tissue tropism and hence disease specificity. To test this possibility, we turned to the strongly disease-specific retroviruses, Moloney and Friend. Moloney leukemia virus induces T cell leukemias, while Friend virus induces erythroleukemia (Friend 1957; Moloney 1960; Oliff et al. 1980). Using molecular clones of these two viruses, Chatis exchanged their LTRs (Chatis et al. 1983, 1984). The results were striking: the resulting recombinants now induced leukemia according to the type of LTR they carried (Table 1). These experiments showed that sequences in the LTR, which encodes just transcriptional signals but no proteins, could determine what type of leukemia would be induced. When one examines the organ tropism of these viruses, it is apparent that the change in LTR is accompanied by a change in ability

Table 1
LTR–A Strong Determinant of Disease Specificity in Recombinants Between Nondefective Friend and Moloney Viruses

Virus	Source of U3 portion of LTR	Source of other viral genes	No. of mice with leukemia[a]		
			T cell	Erythro	Myeloid
Moloney	Moloney	Moloney	15		
Friend	Friend	Friend		50	
FM[b]	Moloney	Friend	46	1	
MF	Friend	Moloney	1	53	3
FMdrR	Moloney	Friend	19	1	

[a]Viruses were injected into NFS mice less than 2 days old. Animals were sacrificed when severely ill. Gross and microscopic pathologic examinations and hematocrits were performed to determine the type of leukemia present (Chatis et al. 1983, 1984).

[b]FM is a recombinant constructed from molecular clones of Friend and Moloney viruses. It contains about 620 nucleotides of Moloney virus. The rest of its genome is derived from Friend. The 620 nucleotides derived from Moloney encompass the U3 region. MF is the reciprocal recombinant to FM. FMdrR is a recombinant with only about 380 nucleotides from Moloney and the rest of its genome from Friend. The sequences from Moloney lie entirely within the U3 and R portions of the LTR. They include the transcriptional enhancer and promoter and the cap site but no protein coding sequences.

Table 2
LTR of Moloney Leukemia Virus Increases the Thymotropism of Nondefective Friend Virus[a]

Virus	Source of U3	Days after inoculation	Thymocytes producing virus ($Log_{10}/10^7$ cells)
Moloney	Moloney	21–36	6.1
Friend	Friend	21–36	4.4
FM[b]	Moloney	21–36	5.2
FMdrR	Moloney	21–36	5.3
MF	Friend	21–36	4.5

[a]Data from J. W. Hartley and N. Hopkins, unpubl. NFS mice are injected with Moloney, Friend, or LTR recombinants between Moloney and Friend. At some time between 21 and 36 days later, individual mice are sacrificed, their thymocytes removed, and the number of cells expressing XC plaque forming (input) virus is determined. The result is expressed as $\log_{10}$ virus producing cells per 10^7 thymocytes. The numbers shown are averages of data obtained from many individual mice.

[b]FM and FMdrR recombinants have U3 regions derived from Moloney virus (see legend to Table 1) whereas MF has its U3 derived from Friend. These and other data (not shown) indicate that quite a small difference in thymotropism as measured by this assay (half a log perhaps) can be correlated with a big difference in whether animals develop T cell or erythroleukemia. More striking differences in organ tropism are seen in recombinants between the thymotropic MCF 247 virus and erythroleukemia-inducing Friend virus (C. A. Holland, N. Hopkins, and J. W. Hartley, unpubl.).

to replicate in the thymus. The Moloney LTR confers about a tenfold increase in the number of thymocytes that become infected following virus injection (Table 2) (J. W. Hartley and N. Hopkins, unpubl.).

At the time these observations were made, we learned that studies from Khoury and co-workers were showing that transcriptional enhancer elements can function better in some cells than others (Laimins et al. 1982). This finding immediately illuminated our own observations, suggesting a mechanism to explain the disease specificity, organ tropism, and in vitro host range that we had mapped to the LTR. As has now been substantiated in many laboratories, transcriptional signals, both promoters and enhancer elements, can be tissue-specific.

It should be noted that although our studies clearly show that the LTR can be a very strong determinant of disease specificity, other studies have shown that other viral genes can also contribute to this phenotype (Robinson et al. 1982; Oliff and Ruscetti 1983).

DISCUSSION

In the experiments described above, we showed that a highly leukemogenic virus, the MCF type of mouse retrovirus, is carefully constructed for maximum speed and efficiency and that tampering with any of its genes is likely to decrease its potency.

However, it should be noted that sometimes tampering can extend leukemogenic potential. For example, neither the MCF or Akv mouse retroviruses used in the studies described above is able to cause leukemia in NFS mice. (The studies were performed in AKR mice in which MCF viruses induce leukemia in essentially 100% of recipients in about 4 months, well before the age at which the mice would normally contract spontaneous disease.) However, recombinants that had the LTR of MCF but the remainder of their genome derived from Akv were able to induce a low incidence (20–35%) of leukemia in NFS mice after a long latent period (Holland et al. 1985a). Thus, these engineered viruses were leukemogenic in a situation where neither parent can cause disease. Other cases help to illustrate the complexity of the situation. Studies by Haseltine and his collaborators showed that the LTR of a potent leukemogenic virus called SL3 is sufficient to confer leukemogenicity on Akv virus (Lenz et al. 1984); and studies by Jolicoeur's group with the strongly leukemogenic Gross Passage A virus yielded similar results (DesGroseillers et al. 1983b). In contrast, when the LTR of the highly leukemogenic Friend virus was joined to Akv or to another nonleukemogenic virus, the amphotropic virus, the resulting recombinants were essentially nonleukemogenic (Oliff et al. 1984; M. Cloyd, pers. comm.). These data sharply contrast to the results discussed in this paper where the Friend LTR was placed in the Moloney virus genome. In our other studies, a recombinant with the LTR of Friend virus and the remainder of its genes from MCF was able to to induce a low incidence of erythroleukemia in NFS mice (where MCF itself is nonleukemogenic but Friend is highly leukemogenic) (C. A. Holland, N. Hopkins, and J. W. Hartley, unpubl.). Perhaps these studies merely emphasize again the fact that multiple viral genes contribute to leukemogenicity and that some combinations are compatible and some are not.

Finally, the studies described here have served to emphasize the critical importance of virus tropism in disease induction and disease specificity and also the fact that many different types of viral genes can determine host range.

ACKNOWLEDGMENTS

I thank my collaborators who participated in the work summarized here. In particular, I acknowledge with pleasure the long-standing collaboration between my laboratory and that of Drs. Janet Hartley and the late Wallace P. Rowe.

REFERENCES

Chatis, P. A., C. A. Holland, J. W. Hartley, W. P. Rowe, and N. Hopkins. 1983. Role for the 3′ end of the genome in determining disease specificity of Friend and Moloney murine leukemia viruses. *Proc. Natl. Acad. Sci. U.S.A.* **80**: 4408.

Chatis, P. A., C. A. Holland, J. E. Silver, T. N. Fredrickson, N. Hopkins, and J. W. Hartley. 1984. A 3′ end fragment encompassing the transcriptional enhancers of nondefective Friend virus confers erythroleukemogenicity on Moloney leukemia virus. *J. Virol.* **52**: 248.

Chattopadhyay, S. K., M. R. Lander, S. Gupta, E. Rands, and D. R. Lowy. 1981. Origin of mink cytopathic focus-forming (MCF) viruses: Comparison with ecotropic and xenotropic murine leukemia virus genomes. *Virology* **113**: 465.

Cloyd, M. W., J. W. Hartley, and W. P. Rowe. 1980. Lymphomagenicity of recombinant mink cell focus-inducing murine leukemia viruses. *J. Exp. Med.* **151**: 542.

Corcoran, L. M., J. M. Adams, A. R. Dunn, and S. Cory. 1984. Murine T lymphomas in which the cellular myc oncogene has been activated by retroviral insertion. *Cell* **37**: 113.

Cuypers, H. T., G. Selten, W. Quint, M. Zijestra, E. R. Maandag, W. Boelens, P. van Wezenbeeck, C. Melief, and A. Berns. 1984. Murine leukemia virus induced T-cell lymphomagenesis: Integration of proviruses in a distinct chromosomal region. *Cell* **37**: 141.

DesGroseillers, L., E. Rassart, and P. Jolicoeur. 1983a. Thymotropism of murine leukemia virus is conferred by its long terminal repeat. *Proc. Natl. Acad. Sci. U.S.A.* **80**: 4203.

DesGroseillers, L., R. Villemur, and P. Jolicoeur. 1983b. The high leukemogenic potential of Gross passage A murine leukemia virus maps in the region of the genome corresponding to the long terminal repeat and to the 3′ end of env. *J. Virol.* **47**: 24.

Fischinger, P. J., S. Nomura, and D. P. Bolognesi. 1975. A novel murine oncornavirus with dual eco- and xenotropic properties. *Proc. Natl. Acad. Sci. U.S.A.* **72**: 5150.

Friend, C. 1957. Cell-free transmission in adult Swiss mice of a disease having the character of a leukemia. *J. Exp. Med.* **105**: 307.

Gross, L. 1951. Spontaneous leukemia developing in C3H mice following inoculation, in infancy, with AK-leukemic extracts, or AK-embryos. *Proc. Soc. Exp. Biol. Med.* **76**: 27.

——— . 1970. *Oncogenic viruses* 2nd ed. Pergamon Press, Oxford.

Hartley, J. W., N. K. Wolford, L. J. Old, and W. P. Rowe. 1977. A new class of murine leukemia virus associated with development of spontaneous lymphomas. *Proc. Natl. Acad. Sci. U.S.A.* **74**: 789.

Hayward, W. S., B. G. Neel, and S. M. Astrin. 1981. Activation of a cellular onc gene by promoter insertion in ALV-induced lymphoid leukosis. *Nature* **290**: 475.

Holland, C. A., J. Wozney, P. A. Chatis, N. Hopkins, and J. W. Hartley. 1985a. At least four viral genes contribute to the leukemogenicity of murine retrovirus MCF 247 in AKR mice. *J. Virol.* **53**: 158.

Holland, C. A., J. Wozney, P. A. Chatis, N. Hopkins, and J. W. Hartley. 1985b. Construction of recombinants between molecular clones of murine retrovirus MCF 247 and Akv: Determinant of an in vitro host range property that maps in the long terminal repeat. *J. Virol.* **53**: 152.

Kelly, M., C. A. Holland, M. L. Lung, S. K. Chattopadhyay, D. R. Lowy, and N. Hopkins. 1983. Nucleotide sequence of the 3′ end of MCF 247 murine leukemia virus. *J. Virol.* **45**: 291.

Laimins, L. A., G. Khoury, C. Gorman, B. Howard, and P. Gruss. 1982. Host-specific activity of transcription by tandem repeats from simian virus 40 and Moloney murine sarcoma virus. *Proc. Natl. Acad. Sci. U.S.A.* **79**: 6453.

Lenz, J. D., D. Celander, R. L. Crowther, R. Pataraca, D. W. Perkins, and W. A. Haseltine. 1984. Determination of the leukemogenicity of a murine retrovirus by sequences within the long terminal repeat. *Nature* **308**: 467.

Li, Y., C. A. Holland, J. W. Hartley, and N. Hopkins. 1984. Viral integration near c-myc in 10–20% of MCF 247-induced AKR lymphomas. *Proc. Natl. Acad. Sci. U.S.A.* **81**: 6808.

Lieberman, M. and H. Kaplan. 1959. Leukemogenic activity of filtrates from radiation-induced lymphoid tumors of mice. *Science* **130**: 387.

Lowy, D. R., E. Rands, S. K. Chattopadhyay, C. F. Ganon, and G. L. Hager. 1980. Molecular cloning of infectious integrated murine leukemia virus DNA from infected mouse cells. *Proc. Natl. Acad. Sci. U.S.A.* **77**: 614.

Lung, M. L., J. W. Hartley, W. P. Rowe, and N. Hopkins. 1983. Large RNase T1-resistant oligonucleotides encoding p15E and the U3 region of the long terminal repeat distinguish two biological classes of mink cell focus-forming type C viruses of inbred mice. *J. Virol.* **45**: 275.

Lung, M. L., C. Hering, J. W. Hartley, W. P. Rowe, and N. Hopkins. 1980. Analysis of the genomes of mink cell focus-inducing murine type C viruses: A progress report. *Cold Spring Harbor Symp. Quant. Biol.* **44**: 1269.

Moloney, J. B. 1960. Biological studies on a lymphoid-leukemia virus extracted from sarcoma 37. I. Origin and introductory investigations. *J. Natl. Cancer Inst.* **24**: 933.

Neel, B. G., W. S. Hayward, H. L. Robinson, J. Fang, and S. M. Astrin. 1981. Avian leukosis virus-induced tumors have common proviral integration sites and synthesize discrete new RNAs: Oncogenesis by promoter insertion. *Cell* **23**: 323.

Oliff, A. and S. Ruscetti. 1983. A 2.4-kilobase-pair fragment of the Friend murine leukemia virus genome contains the sequences responsible for Friend murine leukemia virus-induced erythroleukemia. *J. Virol.* **46**: 718.

Oliff, A., K. Signorelli, and L. Collins. 1984. The envelope gene and long terminal repeat sequences contribute to the pathogenic phenotype of helper-independent Friend viruses. *J. Virol.* **5**: 788.

Oliff, A. I., G. L. Hager, E. H. Chang, E. M. Scolnick, H. W. Chan, and D. R. Lowy. 1980. Transfection of molecularly cloned Friend murine leukemia virus DNA yields a highly leukemogenic helper-independent type C virus. *J. Virol.* **33**: 475.

Payne, G. S., S. A. Courtneidge, L. B. Crittenden, A. M. Fadley, J. M. Bishop, and H. E. Varmus. 1981. Analysis of avian leukosis virus DNA and RNA in bursal tumors: Viral gene expression is not required for maintenance of the tumor state. *Cell* **23**: 311.

Robinson, H. L., B. M. Blais, P. N. Tsichlis, and J. M. Coffin. 1982. At least two regions of the viral genome determine the oncogenic potential of avian leukosis viruses. *Proc. Natl. Acad. Sci. U.S.A.* **79**: 1225.

Rowe, W. P. 1972. Studies of genetic transmission of murine leukemia virus by AKR mice. Crosses with Fv-1^n strains of mice. *J. Exp. Med.* **136**: 1272.

——. 1973. Genetic factors in the natural history of murine leukemia virus infection. *Cancer Res.* **33**: 3061.

Rowe, W. P. and J. W. Hartley. 1972. Studies of genetic transmission of murine leukemia virus by AKR mice II. Crosses with Fv-1^b strains of mice. *J. Exp. Med.* **136**: 1286.

Rowe, W. P. and T. Pincus. 1972. Quantitative studies of naturally occurring murine leukemia virus infection of AKR mice. *J. Exp. Med.* **135**: 429.

Steffen, D. 1984. Proviruses are adjacent to c-myc in some murine leukemia virus-induced lymphomas. *Proc. Natl. Acad. Sci. U.S.A.* **81**: 2097.

Thomas, C. Y. and J. M. Coffin. 1982. Genetic alterations of RNA leukemia viruses associated with development of spontaneous thymic leukemia in AKR mice. *J. Virol.* **43**: 416.

Tsichlis, P. N., P. G. Strauss, and L. F. Hu. 1983. A common region for proviral DNA integration in Mo-MuLV induced rat thymic lymphomas. *Nature (Lond)* **302**: 445.

COMMENTS:

KHOURY: When you reduced the tumorigenicity of the MCF virus, did you destroy any other genes in your reconstruction at the same time you switched LTRs?

HOPKINS: In the case of the MCF, it was a very small piece that was exchanged. It was cut in U3 and in R so it included most of U3 and about half of R.

KHOURY: So it didn't involve any coding regions?

HOPKINS: No. It didn't touch them.

KHOURY: You know from your focus assay that the replication was reasonably high, so the model—that the tumorigenicity is simply related to levels of replicating virus searching around for oncogenes to integrate next to—is not totally correct.

HOPKINS: No, the model is not contradicted by the MCF-Friend recombinant, although this recombinant is rather complicated. In general, however, in this case as well as others, we find that virus replication levels are critical, and the data we have shows that, to a first approximation, organ tropism as determined by the LTR is correlated with disease. However, you can't always predict from the titer you measure in vivo what will happen in terms of disease. The correlation is not perfect, perhaps because the tests are too crude, or perhaps because other factors can intervene. In the case of the MCF virus with the Friend LTR, I wonder if the LTR is dragging the virus towards one target while the other genes are dragging it towards another, so it's torn and can't replicate well enough in either the spleen or thymus to be really effective. We know from other examples that small differences in virus titer between spleen cells and thymus, half a log and even less, may be enough to mean the difference between getting a disease or not getting a disease or slowing it down considerably. Virus replication in this system is so critical to whether you are going to get disease; very small differences in titer might have big effects on disease induction.

KHOURY: I agree. Can you compare those replication levels to the recombinants that you made in your earlier experiments where you weren't using MCF but Moloney and Friend?

HOPKINS: Does it correlate?

KHOURY: Yes.

HOPKINS: It's difficult to make a meaningful comparison of MCF titers with those of Friend and Moloney. But there seems to be another element here and it isn't clear to what extent it depends on the level of replication. It is seen in the case of Moloney, Friend, and recombinants where the Moloney and Friend switch their tropism, switch their disease, and when their LTRs are exchanged. You see a 1 log switch in replication going to the spleen versus thymus, but you see 3 or 4 logs difference in the ability to generate MCF viruses in the thymus; and this correlates with T cell leukemia induction. So it comes back to the issue of whether when you make leukemia in a mouse, you are really retailoring a virus which is just designed for that organ and that cell type; and you have this reservoir of endogenous viruses which you just use as tinker toys to build viruses that can do the job. It could be that there is something more to it that involves MCF generation.

AHMED: Is it known what kind of T cell is infected?

GALLO: It's relatively immature. It is not like the human. It's a more immature cell of the thymus, I think, not the most primitive, but of intermediate maturity from what cell markers I remember.

HOPKINS: Yes, that's my understanding of it.

GALLO: Do you think the total level of virus is equal, though? I mean, if you had to estimate viremia in the animal who gets the T cell leukemia versus when you switch the LTR and it doesn't get a T cell leukemia, do you feel confident (I was thinking along the same lines as George Khoury) that the titer of virus is almost the same? If not, if the replication were different, you could argue that it is just related to varying the amount of replication, that to get to the right T cell simply needs the greater amount of replication, and that you don't do as well when you use the Friend LTR. Generally in mice, in cats, and in chickens, malignancy correlates with the magnitude of viremia. This is, of course, not true with bovine leukemia virus and the human retroviruses, where virus integration is random, replication minimal, and still leukemia occurs. This is probably because of the *trans* mechanism, more like DNA tumor viruses.

HOPKINS: I agree that replication is critical. What the results show is that LTRs confer organ tropism and that this does correlate with the type of disease that will be induced. However, if someone tells you the virus titer in thymus, you can't always predict what will happen. One problem might be that the test of virus titers in splenocytes and thymocytes may be too crude. You might have a big growth preference for replication in the actual target cell, but you might not see it in this crude assay.

MARTIN: Nancy [Hopkins], in your summary, you listed the various viral genes and indicated their functional roles. Can you assign a function to the envelope? Most of these MCF viruses possess unique envelopes. You have discussed that yourself in several papers. Do you think that's important?

HOPKINS: We do not know why the envelope has changed to the MCF type in the leukemogenic virus. The virus that you start out with, which is not leukemogenic, can replicate very well in mouse cells. The simplest argument is that, in fact, the MCF envelope allows you to get into T cells better. There is no evidence for it and there is evidence that eco envelopes can get you into at least some T cells; however, it still might be true. It is just that nobody has ever been able to figure out what the MCF envelope does.

MARTIN: It's probably similar to the feline system too, which has a changing envelope.

HOPKINS: Yes. All sorts of things have been proposed. Thymotropism again, and again it could be just a slight preference. Maybe there is a slightly tighter binding for some receptor on some T cells, but there's no evidence.

KHOURY: What is wrong with AKV? Is AKV viremic at high titers?

HOPKINS: Yes, although not nearly as high as Moloney.

KHOURY: Does it have an enhancer?

GALLO: Yes. It has differences in the enhancer, though, right?

HOPKINS: Yes.

KHOURY: But if you take out its U3 region, will it stimulate heterologous genes? If so, why wouldn't it drive an oncogene next to which it integrates?

HOPKINS: I'm not sure that I completely know the answer to your question, although the level of replication might be too low, and also, I think it fails to generate MCF viruses, except of course in AKR mice. Perhaps Bill Haseltine's experiments bear on this. If he makes a recombinant in which the LTR of a leukemogenic virus, called SL3, is put onto Akv, then the resulting virus is leukemogenic. However, Todd Cloyd tells me that when they repeated our experiment to see if they could change disease specificity with LTRs, they put the Friend LTR onto Akv and the resulting virus was essentially non-leukemogenic. So something else is going on. The experiment works with Moloney genes, but not with Akv genes.

KHOURY: I see.

KAMELY: Do these viruses cross species?

HOPKINS: They go in rat.

GALLO: You mean retroviruses in general or this one?

KAMELY: How about Moloney virus? Is it conceivable that your recombinant could be made to cross species in one direction or another?

HOPKINS: That's a good question. We haven't tried it. I'm not sure. Actually, now that I think about it, I doubt it. MCF viruses only work in certain mouse strains. Most of these viruses really only work in certain mouse strains and often only in very young animals. It's very much a model system, and what is the probability that the model could ever work on anything else? Moloney, however, causes leukemia in rats, and so do several other mouse retroviruses.

KAMELY: What is the answer to retroviruses in general? Are you concerned that if you make enough recombinants, a certain recombination may cross species?

GALLO: Some cross species. Some don't. Some do it very well.

HOPKINS: One thing I didn't say here, and I think I should emphasize, these viruses are really weakly leukemogenic, except in their optimal situations where they are extremely effective. I suspect they wouldn't do much in real life. However, if one put a different envelope gene on the virus, I'm not so sure.

GALLO: In the mouse?

HOPKINS: In the mouse, at least the ones that we've worked on most, the MCFs. They work in a particular strain of mice if you inject them very young.

GALLO: To answer the question, a good example would be bovine leukemia virus. People have induced virtually 100% leukemia in young sheep; when you inoculate a young sheep, it works better than it does in a cow, its natural host. It produces B cell malignancy in cows, but apparently usually an acute T cell leukemia in sheep. Recently, Arsene Burney put it in rabbits and produced an AIDS-like disease.

HOPKINS: Actually, the other one that jumps is feline leukemia virus, which I guess jumps across species and works in adults.

FIELDS: Is there anything known about what determines whether it will jump across species and has it been done enough times so that one could predict anything? You just cited an example.

GALLO: No, I don't think so, at least not to my knowledge. Does anybody else have insight into a molecular determinant's favoring interspecies transmission of a retrovirus?

MARTIN: The xenotropic MLVs produce no disease that anybody knows about in the laboratory on every kind of cell.

GALLO: There is an interesting story that happened naturally, with the Gibbon ape leukemia (GalV) virus. There is evidence that the simian sarcoma virus (SSV) was generated by one interspecies transmission of GaLV from a Gibbon ape to a woolly monkey. Believe it or not, this happened in a house in California. A woman in Sacramento had pet monkeys and a pet Gibbon ape, and she had them cohabitating. GaLV causes leukemia in Gibbon apes. The woolly monkey developed a sarcoma of the neck and a fibrosis of the bone marrow and spleen. SSV was isolated from this woolly monkey. When analyzed, SSV was shown to have a new *onc* gene (now called *sis*, which we showed was derived from a normal gene of the woolly monkey) plus the genome of GaLV.

FIELDS: What effects would enhancers have on their capacity to enter different tumor viruses and alter their host range or environmental stability?

KHOURY: This leads to speculation. The SV-40 enhancer, like that of RSV, is rather promiscuous. There aren't many diffentiated cells in tissue culture cells in which it won't work. If it were introduced into a retrovirus, let's say, could it extend the host range or the tissue tropism? Presumably, yes. But let me say that, while it is promiscuous, it is now clear that other promoter elements contribute to host range. For example, if you take a promiscuous enhancer and insert it into an immunoglobulin gene, the activity is still restricted to B cells, presumably because its promoter is B cell-specific. If you switch the promoter and the enhancer, the gene for some reason may still be restricted to B cells. So you might have to do a little more playing around. I could conceive of a situation in which adding the SV-40 enhancer to another virus would extend its tissue tropism or its host range. Would that be dangerous? If that virus were dangerous, conceivably. I would not hook up the SV-40 enhancer to a toxin and introduce the recombinant into transgenic mice or other organisms. One experiment which shows how complex the situation can be in vivo was performed by Brinster and his colleagues. The SV40 early region and its gene, T-antigen, is expressed in virtually all tissue culture cells. When it was introduced into transgenic mice, some eventually developed choroid plexus adenomas for reasons that are unclear. SV-40 is present in every tissue, integrated, with an enhancer. However, in the animal, it's not making significant levels of T-antigen in tissues other than choroid plexus. When you take other tissues out of the animal and put them in tissue culture, surprisingly you get T-antigen and cell transformation—a real surprise. So, there are many things that we don't understand about the host, which perhaps can override some of these mechanisms that we're looking at in tissue culture. One can't err on the side of caution. But fortunately there are some protective mechanisms in the host that go well beyond what we understand from tissue culture assays. I think that's the "take-home" message.

Penetration of Enveloped Animal Viruses: Relevance in Determining Cell Tropism and Host Range

ARI HELENIUS,* ROBERT DOMS,* JUDY WHITE,† AND MARGARET KIELIAN*
*Department of Cell Biology
Yale University School of Medicine
New Haven, Connecticut 06510
†Department of Pharmacology
University of California
San Francisco, California 94143

OVERVIEW

After adsorbing to the cell surface, animal viruses penetrate into the cell by a variety of different mechanisms. The penetration reactions occur either at the plasma membrane or after endocytosis in endosomes and lysosomes. The surface proteins of the virus are important in all of these events, but in addition, processes and activities of the host cell frequently play a central role.

Enveloped animal viruses penetrate, as a rule, by a membrane fusion reaction between the viral envelope and a cellular membrane. This reaction is catalyzed by the spike glycoproteins of the virus, and in the majority of cases, fusion is triggered by the mildly acidic pH which the virus encounters in endosomes. The low pH appears to act by inducing activating conformational changes in the viral fusion factors. For two viruses, Semliki Forest virus and Influenza virus A, details of the conformational changes and the subsequent fusion reaction are becoming increasingly clear from biochemical, cell biological, immunochemical, and genetic studies. It is as yet too early to decide whether alterations in the viral penetration activities of these and other enveloped viruses could have an impact on their cell tropism and environmental effects. Two hypothetical situations in which such an effect could conceivably occur are discussed.

INTRODUCTION

Enveloped animal viruses transfer their genome between cells and organisms by a vesicle transport mechanism. The viral envelope, which serves as the transport vesicle, is derived by budding from a cellular membrane of the infected cell. The budding process also provides the mature virus with a set of virus-specific spike glycoproteins which play an important role in the entry phase. They are responsible for attachment, membrane fusion and, in some cases, receptor destruction. As they are crucial in the early interactions between virus and host cell, they constitute one of the key factors that determine viral tropism and pathogenicity.

The main stages in the early interactions are attachment, endocytosis (not applicable to all viruses), penetration, and uncoating (Fig. 1). During the penetration step, the viral genome and any accessory proteins must pass through a membrane barrier into the cytoplasmic compartment. In this paper we will briefly describe the role of viral proteins in penetration via membrane fusion. We will discuss possible effects that genetic alterations might have on the properties of these proteins, and an attempt will also be made to predict the environmental impact of such alterations.

Membrane Fusion

The mechanism of virus interaction with the cell surface and membrane fusion are best understood for alphaviruses and myxoviruses, which depend on endocytosis and acid-activated fusion, and for paramyxoviruses, which enter directly by fusion with the plasma membrane. (For recent reviews, see Helenius et al. 1980a; Dimmock 1982; Lenard and Miller 1982; White et al. 1983; Marsh 1984; and Kielian

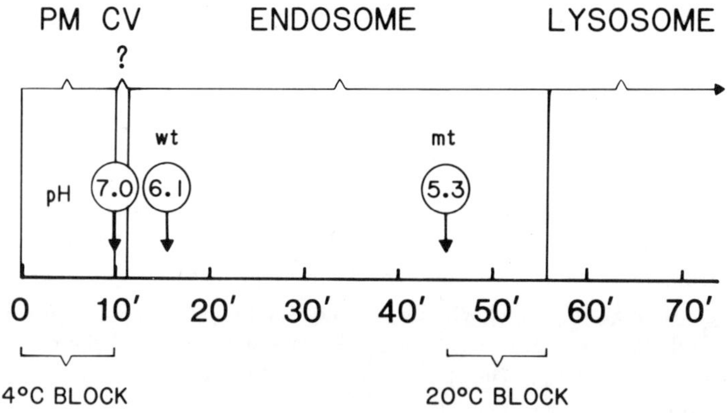

Figure 1
The time course of Semliki Forest virus transport in the endocytic pathway of BHK-21 cells. The figure indicates schematically the average times that the virus spends at the various stations of the pathway. It is based on a number of biochemical and morphological studies (Helenius et al. 1980b; Marsh and Helenius 1980; Marsh et al. 1983; Kielian et al. 1984; M. Kielian and A. Helenius, unpubl.). The half-life of the bound virus on the cell surface is about 10 min; the time spent in a coated vesicle probably only a minute or two; the time in the endosomal compartment about 45 min; and delivery into the lysosomes occurs on the average about 45–50 min after uptake. It is important to emphasize that these are average values; the pathway functions nonsynchronously and therefore individual virus particles reach the various steps at widely different times. The values indicated for the pH of the various stages is based on the timing of average virus penetration using wild-type (wt) or mutant (mt) viruses which fuse at pH values of 6.2 and 5.5, respectively.

and Helenius 1985). In each case, a specific viral glycoprotein has been identified as responsible for the membrane fusion activity. Reconstitution and expression studies have shown that the fusion factors, which are integral membrane proteins, are functional in the absence of other viral gene products. However, to catalyze fusion they must be anchored in a membrane. They display relatively little specificity as to the composition of the target membrane for fusion. Fusion can, for instance, easily be obtained using artificial protein-free liposomes as targets, a finding which has allowed detailed analysis of fusion parameters. Table 1 lists some of the most important properties of Semliki Forest virus and Influenza virus fusion with various artificial liposysomes.

One of the most important conclusions to be drawn from such in vitro results is that, while crucial for attachment and internalization, the cell surface receptors for the viruses are not required during the penetration step. Unlike attachment, which depends on the presence of specific binding sites, penetration is probably not directly involved in determining host cell specificity. The implication is that once internalized by a cell via any functional "receptor," these viruses should in general be able to penetrate.

The threshold pH values at which fusion is activated varies greatly between virus families, between members of virus families, and even between strains. Paramyxoviruses, as shown in Figure 2, display fusion activity over the entire relevant pH range, while the pH thresholds of Influenza viruses vary between pH 5 and 6, and alpha and rhabdoviruses fuse around pH 6. Moreover, the pH threshold of fusion is easily modified by mutations. The first mutant of this type was described in Semliki Forest virus (Kielian et al. 1984). The mutant had a pH of fusion approximately 0.8 pH units lower than the wild-type. Mutants with higher fusion pH

Table 1
Fusion of Semliki Forest Virus and Fowl Plaque Virus with Liposysomes[a]

	Semliki forest virus (alpha virus)	Fowl plague virus (orthomyxovirus)
pH-dependence	pH < 6.2	pH < 5.0–6.0
Efficiency	98%	- 95%
Lytic fusion	< 5%	< 5%
Fusion rate (T1/2)	< 1 min	< 30 sec
Requirement for bivalent cations	None	None
Requirement for cholesterol	Absolute	None
Requirement for specific phospholipids	None	Partial for PE
Receptor requirement	None	None
Requirement for ion gradients	None	—
Temperature dependence	0–40°C	0–40°C

[a]White and Helenius (1980); White et al. (1983); Kielian and Helenius (1984)

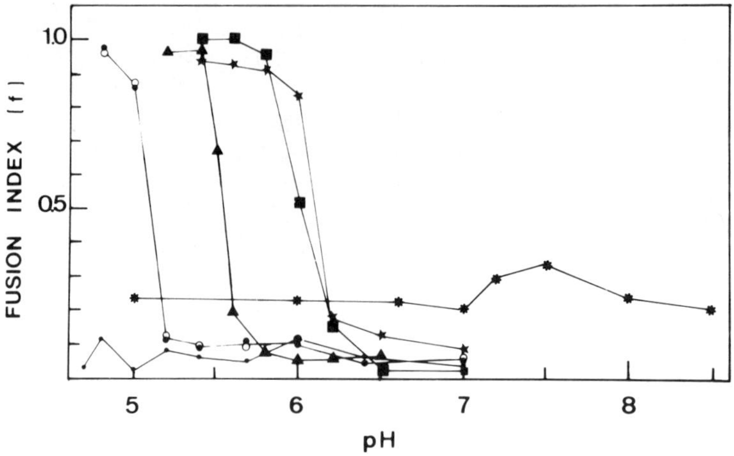

Figure 2
pH-Dependence of cell-cell fusion induced by a variety of enveloped animal viruses. Isolated viruses were allowed to bind to cell monolayers whereafter the pH was adjusted to the indicated values. After neutralization the extent of fusion was scored by counting the number of nuclei per cell. A fusion index of 0.5 denotes a situation in which each cell had an average of two nuclei. It is seen that Sendai virus (*), a paramyxovirus, has essentially pH independent fusion activity whereas Vesicular Stomatitis virus, a rhabdo virus (★), Semliki Forest virus (■), and various strains of Influenza virus: Fowl plague (▲); A/Japan/305/1957 H2N2 (○–○); and A/Hong Kong/1968 H3N2 (○) have fusion activities activated at mildly acid pH. The Influenza virus which does not have the mature HA (●) is not fusogenic at any pH.

threshold than wild-type have subsequently been isolated from various human and avian influenza viruses (Rott et al. 1984; Daniels et al. 1985; R. Doms, in prep.; M.J. Gething et al., in prep.). It is interesting and significant that in many cases single mutations in the fusion factor result in significant changes in fusion pH.

The fusion mutants are all capable of normal infection and replication. The Semliki Forest virus mutant was shown to enter through endosomes as does the wild-type (Kielian et al. 1984). However, after internalization via the endocytic pathway, its penetration was delayed approximately 45 minutes compared to wild-type (Fig. 1). From this, we concluded that incoming viruses encounter decreasing pH during the time they spend in the endosomal compartment. The mutant virus apparently penetrates from a relatively late endosome compartment of lower pH, in contrast to the wild-type virus, which penetrates in an earlier, more peripheral compartment. The pH of fusion may in this way determine the timing and the location of the penetration reaction inside the host cell. (For recent reviews on the cell biology of endosomes, see Brown et al. 1983; Helenius et al. 1983; and Pastan and Willingham 1983.)

In addition to the acid dependence of fusion, the only absolute requirement observed to date in liposome fusion assays of these viruses has been for cholesterol in the target membrane. This requirement has been demonstrated for the alphaviruses and Sendai virus (White and Helenius 1980; Hsu et al. 1983; Kundrot et al. 1983; Kielian and Helenius 1984). In the case of Semliki Forest virus, we have found that the cholesterol (or other three β-hydroxysterols) is needed for the acid-induced conformational change occurring in the E1 subunit of the spike glycoprotein (M.C. Kielian and A. Helenius, in prep.). As a nearly ubiquitous component of the plasma membrane and the membranes of the endocytic compartment, cholesterol hardly qualifies as an important factor for determining tropism among higher vertebrates. However, there is one situation in which cholesterol dependence might be important, i.e., infectivity in insects which cannot synthesize cholesterol and are therefore dependent on dietary sterol. Further studies will be needed to determine whether sterols play any role in determining the tropism among the insect-borne viruses.

Possible Relevance for Cell Tropism

Our understanding of the penetration reactions of most enveloped viruses is still very incomplete. Any speculations on the role of penetration in tropism and pathogenicity (not to mention the environmental effects) remain highly speculative. From the information summarized above, one can, however, identify two hypothetical situations in which modification of penetration, for example by genetic manipulation of the fusion factors, might affect cell tropism and/or the environmental impact.

Scenario 1

Consider a virus being endocytosed by the presynaptic terminae of neuronal cells and carried to the cell body by axonal transport. In this case, the initial uptake and the early endosomes are distant from the cell body in which the perinuclear, mature endosomes are located. If the gradual decrease in vacuolar pH observed for other cell types (Merion et al. 1983; Murphy et al. 1984; Tanasugarn et al. 1984) applies to nerve cells, one would expect that a virus triggered at pH 6 could fuse in the cell processes, whereas a virus modified to have a lower fusion pH might be expected to penetrate in the cell body. By thus altering the site of penetration, the efficiency of infection might be affected. An alteration in a virus which would lower the pH dependence of fusion could possibly make a nonneurotropic virus neurotropic.

Although conceivable in the light of present data on virus penetration, this scenario is only a theoretical possibility. It is already clear that viruses can spontaneously modify their fusion threshold by single amino acid changes, and probably any variants that could be created in the laboratory have already occurred in nature. There is no evidence that these mutations in enveloped viruses have resulted in any sudden changes in tropism. Further, neurotropic rabies virus appears to have a fusion pH threshold of approximately 6 (Mifune et al. 1982).

Scenario 2

Consider a virus that requires cholesterol for fusion activity. This requirement could prevent it from infecting certain insects or insect tissues. Altering it genetically such that it no longer depends on this factor for penetration might alter its insect host range.

Less is known about the cholesterol dependence of fusion than about the pH dependence. It is important to note that although cholesterol dependence has been demonstrated in vitro, there is as yet no evidence that sterols play any role in vivo. Semliki Forest virus, whose fusion activity with liposomes is cholesterol-dependent is, in fact, fully capable of infecting mosquitoes both in nature and in the laboratory. The example once again demonstrates how difficult it is to extrapolate from artificial systems to viral infection at the level of intact organisms, and that it will be difficult to predict the results of modification in viral penetration factors.

REFERENCES

Brown, M.S., R.G.W. Anderson, and J.L. Goldstein. 1983. Recycling receptors: The round-trip itenerary of migrant membrane proteins. *Cell* **32**: 663.

Daniels, R.S., J.C. Downie, A.J. Hay, M. Knossow, J.J. Skehel, M.L. Wang, and D.C. Wiley. 1985. Fusion mutants of the influenza haemagglutinin. *Cell* **40**: 431.

Dimmock, M.J. 1982. Initial stages in infection with animal viruses. *J. Gen. Virol.* **59**: 1.

Helenius, A., M. Marsh, and J. White. 1980a. The entry of viruses into animal cells. *Trends Biochem. Sci.* **5**: 104.

Helenius, A., J. Kartenbeck, K. Simons, and E. Fries. 1980b. On the entry of Semliki Forest virus into BHK-21 cells. *J. Cell Biol.* **84**: 404.

Helenius, A., I. Mellman, D. Wall, and A. Hubbard. 1983. Endosomes. *Trends Biochem. Sci.* **8**: 245.

Hsu, M.-C., A. Scheid, and P.W. Choppin. 1983. Fusion of Sendai virus with liposomes: Dependence on the viral fusion protein (F) and the lipid composition of liposomes. *Virology* **126**: 361.

Kielian, M.C. and A. Helenius. 1984. Role of cholesterol in fusion of Semliki Forest virus with membranes. *J. Virol.* **52**: 281.

———. 1985. Entry of alphaviruses. In *The viruses series*, H. Fraenkel-Conrat and R. Wagner. (In press).

Kielian, M., S. Keränen, L. Kääriänen, and A. Helenius. 1984. Membrane fusion mutants of Semliki Forest virus. *J. Cell Biol.* **98**: 139.

Kundrot, C.E., E.A. Spangler, D.A. Kendall, R.C. MacDonald, and R.I. MacDonald. 1983. Sendai-virus-mediated lysis of liposomes requires cholesterol. *Proc. Natl. Acad. Sci. U.S.A.* **80**: 1608.

Lenard, J. and D. Miller. 1982. Uncoating of enveloped viruses. *Cell* **28**: 5.

Marsh, M. 1984. The entry of enveloped viruses into cells by endocytosis. *Biochem. J.* **218**: 1.

Marsh, M. and A. Helenius. 1980. Adsorptive endocytosis of Semliki Forest virus. *J. Mol. Biol.* **142**: 439.

Marsh, M., E. Bolzau, and A. Helenius. 1983. Penetration of Semliki Forest virus from acidic prelysosomal vacuoles. *Cell* **32**: 931.

Merion, M., P. Schlesinger, R.M. Brooks, J.M. Moehring, T.J. Moehring, and W.S. Sly. 1983. Defective acidification of endosomes in Chinese hamster ovary cell mutants "cross-resistant" to toxins and viruses. *Proc. Natl. Acad. Sci. U.S.A.* **80**: 5315.

Mifune, K., M. Ohuchi, and K. Mannen. 1982. Hemolysis and cell fusion by rhabdo viruses. *Febs. Eur. Biochem. Soc. Lett.* **137**: 293.

Murphy, R.F., S. Powers, and C.R. Cantor. 1984. Endosome pH measured in single cells by dual fluorescence flow cytometry: Rapid acidification of insulin to pH 6. *J. Cell Biol.* **98**: 1757.

Pastan, I. and M.C. Willingham. 1983. Receptor-mediated endocytosis: Coated pits, receptosomes and the Golgi. *Trends Biochem. Sci.* **8**: 250.

Rott, R., M. Orlich, H.-D. Klenk, M.L. Wang, J.J. Skehel, and D.C. Wiley. 1984. Studies on the adaptation of influenza viruses to MDCK cells. *Eur. Mol. Biol. Organ. J.* **3**: 3329.

Tanasugarn, L., P. McNeil, G.T. Reynolds, and D.L. Taylor. 1984. Microspectrofluorometry by digital image processing: Measurement of cytoplasmic pH. *J. Cell Biol.* **98**: 717.

White, J. and A. Helenius. 1980. pH Dependent fusion between the Semliki Forest virus membrane and liposomes. *Proc. Natl. Acad. Sci. U.S.A.* **77**: 3273.

White, J., M. Kielian, and A. Helenius. 1983. Membrane fusion proteins of enveloped animal viruses. *Q. Rev. Biophys.* **16**: 151.

COMMENTS:

KAWANISHI: Is there any evidence of viral products that keep the virus from being shunted into the lysosomal pathway?

HELENIUS: The receptor-mediated pathway works in such a way that anything that has not been sorted out from the endosome will eventually be routed to the lysosomes. UV-inactivated Semliki Forest viruses, which are unable to fuse, will, for instance, all go into lysosomes and get degraded. Of intact active viruses, usually 45-60% deliver their RNA into the cytoplasm from endosomes; the remaining 40-55% get degraded in lysosomes.

KILBOURNE: Is the coating of the pits always necessary? I thought there were exceptions to that.

HELENIUS: In the case of Semliki Forest virus, we can't see entry by any other endocytic structures. In the case of influenza virus, there seem to be two structures. One is the normal coated pit. The other, which Bachi and Dourmashkin and others have reported, corresponds to uncoated pits. In the latter case it may be that the virus particles simply wrap themselves into the plasma membrane. It is not clear whether this will result in delivery to the endosome and penetration.

FIELDS: You described a normal pathway, which many viruses use, that doesn't play much role in specificity. You also described pH mutants, some of them with altered pH. If you were to think of genetically altering this part of the pathway—you described one type of mutation in pH and configuration—do you anticipate that those would, in general, block entry or alter entry? Are there any data dealing with mutants in this part of the pathway that would potentially broaden the host range by allowing viruses to enter the cell that don't ordinarily enter? Or, again from the theme of genetic alterations, that it might? How would these manipulations relate to either broadening or narrowing host range or spread? You described some interesting mutants. What would their impact be? Would they all be attentuating or would some of them really alter the interaction in ways that might cause people to worry about genetically modifying viruses in this part of the pathway?

HELENIUS: As a rule, the pH in the peripheral endosomes which incoming viruses will encounter first, is about 5.5 or higher. Later on, the endosomes are more acidic, probably close to lysosomal pH values, around 5. Depending on the threshold pH, viruses will fuse either in an early or a late endosome. The implications are that the timing and localization of the fusion reaction may differ as a function of pH-threshold, for instance, between nerve terminal and cell body of a nerve cell. If the virus penetrates too early during axonal transport, infection may be aborted. The cell biology of the host cell may thus determine whether a change in pH-threshold will have any effects on tropism. But there are very little data related to this. It is surprising how little detail and quantitative data there are about entry of enveloped viruses in general, not to mention the nonenveloped viruses. Nonenveloped viruses clearly don't have a membrane fusion reaction, but in many instances it is known that they have acid requiring penetration.

CROWELL: Enteroviruses really, by definition, are acid pH-resistant, all the way down to pH 1 or 2. You can also find there are trypsin and chymotrypsin—they're protease-resistant. There has to be something very specific going on in these endosomes, and that is the receptor, and the multivalent binding, presumably of these receptors to the site. For instance, neutralized virus will bind to receptors, but does not generate an infection. The question is why. I don't really know the answer to that, because presumably the antibody molecule is disturbing the fit between the receptors to accomplish a productive uncoating. We have a very different situation in perhaps explaining the specificity of a specific kind of receptor for these viruses.

HELENIUS: The astounding thing with Influenza virus is that we find no specificity in the fusion reaction; any target lipid membrane will do. In Semliki Forest virus and other Toga-viruses there is a requirement for cholesterol in the target membrane. But, beyond that there are no further requirements; no bona fide reception is required.

MARTIN: Is it known what the intercellular status of the latent viruses is? Do they hang around in structures such as those you are discussing.

HELENIUS: It's not inconceivable. One of the functions of endosomes in some cells is storage. The most dramatic example is the egg yolk, which is essentially an endosome. Another is the glucose transport protein which is assumed to be located in some sort of storage organelles. But again, the cell biology of such endosomal functions is not clear, nor is it known whether they are important for any virus infection.

FOWLE: I was wondering, what might lead to retroviruses' becoming transmitted from one species to another and becoming incorporated as endogenous viruses in the genome of the recipient species?

MARTIN: In the mouse system, for example, we know that over an 18-month period the AKR mice purchased from Jackson Laboratories contained three, four, or five endogenous ecotropic proviruses. In mice, at least, retroviral copies are constantly being acquired and lost. In other species, such as man, the endogenous retroviral "load" is pretty constant; there doesn't seem to be much in the way of provirus gains and losses. Does anybody know what goes on in cats?

GALLO: RD114.

MARTIN: I think it's pretty stable, but there are not much good data on this point.

FOWLE: Has work been done on factors leading to the acquisition of endogenous retroviruses?

GALLO: Sometime in the course of the animals' evolutionary history infection of the germline occurred.

MARTIN: I think in most cases the idea is that they were acquired as a result of exogenous infection–a million years ago in the case of man or last week in the mouse.

Session 4:
Host Interactions

Viral Persistence: Role of Virus Variants and T Cell Immunity

RAFI AHMED* AND MICHAEL B. A. OLDSTONE†
*Department of Microbiology and Immunology
UCLA School of Medicine
Los Angeles, California 90024
†Department of Immunology
Scripps Clinic and Research Foundation
La Jolla, California 92037

OVERVIEW

We have studied the mechanism of viral persistence and the potential of specific immune therapy to clear virus from a persistently infected host. Using infection of mice with lymphocytic choriomeningitis virus (LCMV) as a model system we have found the following: (1) LCMV undergoes mutation during infection in its natural host, and there is organ-specific selection of viral mutants. LCMV isolates with different biological properties are present in brain and spleen of persistently infected mice. (2) Viral variants isolated from the spleen suppress LCMV-specific cytotoxic T lymphocyte (CTL) response of immunocompetent mice and cause persistent infections. In striking contrast, LCMV clones isolated from the brain induce potent virus-specific CTL responses and the infection is rapidly cleared. (3) The spleen variants selectively suppress the induction of primary CTL responses and LCMV-specific antibody responses are not affected. (4) Immune therapy of persistently infected mice with LCMV-specific memory cells is effective in reducing both infectious virus and viral genetic material. Viral clearance is mediated by major histocompatibility complex (MHC)-restricted virus-specific T cells.

INTRODUCTION

The immune system protects vertebrates against viral diseases and enables the host to clear the viral infection. The ability of viruses to escape or suppress the immune surveillance mechanisms and persist in the host is one of the most intriguing problems in virology. A variety of viruses, such as herpes simplex, cytomegalovirus, papilloma virus, human T cell leukemia/lymphoma virus, Epstein-Barr virus, measles, and hepatitis B, are known to cause persistent infections in humans (Wolinsky and Johnson 1980; Notkins and Oldstone 1984). Persistent viral infections are medically significant due to their association with a variety of diseases including neoplasms. In recent years, it has become evident that a number of chronic human diseases such as subacute sclerosing panencephalitis, progressive multifocal leukoencephalopathy, and progressive rubella panencephalitis are caused by persistent viral infections. A viral etiology is also suspected in several other chronic human illnesses. Thus, there is a need to study the host and viral factors involved in viral persistence, to identify the host effector mechanisms responsible for viral clearance, to understand the strategies used by the virus to escape the host immune response, and to develop methods to cure the infection.

Infection of mice with lymphocytic choriomeningitis virus (LCMV) is an excellent experimental model for studying the interaction between the virus and the host's immune response, and in defining the conditions that lead to viral clearance or persistence. This classic model of viral persistence has been extensively studied during the past 50 years and has provided significant information about virus-host interactions (Lehmann-Grube et al. 1983; Oldstone et al. 1985). However, a major unresolved question in the immunobiology of LCMV infection is the mechanism of suppression of cell-mediated immunity and its role in persistence. In this paper we describe our studies on the mechanism of LCMV persistence and suppression of T cell responses in BALB/c WEHI mice persistently infected with LCMV Armstrong strain.

RESULTS

When adult BALB/c mice are infected intravenously (i.v.) or intraperitoneally (i.p.) with wild-type (wt) LCMV Armstrong, they mount a vigorous primary cellular and humoral response against the virus and clear the infection within 8-10 days. After the primary immune response has subsided, these mice develop immunity and contain memory B and T cells against LCMV. Potent secondary responses can be generated following rechallenge with LCMV in vivo, or by culturing lymphocytes from immune mice with LCMV in vitro. In striking contrast to the acute infection seen in adult mice, when BALB/c mice are infected with LCMV at birth or in utero they become persistently infected with lifelong viremia and all their major organs contain high levels of infectious LCMV. Such carrier mice make reduced antibody response against the virus, and contain no detectable LCMV-specific cytotoxic T cells (CTL). Numerous attempts to induce CTL activity in carrier mice following rechallenge with LCMV in vivo or after stimulation of carrier lymphocytes with LCMV in vitro have also been unsuccessful. LCMV carrier mice do not have a generalized defect in making CTL responses since spleen cells from carrier mice are capable of generating hapten-specific and allogeneic CTL responses in vitro.

Using adoptive transfer experiments, we have shown that spleen and lymph node (LN) cells from persistently infected (carrier) mice actively suppress the expected LCMV-specific CTL response of spleen cells from normal adult mice (Ahmed et al. 1984). The results of these experiments are summarized in Table 1. There are four major points: (1) Carrier spleen cells suppress the induction of primary LCMV-specific CTL response of normal adult spleen cells. (2) The suppression is specific for the CTL response and has no effect on LCMV-specific antibody response. (3)

Table 1
Summary of Transfer Experiments

Source of BALB/c spleen cells transferred[a]	LCMV specific immune response[b]		Outcome of infection[c]
	CTL	Antibody	
Normal (N)	++++	++++	Clearance
Persistently infected carrier (C)	–	±	Persistence
C + N	±	++++	Persistence
C + Immune	++++	++++	Clearance

[a] 5×10^7 cells from 6–10-week-old BALB/c (normal, carrier, or immune) mice were transferred intravenously into 2–4-month-old normal BALB/c WEHI mice irradiated (600 rad) 1 day previously. Mice reconstituted with lymphoid cells from two sources received 5×10^7 cells from each source. At the time of the cell transfer, all recipient mice were infected intravenously with 2×10^5 pfu of wt LCMV Armstrong.

[b] At various times postinfection (days 3, 6, 8, 10, and 15) mice were checked for (1) LCMV-specific CTL activity in spleen and lymph nodes (LN) by ^{51}Cr-release assay, and for (2) LCMV-specific antibody by an ELISA using purified virus as antigen.

[c] Mice were checked for presence of infectious LCMV in serum, spleen, LN, thymus, lung, liver, brain, and kidney for a period of 6–9 months postinfection. The titer of infectious LCMV was determined by plague assay on Vero cell monolayers.

Suppression of the CTL response is associated with the establishment of persistent infection. (4) Carrier spleen cells are unable to suppress LCMV-specific secondary CTL responses; transfer of immune cells containing memory T cells results in a potent CTL response and subsequent viral clearance. We have shown that the active suppression of CTL responses by carrier spleen cells and the associated persistent infection are due to genetic variants of LCMV in spleens of carrier mice (Ahmed et al. 1984).

Organ-Specific Selection of Variants: Selection of LCMV Variants with Different Biological Properties in Brain and Spleen of Persistently Infected Mice

Five 7-week-old BALB/c WEHI carriers, infected at birth with wt LCMV Armstrong, were sacrificed and their spleens and brains harvested. The organs were homogenized individually and titrated on Vero cell monolayers. Well-isolated plaques were picked from each sample, recloned twice, and then stocks grown in BHK-21 cells. Six LCMV clones were isolated from each carrier mouse; three clones from the brain (brain isolates) and three clones from the spleen (spleen isolates). These spleen and brain isolates were authenticated as LCMV by their positive reactivity with monoclonal antibodies specific for the nucleoprotein and glycoprotein of wt LCMV Armstrong. The brain and spleen isolates were tested for their ability to induce LCMV-specific CTL and antibody responses in adult BALB/c mice. All 30 LCMV isolates, irrespective of their origin, induced potent antibody responses. In striking contrast, dramatic differences were observed between the brain and spleen isolates in their ability to induce CTL responses and to cause acute or persistent infection in adult immunocompetent mice (Fig. 1). The majority (12/15) of the brain isolates induced good CTL responses and the virus was cleared within 8–10 days. However, greater than 80% (13/15) of the spleen isolates induced poor LCMV-specific CTL responses and were present in high titers in the spleens of infected mice at 8 days postinfection. Adult mice infected with the spleen isolates remain persistently infected for at least 2 months. These spleen isolates were derived from 7-week-old carrier mice that were infected at birth with wt LCMV Armstrong. Ten LCMV clones derived from the original stock of wt LCMV were tested for their ability to induce CTL responses in adult mice. All ten clones induced potent CTL responses and the infection was cleared within 8 days (Fig. 1). This result shows that the original inoculum was reasonably homogeneous and the LCMV variants that suppress CTL responses emerged during the persistent infection.

These studies demonstrate that LCMV undergoes mutation during persistent infection in its natural host and there is organ-specific selection of viral mutants. LCMV isolates with different biological properties are present in brain and spleen of carrier mice. LCMV clones isolated from brains of carrier mice induce strong CTL responses in adult mice and the infection is rapidly cleared, whereas LCMV clones derived from spleens of carrier mice induce poor CTL responses and cause persistent infections in adult immunocompetent mice (Fig. 2).

Immune Therapy of Persistently Infected Mice: Reduction of Infectious LCMV and Viral Genetic Material

Experiments were done to determine whether virus can be cleared from persistently infected mice upon transfer of LCMV-specific immune cells. Adult (6–8 weeks old) BALB/c mice were infected intraperitoneally with 10^4 pfu of LCMV. The mice were sacrificed 30–60 days later, and spleen and lymph node cells from these mice were used as LCMV-specific immune cells. Four to six-week-old BALB/c carrier mice infected at birth with LCMV were treated with the immune lymphocytes (5×10^7 cells/mouse given intravenously). As previously reported (Volkert 1963), a reduction in the level of infectious LCMV was noted in the majority (23/30) of the treated carriers. The treated carriers showed a gradual decrease in the level of infectious LCMV in the serum and contained < 50 pfu/ml 3–4 months after the transfer of immune cells. In contrast, age- and sex-matched untreated carrier mice contained $> 10^4$ pfu/ml in the serum. The treated carriers contained no detectable or trace amounts of infectious LCMV in the spleen, lymph nodes, thymus, liver, lung, and brain whereas untreated carriers contained between 10^5 to 10^7 pfu per organ (Fig. 3). The kidney was the only organ where high levels of LCMV were still present at 3–4 months after the transfer of immune cells. However, there was ~ 500-fold reduction in the amount of LCMV present in the kidneys of treated carriers compared to untreated carriers. Using in situ hybridization of whole mouse sections with ^{32}P-labeled LCMV-specific cDNA probes, we found that LCMV

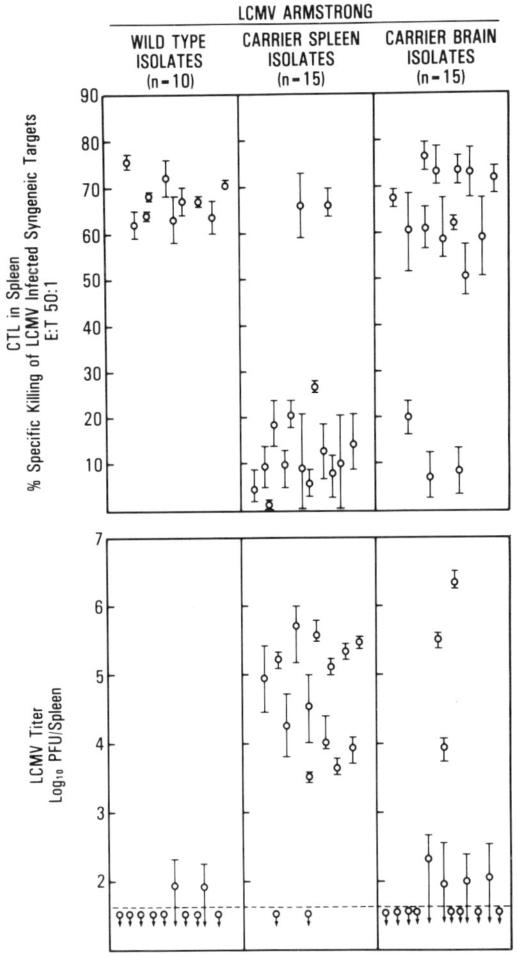

Figure 1
LCMV variants with different biological properties are present in brain and spleen of persistently infected mice. Six to ten-week-old BALB/c WEHI mice were infected i.v. with 2×10^5 pfu of LCMV isolates derived from the indicated source. LCMV-specific CTL activity and virus titer were determined 8 days postinfection. Each point represents the mean value of three mice infected with a particular LCMV isolate. The bars indicate the range.

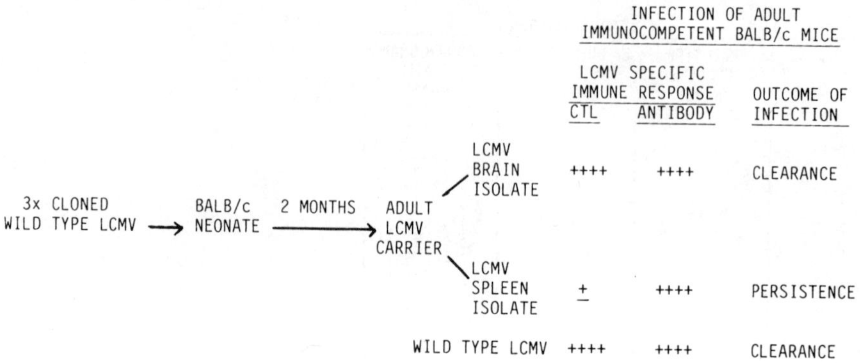

Figure 2
Organ-specific selection of viral mutants during persistent infection

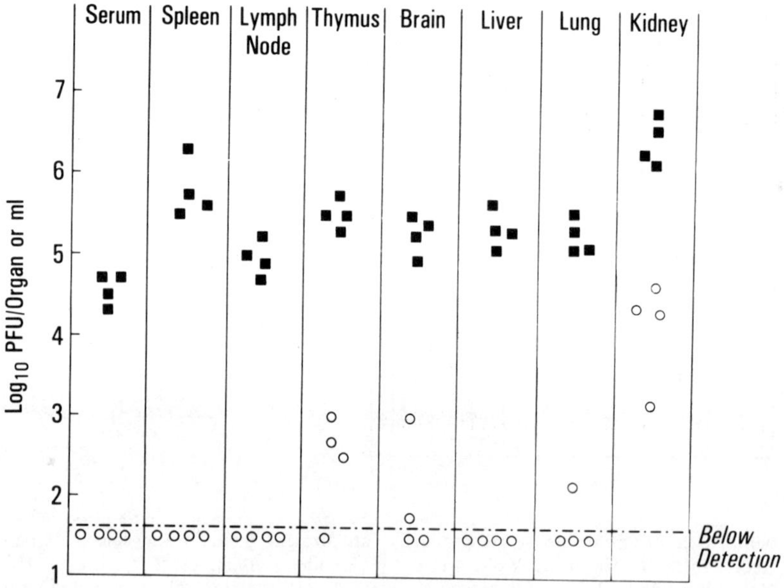

Figure 3
Immune therapy of persistently infected mice: Clearance of infectious virus from the organs. Four to six-week-old BALB/c WEHI carrier mice were treated with LCMV-specific immune lymphocytes (5×10^7 cells/mouse given intravenously). Mice were sacrificed 120 days after the cell transfer and the amount of infectious LCMV was determined by a plaque assay on Vero cells. (○) Treated carriers; (■) untreated carriers.

genetic material was dramatically reduced in persistently infected mice receiving immune cells (Ahmed et al. 1985). Thus, our studies show that immune cells can reduce both infectious LCMV and viral genetic material from persistently infected mice.

Immune Therapy of Persistently Infected Mice: MHC-restricted Virus-specific T cells Mediate Viral Clearance

LCMV-specific immune cells treated with anti-theta plus complement prior to transfer were unable to clear virus from persistently infected mice. Transfer of immune lymphocytes into non-H-2 matched persistently infected mice had no effect on the virus level. In these experiments the carrier recipients were irradiated prior to cell transfer to prevent rejection of the allogeneic immune cells. Syngeneic transfer of Pichinde virus-specific immune lymphocytes was also ineffective in reducing the virus titer of mice persistently infected with LCMV. Taken together, these three experiments show that viral clearance of persistently infected host is mediated by H-2 restricted virus-specific T cells. Experiments are currently in progress to identify the subsets of T cells required for viral clearance.

DISCUSSION

We have studied the interaction of a virus with its natural host and defined conditions that lead to viral clearance or persistence. Our study establishes three major points: (1) There is organ-specific selection of viral mutants during persistent infection. (2) Suppression of virus-specific T cell response is associated with the establishment of persistent infection. (3) H-2 restricted virus-specific T cells are effective in clearing virus from a persistently infected host.

The ability of several viruses, especially those with RNA genomes, to undergo rapid mutation during both acute and persistent infections has been amply documented (Ahmed et al. 1980; Youngner and Preble 1980; Holland et al. 1982; Ahmed et al. 1983). The majority of these studies have examined the evolution of viruses in tissue culture cells, and there have been relatively few studies analyzing genetic variation in viruses during replication in their natural hosts (Kono et al. 1973; Narayan et al. 1977). Also, little is known about the selective pressures involved in the emergence of viral variants and the role of such variants in persistence and pathogenesis. The generation of LCMV variants that suppress CTL responses represents an example of a biologically meaningful selection. Such variant viruses are capable of causing a persistent infection in an adult immunocompetent host. Our studies also demonstrate the importance of the host organ in selection of viral mutants. Work is in progress to identify the cell type in which variants are generated and to analyze the genetic differences between the viral variants.

Cell-mediated immunity is an important factor in the control and elimination of many viral infections, and several studies have shown that major histocompatibility complex (MHC) restricted virus-specific CTL play a crucial role in viral clearance (Blanden 1974; Zinkernagel and Doherty 1979; Finberg and Benacerraf 1981). The availability of LCMV variants that specifically suppress CTL responses should allow us to study the mechanism of this suppression and provide basic information about virus-lymphocyte interaction. We have also shown that H-2-restricted T cells can effectively reduce both infectious virus and total viral genetic material. Thus, specific immune therapy has the potential to clear virus from a persistently infected host.

ACKNOWLEDGMENTS

We thank Rita J. Concepcion for excellent technical assistance. This work was supported by grants AI-09484 and NS-12428 from the U.S. Public Health Service to M.B.A.O., and by grants from the Keck Foundation and UCLA Medical School BRSG start-up funds to R.A.

REFERENCES

Ahmed, R., P.R. Chakraborty, and B.N. Fields. 1980. Genetic variation during lytic reovirus infection: High-passage stocks of wild-type reovirus contain temperature-sensitive mutants. *J. Virol.* **34**: 285.

Ahmed, R., R.S. Kauffman, and B.N. Fields. 1983. Genetic variation during persistent reovirus infection: Isolation of cold-sensitive and temperature sensitive mutants from persistently infected L cells. *Virology* **130**: 71.

Ahmed, R., A. Salmi, L.D. Butler, J.M. Chiller, and M.B.A. Oldstone. 1984. Selection of genetic variants of lymphocytic choriomeningitis virus in spleens of persistently infected mice. *J. Exp. Med.* **60**: 521.

Ahmed, R., P. Southern, P. Blount, J. Byrne, and M.B.A. Oldstone. 1985. Viral genes, cytotoxic T lymphocytes, and immunity. In *Vaccines 85* (ed. R.A. Lerner, R.M. Chanock, and F. Brown), p. 125. Cold Spring Harbor Laboratory, Cold Spring Harbor, New York.

Blanden, R.V. 1974. T cell response to viral and bacterial antigens. *Transplant. Rev.* **19**: 56.

Finberg, R. and B. Benacerraf. 1981. Induction, control and consequences of virus specific cytotoxic T cells. *Immunol. Rev.* **58**: 157.

Holland, J.J., K. Spindler, F. Horodyski, E. Graham, S. Nichol, and S. Vendepol. 1982. Rapid evolution of RNA genomes. *Science (Wash. D.C.)* **215**: 1577.

Kono, Y., K. Kobayashi, and Y. Fukunaga. 1973. Antigenic drift of equine infectious anemia virus in chronically infected horses. *Arch. Gesamte Virusforsch.* **41**: 1.

Lehmann-Grube, F., L.M. Peralta, M. Bruns, and J. Lohler. 1983. Persistent infection of mice with the lymphocytic choriomeningitis virus. In *Comprehensive virology,* (ed. H. Fraenkel-Conrat and R. Wagner), vol. 18, p. 43. Plenum Publishing, New York.

Narayan, O., D.E. Griffen, and J. Chase. 1977. Antigenic shift of visna virus in persistently infected sheep. *Science (Wash. D.C.)* **199**: 376.

Notkins, A.L., and M.B.A. Oldstone. 1984. *Concepts in viral pathogenesis.* Springer-Verlag, New York.

Oldstone, M.B.A., R. Ahmed, J. Byrne, M.J. Buchmeier, Y. Riviere, and P. Southern. 1985. Viruses and immune responses: LCMV as a prototype model of viral pathogenesis. *Br. Med. Bull.* **41**: 70.

Volkert, M. 1963. Studies on immunological tolerance to LCM virus. 2. Treatment of virus carrier mice by adoptive immunization. *Acta Pathol. Microbiol. Scand.* **57**: 465.

Wolinsky, J.S., and R.T. Johnson. 1980. Role of viruses in chronic neurological diseases. In *Comprehensive virology.* (ed. H. Fraenkel-Conrat and R. Wagner), vol. 16, p. 257. Plenum Publishing, New York.

Youngner, J.S. and O.T. Preble. 1980. Viral persistence: Evolution of viral populations. In *Comprehensive virology.* (ed. H. Fraenkel-Conrat and R. Wagner), vol. 16, p. 73. Plenum Publishing, New York.

Zinkernagel, R.M. and P.C. Doherty. 1979. MHC-restricted cytotoxic T cells: Studies on the biological role of polymorphic major transplantation antigens determining T-cell restriction-specificity, function, and responsiveness. *Adv. Immunol.* **77**: 51.

COMMENTS

CROWELL: What do you think causes the tropism for the spleen or the brain? Do you think that it happens ahead of time or after the virus is there? In other words, is the virus population being modified in these sites or is it a selection of a preexisting mutant?

AHMED: There is certainly organ-specific selection. Although presence of a preexisting mutant in the original inoculum cannot be completely ruled out, it is likely that the mutants are generated in vivo since we have worked with plaque-purified populations and, in some cases, have made virus carriers by injecting neonatal mice with an individual plaque of LCMV and still observed the selection of mutants in the different organs. At the moment, we do not know the molecular basis for the tissue-specific selection. This could be operating at the level of virus adsorption/penetration (i.e., cellular receptors) or viral transcription/replication involving specific host factors.

MARTIN: Has the level of the defect been determined? Do you have any ideas why replication is inhibited in the spleen type virus? Why does it persist?

AHMED: The ability of the LCMV variants to cause a disseminated infection and the suppression of virus-specific T cell responses are the primary factors resulting in viral persistence. Regarding the mechanism of T cell suppression, the information we have at present is that an $L3T4^+$ $Lyt2^-$ T cell is infected in carrier mice. Thus, a T cell of the helper-induced phenotype is infected with LCMV. We are trying to clone these T cells to further characterize them and are also investigating whether T helper cells are required for the generation of LCMV-specific cytotoxic T cell responses.

MERIGAN: What other strategies have been used to try and eradicate this chronic infection? Is this the first time immunotherapy has been used?

AHMED: I am not aware of any other strategies (e.g., anti-viral drugs or interferon) being used to treat the LCMV chronic infection. The ability of LCMV-specific immune cells to eliminate virus from persistently infected mice was first shown by Volkert and his associates. The results I presented are the first identification of the type of T cells required for viral clearance.

MERIGAN: Can you tell us which viral gene may have been altered?

AHMED: We have made genetic reassortments between the spleen variant and wild-type LCMV to map the viral genes responsible for persistence. Our preliminary data indicates that the ability of the spleen variants to cause persistent infection in adult mice is due to mutations in both the large and small genomic segments.

FIELDS: You have examined mutants, selected in vivo, that have a different impact on host immune response. Are there clear examples of in vitro selected mutants?

AHMED: I am not aware of any examples using in vitro selected viral mutants showing such dramatic differences on host immune responses.

FIELDS: One of the issues, again, is how the host would respond to a genetically altered virus. Would variants of HTLV, altered by in vitro mutagenesis lead to changes in the host response? What you have cited is an example where the host has selected the mutants. Could you reproduce this in cell culture?

AHMED: Several investigators including myself have tried unsuccessfully to get a productive LCMV infection of T cells in vitro. During persistent infection in vivo about 0.1-1% of peripheral T cells, as well as thymocytes, are infected with LCMV. It is possible that T cells are susceptible to LCMV only at an early stage in the differentiation pathway. This would explain the unsuccessful attempts at infecting mature T cells in vitro.

MERIGAN: How about the role of the cytotoxic cell in the acute infection?

AHMED: The role of cytotoxic T cells in controlling acute LCMV infection was first shown by Zinkernagel and Welsh, and more recently by Byrne and Oldstone, using clones of LCMV-specific CTL.

MERIGAN: But as far as the establishment of the chronic infection, can you tell us whether there is a requirement for a deficiency in certain immune cells?

AHMED: I didn't show you all the data, but we have recently published a paper demonstrating a striking correlation between suppression of LCMV-specific CTL response and the establishment of persistent infection.

Genomic Variation of HTLV-III/LAV, the Retrovirus of AIDS

BEATRICE H. HAHN,* GEORGE M. SHAW,* FLOSSIE WONG-STAAL, AND ROBERT C. GALLO
Laboratory of Tumor Cell Biology
Division of Cancer Treatment
National Cancer Institute
Bethesda, Maryland 20205

OVERVIEW

A novel human retrovirus, HTLV-III/LAV, with T-lymphotropic and cytopathic properties has been identified as the causative agent of the acquired immunodeficiency syndrome (AIDS). In this study, we examined the restriction enzyme patterns and nucleotide sequence homologies of 18 consecutive isolates of HTLV-III/LAV and found that a spectrum of variation, or diversity, exists among them. No two viral isolates were identical in their restriction enzyme maps but all hybridized throughout their genomes to a full-length, cloned probe derived from the prototype AIDS virus, HTLV-IIIB. We also cloned the full-length HTLV-III/LAV proviral genome from a Haitian man with AIDS, analyzed it by heteroduplex and nucleotide sequence analysis, and compared it to three other similarly characterized HTLV-III/LAV viruses. Results of this analysis confirmed that genomic diversity is a prominent feature of HTLV-III/LAV and that it is the envelope gene which varies most among different viral isolates. Finally, it was found that changes within the viral envelope are not uniform in distribution; instead, genomic changes cluster in the exterior portion of the envelope glycoprotein and coincide with regions that, based on their secondary structure, hydrophilicity and glycosylation pattern, represent predicted antigenic sites. The implications of these findings with respect to the biology of HTLV-III/LAV, its evolution, and the development of diagnostic reagents and vaccines are discussed.

INTRODUCTION

The clear demonstration that a novel human retrovirus is the causative agent of the acquired immunodeficiency syndrome (AIDS) was reported by Gallo and coworkers in 1984 (Gallo et al. 1984; Popovic et al. 1984; Sarngadharan et al. 1984; Schupbach et al. 1984). This virus, called HTLV-III for human T-lymphotropic virus type III, was subsequently shown to be the same virus as that detected in one patient with AIDS-related complex (ARC) by French investigators and called LAV (Barré-Sinoussi et al. 1983), as well as that reported much later by Levy and col-

**Current address*: Division of Hematology and Oncology, University of Alabama at Birmingham, Birmingham, Alabama 35294 leagues and termed ARV (Levy et al. 1984). Since all of these human T-lymphotropic viruses are isolates of the same virus (Gallo et al. 1985; Ratner et al. 1985a), we refer to them here as HTLV-III/LAV.

Although the cause of AIDS has thus now been identified, the mechanisms by which the virus exerts its cytopathic effects, the natural history of viral infection, and the contributing factors responsible for the broad clinical spectrum associated with HTLV-III/LAV infection are just beginning to be understood. Several lines of investigation, however, point toward diversity, or variation, in the HTLV-III/LAV genome as a viral characteristic fundamentally important to its biological behavior and pathogenicity. That genomic variation is a prominent feature of this virus was first recognized by Southern blot analysis of a limited number of virally infected cell lines and fresh uncultured tissues derived from patients with AIDS (Shaw et al. 1984, 1985). At a nucleotide sequence level, it was also shown that genomic variability of different degrees occurs among different viral isolates and that the envelope gene, in particular, may be highly divergent (Ratner et al. 1985a). This latter observation and the recent demonstration of ultrastructural and genetic relatedness between HTLV-III and visna virus (Gonda et al. 1985), which is known to undergo rapid genomic variation in its envelope gene as a consequence of immune selection (Clements et al. 1980), further point toward genomic diversity as a property likely to be important in the biology of the AIDS virus.

In this study, we have analyzed by restriction enzyme mapping and Southern hybridization 18 consecutive isolates of HTLV-III/LAV from patients with AIDS, ARC, or no clinical disease in order to define the nature and spectrum of variation among different viruses. We have also molecularly cloned and sequenced an isolate from a Haitian man with AIDS and compared this virus to prototype HTLV-III/LAV isolates. This particular virus isolate, HTLV-III_{RF}, was chosen for detailed analysis since it was derived from a restricted geographic region relatively early in the AIDS epidemic (Popovic et al. 1984) and since preliminary analysis suggested that it was not closely related to other HTLV-III/LAV viruses (Shaw et al. 1984). The results verify our earliest indications that genomic variation is a prominent feature of HTLV-III/LAV viruses and that it is the envelope gene, especially its extracellular portion, which varies most.

RESULTS

HTLV-III/LAV—A Spectrum of Highly Related but Distinguishable Viruses

The HTLV-III/LAV viral genomes from 18 individuals infected with this virus were evaluated for evidence of genomic variation by detailed restriction enzyme analysis (Wong-Staal et al. 1985). Virus from nine patients with AIDS or AIDS-related complex and one healthy homosexual man was transmitted to established neoplastic T-cell lines or to normal peripheral blood lymphocytes (Popovic et al. 1984; Salahuddin et al. 1985). Primary tissues (lymph node or brain) from eight other patients were examined directly (Table 1). High molecular weight DNA was digested with a panel of restriction enzymes and analyzed by Southern blot analysis using a 9-kb cloned HTLV-III probe (Hahn et al. 1984). All of the 18 viral genomes, including isolates from Haiti, Europe, and Africa, hybridized throughout their entire genome under high stringency conditions indicating a high degree of homology (Fig. 1 and Table 1). However, each of the viral genomes exhibited a distinct restriction enzyme pattern that was distinguishable from all others by at least one, and usually many, restriction site differences.

No particular restriction enzyme pattern could be identified for a particular disease state such as AIDS, ARC, or asymptomatic carrier; and there was no correlation between certain enzyme site changes and the type of tissue from which the isolate was derived. Viral sequences in brain tissue from patients with AIDS encephalopathy appeared to vary to the same degree as did viral sequences derived from lymphoid tissues. Viruses detected in different tissues within the same patient, at least at a Southern blot level, were found to be identical (B. Hahn unpubl.).

Although most of the 18 viral isolates were found to differ substantially from each other, we identified two viruses that were highly related. These isolates, MN and SL, differed from each other only in a single enzyme site, yet they differed from HTLV-IIIB in more than 50% of the sites tested (Fig. 2). MN was derived in 1984 from a child with AIDS, and SL was obtained at this same time from a homosexual man with ARC. Both patients were from the New York/New Jersey area but had had no known direct or indirect contact with each other (Salahuddin et al. 1985).

Genomic Variation in the Envelope Genes of Different Viral Isolates

Southern blot analysis had shown that genomic diversity is a characteristic and prominent feature of AIDS virus isolates and that the variation ranges from slight to rather extensive changes. At a nucleotide sequence level, it was also demonstrated that genomic variability of different degrees occurs among HTLV-III/LAV viruses and that the envelope gene may be highly divergent (Ratner et al. 1985a,b; Sanchez-Pescador et al. 1985; Wain-Hobson et al. 1985). Since these data were based on the comparison of only three HTLV-III/LAV viruses, two of which (HTLV-IIIB and LAV) were highly related to each other, sequence analysis of an additional variant virus was needed to identify the regions of high sequence variability and conservation. We therefore selected an isolate from a Haitian man with AIDS, HTLV-III$_{RF}$, which differed considerably in its restriction enzyme pattern from all other isolates characterized so far (Popovic et al. 1984; Shaw et al. 1984).

A comparison of the molecularly cloned genome of HTLV-III$_{RF}$ (λHAT-3) to representative clones of HTLV-IIIB, LAV and ARV demonstrated up to 50%

Table 1
Patients Evaluated for HTLV-III DNA

Isolate	DNA source	Patient diagnosis	Risk factor	Geographic location	Year of isolation
H9/HTLV-IIIB	PBL/H9[a]	Pooled from AIDS and ARC	Homosexuals	New York and New Jersey	1983
RF	PBL/H4[a]	AIDS	Haitian	Haiti and Philadelphia	1983
RH	PBL[b]	Healthy	Homosexual	Washington, D.C.	1984
TM	PBL/N-PBL[a]	AIDS	Homosexual	Boston	1984
HW	PBL/N-PBL[a]	AIDS	Heterosexual, promiscuity	Europe and Africa	1984
MN	PBL/JM[a]	ARC	Child of IV drug user mother	New York and New Jersey	1984
SL	PBL/N-PBL[a]	ARC	Homosexual	New York and New Jersey	1984
SC	PBL/N-PBL[a]	ARC	Homosexual	California	1984
JS	PBL/N-PBL[a]	ARC	Homosexual	California	1984
MJ	PBL/H9[a]	AIDS	Child of Haitian mother	Haiti and Miami	1985
RJ	PBL/N-PBL[a]	AIDS	Homosexual	Boston	1985
FO	Lymph node[c]	AIDS	Homosexual	New York and New Jersey	1984
KC	Lymph node[c]	ARC	IV drug user	New York and New Jersey	1984
JR 1	Brain[c]	AIDS	Child of IV drug user mother	New York and New Jersey	1984
JR 2	Brain[c]	AIDS	Homosexual	New York and New Jersey	1984
LS	Brain[c]	AIDS	Child of ARC mother	New York and New Jersey	1984
JT	Brain[c]	AIDS	Homosexual	New York and New Jersey	1984
RC	Brain[c]	AIDS	Homosexual	New York and New Jersey	1984
RR	Brain[c]	AIDS	Homosexual	New York and New Jersey	1984

[a]Virus transmitted from patients' PBL into H9, H4, or JM cell lines, or into normal allogeneic PBL.
[b]Virus grown directly in patients' own PBL without transmission.
[c]Viral DNA detected and analyzed in uncultured tissue.
Reprinted, with permission, from Wong-Staal et al. (1985).

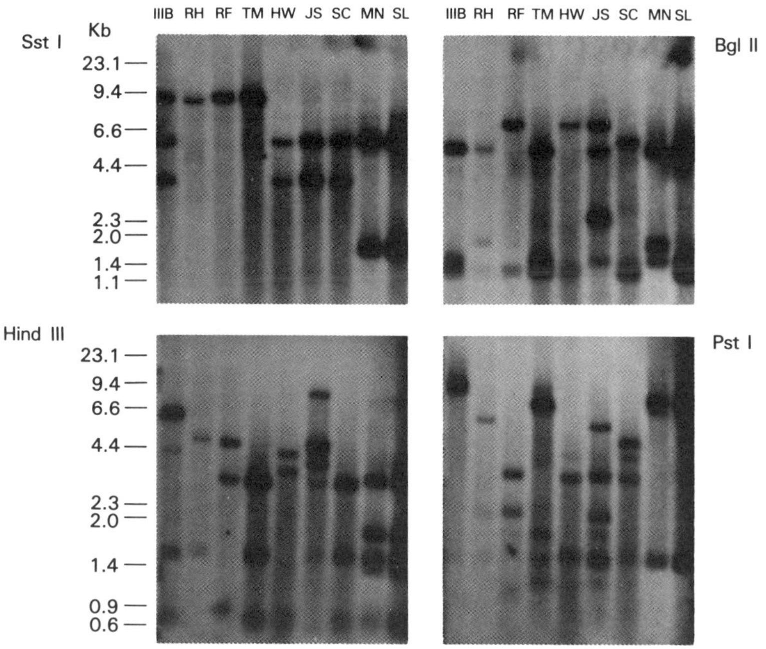

Figure 1
Genomic variation in different HTLV-III/LAV isolates. The restriction enzyme pattern of eight independent HTLV-III/LAV isolates is compared to the prototype HTLV-IIIB (first lane of each panel). A HTLV-III clone (BH-10) containing the entire HTLV-III genome less 180 bps of LTR was used for probe. Hybridization was carried out under high stringency conditions (Tm-25°C). A detailed description of each isolate is summarized in Table 1. Reprinted, with permission, from Wong-Staal et al. (1985).

restriction enzyme site differences scattered throughout the entire length of the molecules (Fig. 3). Heteroduplex thermomelt analysis, however, indicated genomic changes primarily at the 3′ end of the virus (Fig. 4) (Hahn et al. 1985). We have subsequently determined the nucleotide sequence of λHAT-3 and have shown that the envelope gene is the most variable part (B. Starcich et al., in prep.). In this region, HTLV-IIIB (BH-10) differs from HTLV-III$_{RF}$ in 13% of nucleotides (19% amino acids) and from ARV (ARV-2) in 10% of nucleotides (16% amino acids). In the same region, HTLV-III$_{RF}$ differs from ARV in 15% of nucleotides (10% amino acids) (B. Starcich et al., in prep.).

The most conserved areas in the genomes of HTLV-III$_{RF}$, LAV, HTLV-IIIB and ARV were the *gag* and *pol* genes which differed among the viruses in less than 6% of nucleotide sequence and less than 7% of amino acid sequence. In these regions, nucleotide sequence changes were almost exclusively due to point mutations and

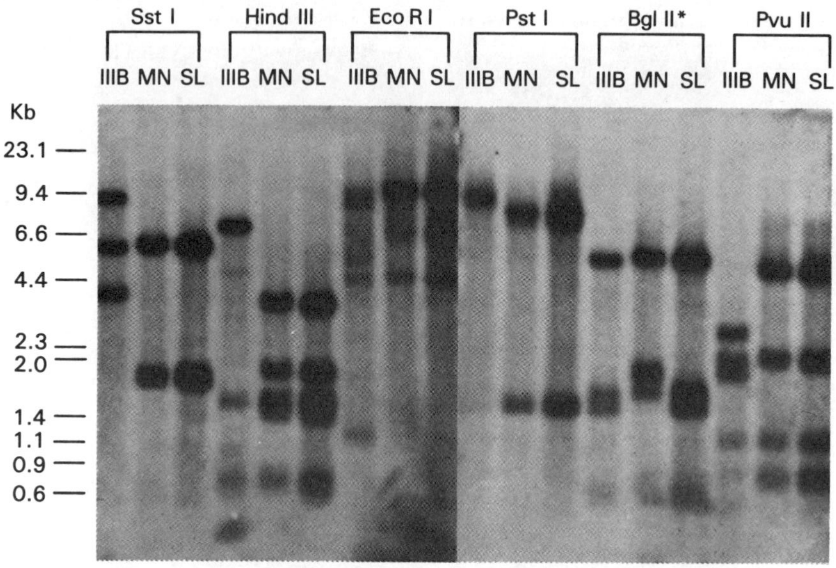

Figure 2

Comparison of two highly related isolates of HTLV-III/LAV to the prototype HTLV-IIIB. As shown, *MN* and *SL* differ from each other only in the size of one *Bgl*II fragment while they differ from HTLV-IIIB in more than 50% of the enzyme sites tested. Reprinted, with permission, from Wong-Stahl et al. (1985).

more than half of these occurred in the third base of a codon, which resulted in no change in the encoded amino acid. In addition, more than half of the actual amino acid changes were conservative. This sequence conservation in *gag* and *pol* contrasts with the substantial divergence in the envelope where clustered nonconservative changes involving in-frame deletions, insertions, and substitutions were common. Furthermore, within the envelope gene the extracellular region was found to vary considerably more than the transmembrane region and to contain localized regions of high sequence variability as well as high sequence conservation.

Prediction of Antigenic Determinants in Different HTLV-III/LAV Envelope Glycoproteins

Since the exterior portion of the envelope glycoprotein constitutes a major immunologic stimulus, we have recently examined this region for predicted antigenic epitopes using a computer program that predicts the secondary structure of proteins superimposed with values for hydrophilicity (Chou and Fasman 1974; Hopp and Woods 1981). Such an analysis of other proteins, including viral envelopes, has shown that continuous antigenic epitopes are often associated with hydrophilic

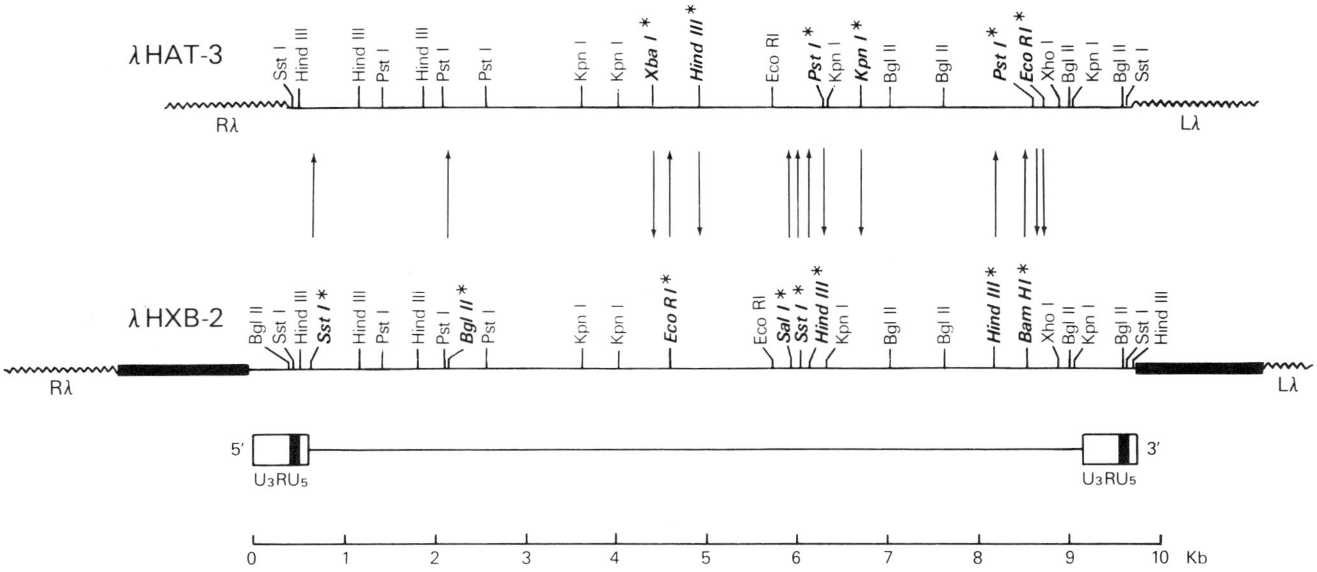

Figure 3
Comparison of the molecular cloned genomes of HTLV-III$_{RF}$ (λHAT-3) and HTLV-IIIB (λHXB-2). Differences in the restriction enzyme maps of both clones are indicated by arrows pointing to missing restriction enzyme sites and by bold letters with asterisks. Reprinted, with permission, from Hahn et al. (1985).

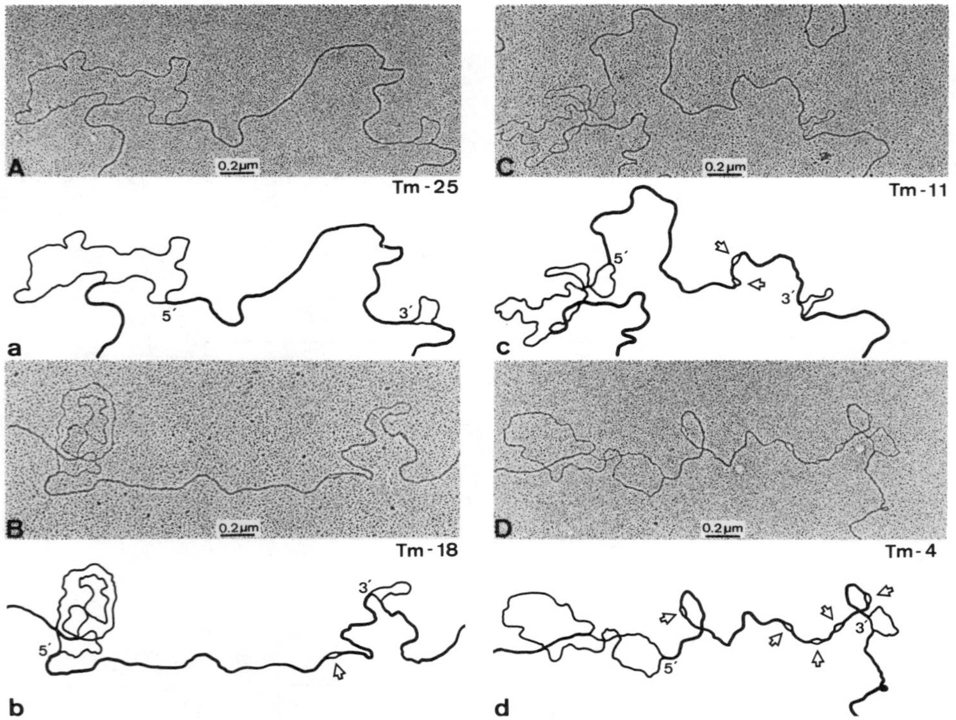

Figure 4
Heteroduplex analysis of HTLV-III$_{RF}$ (λHAT-3) and HTLV-IIIB (λHXB-2) at varied stringencies. Heteroduplex analysis was performed with cloned inserts still in their bacteriophage vectors. Actual heteroduplexes are shown in *A*, *B*, *C*, and *D* and interpretive drawings in *a*, *b*, *c*, and *d*, respectively. (*A*) 50% Formamide (Tm-25°C); (*B*) 60% formamide (Tm-18°C); (*C*) 70% formamide (Tm-11°C); and (*D*) 80% formamide (Tm-4°C). Open arrows (*b*, *c*, and *d*) indicate substitutions as they appear with increasing stringencies. Reprinted, with permission, from Hahn et al. (1985).

protein domains containing β turns (Cohen et al. 1984; Eisenberg et al. 1985; Gunn et al. 1985; Pellett et al. 1985). Continuous epitopes are defined as those peptides whose antigenicity is reflected in the secondary structure of the primary amino acid sequence, whereas discontinuous epitopes are antigenic loci formed by the tertiary folds of the native protein.

Such an analysis shows that the exterior envelope proteins of the four retroviruses, HTLV-III_{RF}, HTLV-IIIB, LAV, and ARV, each contain a number of sites meeting the above criteria for likely antigenic epitopes and that these regions generally coincide with variable regions identified independently by amino acid sequence comparisons (B. Starcich et al., in prep.). In each of the variable regions, differences exist in hydrophilicity, secondary structure and potential glycosylation pattern. Because of this variability, it is likely that the antigenic profiles of the corresponding envelope proteins of these viruses differ considerably.

Interspersed with the variable regions of the exterior envelope protein were areas highly conserved among all four isolates analyzed. With one exception, all of these regions were primarily hydrophobic and contained only few or no β turns; accordingly, they are much less likely to represent antigenic epitopes. One major exception was a stretch of 45 amino acid residues immediately adjacent to the processing site of the envelope precursor. This area was found to be highly conserved among all four viruses, yet it contained numerous β turns and was hydrophilic. Thus, this region of the exterior envelope glycoprotein, more so than other regions, would be anticipated to be both immunogenic and cross-reactive among different viral strains.

The transmembrane region of the envelope glycoprotein was considerably more conserved in its secondary structure than was the exterior portion. This seems to be a common feature among retroviruses and presumably reflects, in part, structural and functional constraints.

DISCUSSION

In this paper, we have summarized a body of evidence obtained by our laboratory over the past year relevant to the issue of genome variation of HTLV-III/LAV. It has become evident that a spectrum of diversity exists among isolates and that this probably reflects progressive changes in the viral genome over time. The existence of this genomic variation has been demonstrated by differences in restriction enzyme patterns and by nucleotide sequence data showing differences up to 20% in certain genomic regions. However, although the extent of the restriction site differences between isolates generally correlates with their degree of nucleotide sequence divergence, it should be noted that restriction enzyme mapping does not adequately reflect the clustered nucleotide sequence changes present in certain regions of the viral genome, especially the viral envelope gene.

The finding of nonuniform genomic changes in HTLV-III/LAV with greatest divergence in the extracellular envelope raises the possibility that changes within this gene could result from selective pressures existent in vivo. Such selection processes could be immune and/or nonimmune in nature. Examples of immune selection are the envelope changes of visna virus and equine infectious anemia virus resulting from the production of high titers of neutralizing antibodies in infected animals (Clements et al. 1980; Montelaro et al. 1984). In this regard, it is particularly noteworthy that HTLV-III/LAV has been shown to be highly related to visna virus in its morphology and genomic organization (Gonda et al. 1985). Studies are now underway to evaluate the nature of the immunologic response to HTLV-III/LAV in man (Robert-Guroff et al. 1985; Weiss et al. 1985).

Another third major conclusion from our recent studies is that even within the envelope gene the nature of genomic variation is nonuniform and different from changes elsewhere in the HTLV-III/LAV genome (B. Starcich et al., in prep.). That is, within the exterior envelope, there are clusters of nucleotide changes involving in-frame deletions, insertions, and substitutions which dramatically alter the protein's predicted hydrophilicity and secondary structure. Conversely, in the *gag* and *pol* genes, changes are almost exclusively due to point mutations, and more than half of these result in no change in the encoded amino acid. The existence of these clustered changes within the exterior envelope gene has allowed us to subdivide this region into variable and constant domains (B. Starcich et al., in prep.). When the deduced amino acid sequences of the *env* genes are analyzed for hydrophilicity and secondary structure, it is apparent that each of the areas of high genomic variability corresponds to a region possessing features characteristic of antigenic epitopes. These findings further suggest that the changes in the HTLV-III/LAV envelope gene result, at least in part, from selective pressures present in vivo and that they are biologically meaningful.

Although much has already been learned about genomic variation in HTLV-III/LAV, important questions remain relevant to the biology, the evolution, and the recent geographic spread of this virus. The discovery of STLV-III (Daniel et al. 1985; Kanki et al. 1985a,b) and ongoing molecular epidemiological studies of HTLV-III/LAV promise to yield insights into these questions. With regard to vaccine development, the findings described here indicating that the exterior HTLV-III/LAV envelope glycoprotein is highly variable, especially in regions of predicted antigenicity, deserve special consideration. In this regard, it is notable that we have also identified other regions which appear to be quite conserved among virus isolates. It is possible that such conserved regions may possess essential biologic functions and therefore serve as important targets for recombinant DNA-based vaccines.

CONCLUSIONS

Different isolates of HTLV-III/LAV exhibit substantial genomic diversity as demonstrated by restriction enzyme mapping, heteroduplex analysis, and nucleic acid sequencing. The region of greatest divergence is located in the envelope gene, especially in the exterior portion of the envelope glycoprotein, as determined by nucleotide sequence comparison of four different viral isolates. These findings raise the possibility that viral isolates from different individuals could have important biological differences in their envelope antigens, a consideration relevant to ongoing attempts to develop diagnostic, therapeutic, and preventive measures against this virus.

REFERENCES

Barré-Sinoussi, F., J.-C. Chermann, F. Rey, M.T. Nugeyre, S. Chamaret, J. Gruest, C. Dauguet, C. Axler-Blin, F. Vezinet-Brun, C. Rouzioux, W. Rozenbaum, and L. Montagnier. 1983. Isolation of a T-lymphotropic retrovirus from a patient at risk of acquired immune deficiency syndrome (AIDS). *Science* **220**: 868.

Chou, P.Y. and G.D. Fasman. 1974. Prediction of protein conformation. *Biochemistry* **13**: 222.

Clements, J.E., F.S. Pedersen, O. Narayan, and W.A. Haseltine. 1980. Genomic changes associated with antigenic variation of visna virus during persistent infection. *Proc. Natl. Acad. Sci. U.S.A.* **77**: 4454.

Cohen, G.H., B. Dietzschold, M. Ponce de Leon, D. Long, E. Golub, A. Varrichio, L. Pereira, and R.J. Eisenberg. 1984. Localization and synthesis of an antigenic determinant of *Herpes simplex* virus glycoprotein D that stimulates the production of neutralizing antibody. *J. Virol.* **49**: 102.

Daniel, M.D., N.L. Letvin, N.W. King, M. Kannagi, P.K. Sehgal, R.D. Hunt, P.J. Kanki, M. Essex, and R.C. Desrosiers. 1985. Isolation of T-cell tropic HTLV-III-like retrovirus from macaques. *Science* **228**: 1201.

Eisenberg, R.J., D. Long, M. Ponce de Leon, J.T. Matthews, P.G. Spear, M.G. Gibson, L.A. Lasky, P. Berman, E. Golub, and G.H. Cohen. 1985. Localization of epitopes of *Herpes simplex* virus type I glycoprotein D. *J. Virol.* **53**: 634.

Gallo, R.C., G.M. Shaw, and P.D. Markham. 1985. The etiology of AIDS. In *AIDS: Etiology, diagnosis, treatment, and prevention* (ed. V.T. De Vita et al.), p. 31. J.B. Lippincott, Philadelphia.

Gallo, R.C., S.Z. Salahuddin, M. Popovic, G.M. Shearer, M. Kaplan, B.F. Haynes, T.J. Palker, R. Redfield, J. Oleske, B. Safai, G. White, P. Foster, and P.D. Markham. 1984. Frequent detection and isolation of cytopathic retroviruses (HTLV-III) from patients with AIDS at risk for AIDS. *Science* **224**: 500.

Gonda, M.A., F. Wong-Staal, R.C. Gallo, J.E. Clements, O. Narayan, and R.V. Gilden. 1985. Sequence homology and morphologic similarities of HTLV-III and visna virus, a pathogenic lentivirus. *Science* **227**: 173.

Gunn, P.R., F. Sato, K.F.H. Powell, A.R. Bellamy, J.R. Napier, D.R.K. Harding, W.S. Hancock, L.J. Siegman, and G.W. Both. 1985. Rotavirus neutralizing protein VP7: Antigenic determinants investigated by sequence analysis and peptide synthesis. *J. Virol.* **54**: 791.

Hahn, B.H., G.M. Shaw, S.K. Arya, M. Popovic, R.C. Gallo, and F. Wong-Staal.

1984. Molecular cloning and characterization of the HTLV-III virus associated with AIDS. *Nature* **312**: 166.

Hahn, B.H., M.A. Gonda, G.M. Shaw, M. Popovic, J. Hoxie, R.C. Gallo, and F. Wong-Staal. 1985. Genomic diversity of the AIDS virus HTLV-III: Different viruses exhibit greatest divergence in their envelope genes. *Proc. Natl. Acad. Sci. U.S.A.* 82: 4813.

Hopp, T.P. and K.R. Woods. 1981. Prediction of protein antigenic determinants from amino acid sequences. *Proc. Natl. Acad. Sci. U.S.A.* **78**: 3824.

Kanki, P.J., R. Kurth, W. Becker, G. Dreesman, M.F. McLane, and M. Esses. 1985a. Antibodies to simian T-lymphotropic retrovirus type III in African green monkeys and recognition of STLV-III viral proteins by AIDS and related sera. *Lancet* **i**: 1330.

Kanki, P.J., M.F. McLane, N.W. King, N.L. Letvin, R.D. Hunt, P. Sehgal, M.D. Daniel, R.C. Desrosier, and M. Essex. 1985b. Serologic identification and characterization of a macaque T-lymphotropic retrovirus closely related to HTLV-III. *Science* **228**: 1199.

Levy, J.A., A.D. Hoffman, S.M. Kramer, J.A. Lanois, J.M. Shimabukuro, and L.S. Oskiro. 1984. Isolation of lymphocytopathic retroviruses from San Francisco patients with AIDS. *Science* **225**: 840.

Montelaro, R.C., B. Parekh, A. Orrego, and C.J. Issel. 1984. Antigenic variation during persistent infection by equine infectious anemia virus, a retrovirus. *J. Biol. Chem.* **259**: 10539.

Pellett, P.E., K.G. Kousoulas, L. Pereira, and B. Roizman. 1985. Anatomy of the *Herpes simplex* virus 1 strain F glycoprotein B gene: Primary sequence and predicted protein structure of the wild type and of monoclonal antibody-resistant mutants. *J. Virol.* **53**: 243.

Popovic, M., M.G. Sarngadharan, E. Read, and R.C. Gallo. 1984. Detection, isolation and continuous production of cytopathic retroviruses (HTLV-III) from patients with AIDS and pre-AIDS. *Science* **224**: 497.

Ratner, L., R.C. Gallo, and F. Wong-Staal. 1985a. HTLV-III, LAV and ARV are variants of the same AIDS virus. *Nature* **313**: 636.

Ratner, L., W. Haseltine, R. Patarca, K. Livak, B. Starcich, S. Josephs, E.R. Doran, J.A. Rafalski, E.A. Whitehorn, K. Baumeister, L. Ivanoff, S.R. Petteway, Jr., M.L. Pearson, J.A. Lautenberger, T.S. Papas, J. Ghrayeb, N.T. Chang, R.C. Gallo, and F. Wong-Staal. 1985b. Complete nucleotide sequence of the AIDS virus, HTLV-III. *Nature* **313**: 277.

Robert-Guroff, M., M. Brown, and R.C. Gallo. 1985. HTLV-III-neutralizing antibodies in patients with AIDS and AIDS-related complex. *Nature* **316**: 72.

Salahuddin, S.Z., P.D. Markham, M. Popovic, M.G. Sarngadharan, S. Orndorff, A. Fladagar, A. Patel, J. Gold, and R.C. Gallo. 1985. Isolation of infectious human leukemia/lymphoma virus type III (HTLV-III) from patients with acquired immunodeficiency syndrome (AIDS) or AIDS-related complex (ARC) and from healthy carriers: A study of risk groups and tissue sources. *Proc. Natl. Acad. Sci. U.S.A.* 82: 5530.

Sanchez-Pescador, R., M.D. Power, P.J. Barr, K.S. Steimer, M.M. Stempien, S.L. Brown-Shimer, W.W. Gee, A. Renard, A. Randolph, J.A. Levy, D. Dina, and

P.A. Luciw. 1985. Nucleotide sequence and expression of an AIDS-associated retrovirus (ARV-2). *Science* **227**: 484.

Sarngadharan, M.G., M. Popovic, L. Bruch, J. Schupbach, and R.C. Gallo. 1984. Antibodies reactive with human T-lymphotropic retroviruses (HTLV-III) in the serum of patients with AIDS. *Science* **224**: 506.

Schupbach, J., M. Popovic, and R.V. Gilden. 1984. Serological analysis of a subgroup of human T-lymphotropic retroviruses (HTLV-III) associated with AIDS. *Science* **224**: 503.

Shaw, G.M., B.H. Hahn, S.K. Arya, J.E. Groopman, R.C. Gallo, and F. Wong-Staal. 1984. Molecular characterization of human T-cell leukemia (lymphoma) virus type III in the acquired immune deficiency syndrome. *Science* **226**: 1165.

Shaw, G.M., M.E. Harper, B.H. Hahn, L.G. Epstein, D.C. Gajdusek, R.W. Price, B.A. Navia, C.K. Petito, C.J. O'Hara, E.-S. Cho, J.M. Oleske, F. Wong-Staal, and R.C. Gallo. 1985. HTLV-III infection in brains of children and adults with AIDS encephalopathy. *Science* **227**: 177.

Wain-Hobson, S., P. Sonigo, O. Danos, S. Cole, and M. Alizon. 1985. Nucleotide sequence of the AIDS virus, LAV. *Cell* **40**: 9.

Weiss, R.A., P.R. Clapham, R. Cheingsong-Popov, A.G. Dalgleish, C.A. Carne, I.V.D. Weller, and R.S. Tedder. 1985. Neutralization of human T-lymphotropic virus type III by sera of AIDS and AIDS-risk patients. *Nature* **316**: 69.

Wong-Staal, F., G.M. Shaw, B.H. Hahn, S.Z. Salahuddin, M. Popovic, P.D. Markham, R. Redfield, and R.C. Gallo. 1985. Genomic diversity of human T-lymphotropic virus type III (HTLV-III). *Science* **229**: 759.

COMMENTS:

KILBOURNE: Have you seen antigenic variability correlated with the nucleotide changes in New York isolates?

GALLO: No, we have not. This is a major part of what we're trying to do now, to test with different variants we have isolated. Results to date with tests of two very different variants, namely a Haitian isolate we call HTLV-III_{RF} and the prototype we call HTLV-III B (BH-10), have not shown significant differences. What we have yet to do is the testing of neutralizing antibody among the various different strains. In other words, we still do not know whether neutralizing antibody neutralizes each of the variants equally. We only know that we can detect antibodies of unknown biological activity reacting equally with two very different strains of HTLV-III.

AHMED: Do you think that the severe immunosuppression in AIDS patients can be accounted for just by direct infection of T4 cells?

GALLO: I assume the question concerns whether the disease is entirely a consequence of a direct killing of the T4 cells by HTLV-III or whether other, perhaps indirect mechanisms, are also involved. There is, in fact, evidence that suppressor factors are liberated from infected cells which can inhibit T cell proliferation. In addition, some investigators have suggested that autoimmunity may be involved, i.e., an antibody induced against one of the viral proteins also reacts against T cells. We have evidence with Jeff Lawrence of Rockefeller that infected cells release some suppressor factors that may inhibit growth of immature T cells. Nonetheless, I think the major effect is a direct cytopathic effect of HTLV-III on mature T4 cells causing their premature death.

FIELDS: Do you think you could reproduce AIDS if you took the recognition portion of the envelope that interacts with the T4 receptor and used it to envelope another virus? Do you think that the killing of the T cell by itself is sufficient or are there any other virulence factors that have been recognized? Clearly, replication is another factor.

GALLO: No, I do not. The extent of replication is, of course, a factor. However, I believe there is something very special about this virus. Like HTLV-I and II, it exhibits the transactivation phenomenon but to an extraordinary degree. Bill Haseltine's data show 1000–5000 times better activity in the LTR-CAT assay than with SV-40 if the HTLV-III LTR-CAT constructs are transfected into HTLV-III pre-infected cells. That must be important; and I expect that it must be activating some viral or cell genes in a way which leads to the premature death of the cell. In addition to the gene for *trans*-acting transcriptional activation, this virus also has at least one more extra gene (extra to the usual three genes of a retrovirus). The second extra gene may code for a cytopathic protein.

FIELDS: If you took the HTLV LTRs or the region involving transactivation and put it into other retroviruses or other viruses, would you have any concern about that in terms of the hosts or the environment?

GALLO: In fact, it's being done.

FIELDS: What would be the nature of your concern, considering everything you've done?

GALLO: I am not overly concerned about attaching the HTLV-III LTR to the genome of another retrovirus because there is much evidence that the cytopathic effect of HTLV-III does involve and need specific regions of the HTLV-III genome. I would be concerned about doing things that broaden the range of target cells for infection; or if something we've done to alter the envelope gives it great capacity for survival, I would be concerned, Incidentally, even HTLV-I is spreading in parts of the world because it has a very long latency period and only a 1:100 disease-to-virus infection ratio. It is much less evident, but I suspect HTLV-I is going to be an increasing problem 10–30 years from now. In summary, I would evaluate each planned alteration carefully. I would not be highly concerned about the modification you mentioned. I would not be surprised if the Japanese did the first vaccination, i.e., vaccination for HTLV-I, before we develop an HTLV-III vaccine.

Genetically Engineered Genomes of Herpes Simplex Virus 1: Structure and Biological Properties

BERNARD ROIZMAN,* AMY E. SEARS,* BERNARD MEIGNIER,† AND MINAS ARSENAKIS*
*The Marjorie B. Kovler Viral Oncology Laboratories
University of Chicago
Chicago, Illinois 60637
†Institut Merieux
Lyon, France

INTRODUCTION

In individuals exposed to herpes simplex viruses 1 and 2 (HSV-1 and HSV-2), the initial virus multiplication occurs at the portal of entry and, in most individuals, this is also the target organ. At the portal of entry, the virus infects sensory nerve fibers and is transported in an as yet undefined form to the cell body (nucleus) of the neurons in the trigeminal ganglia or to the afferent dorsal root ganglia where the virus can remain latent for the life of the host. In a fraction of individuals carrying latent virus, stress resulting from physical, emotional, hormonal, or immune disorders results in activation of the latent virus. The virus is transported by way of the axon to a site at or near the portal of entry where it may cause the formation of herpetic lesions. As a general rule, the formation of recurrent lesions is dependent on the ability of the virus to multiply and on the immunologic status of the host. Thus recurrent lesions are likely to be larger and more severe in immunocompromised than in normal individuals. Little or nothing is known of the molecular events that trigger the induction of virus multiplication and transport of the virus to the portal of entry except that the triggering event may be initiated at sites distal from the ganglion. It may be assumed that in normal individuals the induction of virus multiplication occurs more frequently than the recurrent lesions that may follow.

The differences in the severity of lesions in normal and immunocompromised individuals suggest that the host contributes a major determinant of the virulence of the virus. There is, however, compelling evidence that virulence is also determined by the virus. Thus it has been shown that viruses differ in their capacity to cause stromal disease in the cornea and that the capacity to induce stromal lesions in the eye can be transferred with a viral DNA fragment to a virus producing only epithelial lesions on the cornea (Centifanto-Fitzgerald et al. 1982). The operational assessment of virulence—the capacity to produce disease—is frequently done in susceptible animals and scored on the basis of mortality resulting from central nervous system (CNS) disease. In susceptible animals CNS invasion is common and involves two distinct viral characteristics, i.e., ability to ascend from the periphery to the CNS and ability to grow in the CNS tissues. These operational components of virulence are virus determined. In humans invasion of the CNS is rare. Although there is clear evidence that invasion of the CNS may be facilitated by physiologic events in the host, there is little doubt that fresh human isolates differ in their capacity to ascend and grow in the CNS of experimental animals. Thus, viruses isolated from CNS tissues are uniformly neurovirulent in experimental systems whereas random isolates from other body tissues vary in their capacity to produce CNS disease.

The purpose of this paper is to review the capacity of genetically engineered strains of HSV, carrying specific insertions or deletions, to establish latency and cause disease.

GENETIC ENGINEERING OF HSV

All of the virus constructions to date have involved the HSV-1 thymidine kinase (*tk*) gene using the technique described by Mocarski et al. (1980) and Post and Roizman (1981). This technique is based on three factors:

(1) In eukaryotic cells the function of the *tk* gene is to convert thymidine (T) to thymidine monophosphate (TMP). The main pathway for TMP synthesis, however, is conversions of deoxyuridine monophosphate (dUMP) to TMP by thymidilate synthetase (TS). As long as that pathway is available, cell mutants deficient in the TK enzyme are readily selected in the presence of 5′ bromodeoxyuridine (BUdR) which is phosphorylated by TK but not by TS.

(2) The function of the viral *tk* appears to be broader than its host counterpart inasmuch as the viral *tk* phosphorylates both purines and pyrimidines. Nevertheless, the viral *tk* gene is not essential for viral multiplication in growing, metabolizing cells capable of converting dUMP to TMP, and TK^- viruses are readily selected and grow well in TK^- cells in the presence of BUdR. A characteristic of the viral *tk* is that it phosphorylates a wide variety of analogs that bear little resemblance to their natural counterparts. Many of these analogs are nontoxic to uninfected cells because they cannot serve as substrates for the host enzymes. In HSV-infected cells, however, these analogs become phosphorylated and incorporated into DNA leading to the destruction of the host cell and of the virus. Such analogs (e.g., thymine arabinoside (AraT), acyclovir, BUdR) are suitable for selection of HSV-1 mutants lacking TK activity (TK^-) in TK^- cells.

(3) A powerful factor in the use of the *tk* gene for construction of genetically engineered mutants carrying specific insertions or deletions is the availability of techniques for selection of TK^+ strains. The technique is based on the observation that in TK^- cells conversion of dUMP to TMP by the host TS constitutes the sole and essential source of TMP. Inhibition of this pathway by methotrexate or aminopterin selectively inhibits TK^- mutants and favors the growth of TK^+ viruses.

It is convenient to divide the genetically engineered strains constructed to date into two classes (Fig. 1). The first involves the insertion of HSV or foreign DNA sequences into the transcribed domain of a cloned copy of the HSV *tk* gene. The cloned chimeric fragment is then amplified, cotransfected into appropriate cells with intact wild-type (TK^+) HSV DNA, and recombinant TK^- virus mutants are selected. The recombination event is facilitated by the homology of the TK sequences flanking the insert with sequences in the HSV DNA. Because the recombination frequency is greater than the rate of emergence of spontaneous TK^- mutants, a large fraction of the TK^- progeny carry the insert. This type of construction has been used to confer α or γ_2 gene regulation to the *tk* gene, which is normally regulated as a β gene (Mackem and Roizman 1982; Kristie and Roizman 1984; Sears et al. 1985a; Silver and Roizman 1985); to insert a second copy of an HSV gene into the genome (Post and Roizman 1981; Gibson and Spear 1983; Hubenthal-Voss and Roizman 1985); to identify the *cis*-acting site for inversions by inserting candidate sequences into the *tk* gene (Mocarski et al. 1980); and to test the expression of foreign genes in HSV-1 DNA (Shih et al. 1984).

The second class of recombinants generated by the procedure of Post and Roizman (1981) permits the construction of viable mutants carrying insertions and deletions at any site within the genome, provided the deletion and/or insertion does

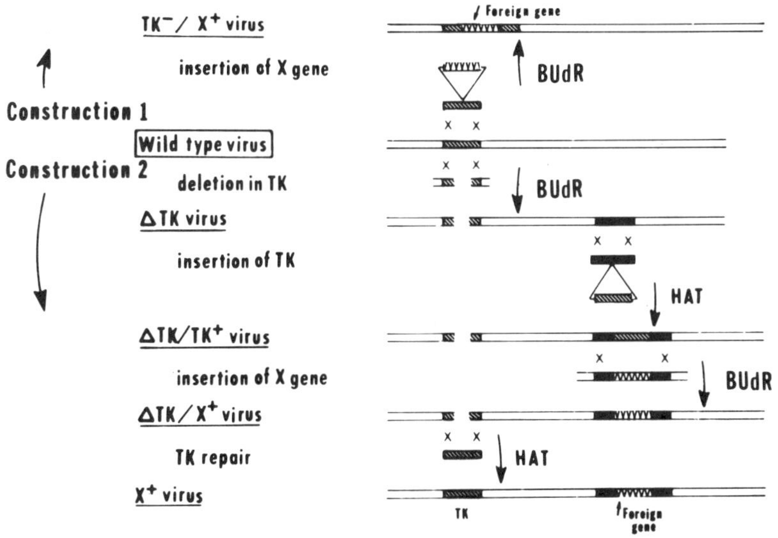

Figure 1

Flow diagram for the construction of class 1 and class 2 recombinant HSV genomes. HSV-1 DNA is represented by open double lines. The *tk* gene is represented by the hatched area. The target gene to be deleted is represented by the solid area and the foreign gene X is represented by the triangled area. The symbol delta indicates deletions. *HAT* and *BudR* refer to growth media containing hypoxanthine, aminopterin thymidine, or 5′-bromodeoxyuridine respectively.

not affect the growth of the virus in the system in which it is selected. This procedure is applicable to large genomes such as those of herpesviruses or poxviruses (Smith et al. 1983; Paoletti et al. 1984) and consists of the following steps: In the first step, a deletion is made in the coding sequences of the *tk* gene carried in a plasmid. The amplified DNA carrying the deletion is then cotransfected with wild-type virus and the TK^- progeny carrying the deletion is selected. The HSV-1(F)TK^- mutant virus selected by this procedure has been designated as HSV-1(F)Δ305, and carries a 700-bp deletion in the *tk* gene (Post and Roizman 1981). In the second step, a native or modified *tk* gene is inserted into the desired site of the target sequence carried on a plasmid. The chimeric plasmid is then amplified and cotransfected with intact HSV-1(F)Δ305 DNA, and TK^+ mutants carrying the insertion are selected. In the last step, either a deletion is made in the target sequence or a foreign gene is inserted in place of the *tk* gene at the previously selected site. The plasmid carrying the target sequence with the deletion or insertion is then amplified and cotransfected with intact viral DNA from the TK^+ recombinant obtained in the previous step. The TK^- progeny in which the inserted *tk* gene is replaced by the new insert or the deletion is then selected. If necessary, the natural *tk* gene can be restored by cotransfection with a plasmid carrying the intact *tk* gene and TK^+ selection of the transfection progeny. This procedure has been used to construct mutants carrying 500-bp and 100-bp deletions in the α22 gene (Post and Roizman 1981), to delete sequences at or near the junction between the L and S components, (Poffenberger et al. 1983), and to insert the chick ovalbumin gene in place of the internal inverted repeats of the HSV-1 genome (M. Arsenakis et al., in prep.).

Both classes of constructs comprise recombinants carrying foreign genes, i.e., the S gene of hepatitis B virus (Shih et al. 1984), the EBNA 1 and EBNA 2 of Epstein-Barr virus in class 1 constructs, and chick ovalbumin in class 2 constructs. (M. Arsenakis and B. Roizman, in prep.). As a rule, foreign genes are expressed readily and efficiently if linked to a HSV promoter regulatory sequence. The use of HSV as a vector for nonherpesvirus genes has many advantages. Foremost, it permits the expression of viral gene products in cells which are not normally infected by the donor viruses. It eliminates the problems associated with large-scale production of potentially hazardous infectious agents, and the gene product can be made in relatively large amounts.

BIOLOGIC PROPERTIES OF GENETICALLY ENGINEERED HSV STRAINS

Although in the last few years this laboratory has produced a large number of diverse recombinants, the biologic properties of the recombinants and particularly data concerning growth, genetic stability, virulence, and capacity to establish latent infections is available on a limited number of the recombinants tested to date.

Virus Growth in Cell Culture

As a general rule, insertions or deletions that affect the expression of a gene may affect the ability of the recombinant to grow. The effect may be minimal or cell line-dependent. For example, the HSV-1(F)Δ305 recombinant carrying a 700-bp deletion in the *tk* gene yields less virus than the wild-type HSV-1(F) virus although the impairment of growth is minimal (less than tenfold) and not cell line-dependent (B. Roizman et al., unpubl.). In contrast the R325 and R328 viruses carrying 500- and 100-bp deletions, respectively, in the α22 gene (Post and Roizman 1981) grow as well as the wild-type parent in HEp-2 and Vero cells but poorly in cell lines derived from rodents or in human embryonic lung cells (Sears et al. 1985a). These studies suggest that mutants in the α22 gene are complemented in some cell lines but not in others and that in both permissive and restrictive cell lines the expression of γ_2 genes may be affected by the deletion in the α22 gene (Sears et al. 1985a).

Genetic Stability

When an insertion or a deletion impair the capacity of a virus to grow, mutations which delete the insert or compensate for the deletion introduced by the insertion or deletion have a selective advantage. The genetic stability of the virus is best tested under conditions in which there is a strong positive pressure for the selection of mutants with increased capacity to grow. The positive pressure can be applied in two systems, i.e., in cell culture or in experimental animals.

Selective pressure in cell culture is exemplified by studies with recombinants R316, R314, R3112, and R3104 (Fig. 2). All of these recombinants carry *tk* genes linked to either α (R314, R316, and R3104) or γ_2 promoter-regulatory sequences (R3112). To effect the conversion of the natural *tk* gene into α genes, *Bam*HI fragment N (R316) or Z (R314) were inserted into the 5′ transcribed noncoding sequence of the *tk* gene (Post et al. 1981). In R3104 the α27 promoter regulatory sequence was fused to the *tk* gene and inserted into the HSV-1 genome between the terminal 500 bp α sequence and the 3′ terminus of the genome (Hubenthal-Voss and Roizman 1985). To construct the recombinant R3112 carrying the γ_2-TK chimera, a HSV-1 DNA fragment (*Bam*HI D′) carrying the γ_2 promoter-regulatory sequence was inserted in the proper transcriptional orientation into the 5′ transcribed noncoding domain of the *tk* gene (Silver and Roizman 1985). All of these viruses appear to be stable on passage in Vero cells. However, in other cell lines, notably HEp-2 cells and human embryonic lung cells, the R316 virus carrying an insertion of the *Bam*HI N fragment, yielded a high frequency of deletions contained entirely within the domain of the inserted *Bam*HI N fragment (Sears and Roizman 1985). Similar deletions in the progeny of R316 virus were obtained upon inoculation into the mouse cornea (Sears et al. 1985b). Because the left and right borders of the deletions appear to vary and because we have isolated a slow-growing

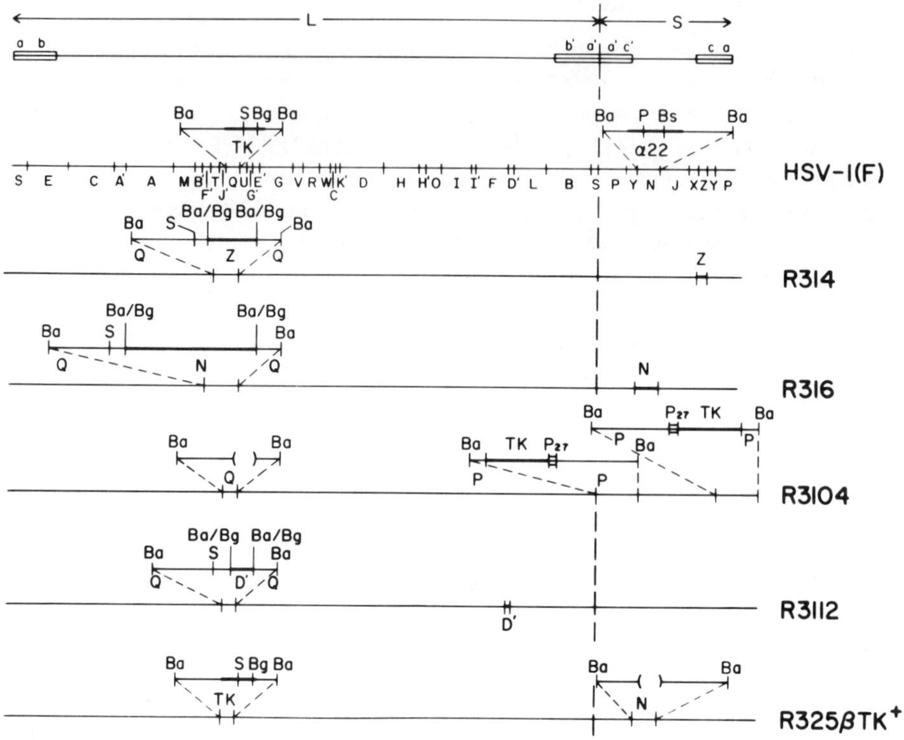

Figure 2

Representation of DNA sequence arrangements of parental and recombinant viruses. The top two lines show the sequence arrangement of HSV-1 DNA, detailing the L and S components and terminal reiterated sequences. The letter L at S refers to L at S covalently linked components represented by lines (unique sequences) flanked by invested repeats (boxes) ab and b'a' for the L component and a'c' of ca for the S component. HSV-1 (F) is the parent virus of all of the recombinants; *Bam*HI restriction fragments are labeled and the sequence arrangement of the *Bam*HI Q fragment is expanded. R314 carries the *Bam*HI Z fragment containing the α4 promoter inserted into the *Bgl*II site of the *tk* leader sequence such that the α promoter and TK are in the same transcriptional orientation; *tk* is regulated as an α gene (Post et al. 1981). R316 is a similar construction, with the *Bam*HI N fragment containing the other copy of the α4 promoter inserted into the *Bgl*II site of *tk* (Post et al. 1981). R3104 contains two copies of an α27 chimeric *tk* gene inserted into the reiterated sequences of the S component in *Bam*HI P between the *a* sequence and the 3′ end of the ICP4 gene, as well as a 700-bp deletion in the original *tk* gene (Hubenthal-Voss and Roizman 1985). R3112 carries the *Bam*HI D′ fragment containing a γ_2 promoter inserted into the *Bgl*II site of the *tk* gene; *tk* is regulated as a γ_2 gene (Silver and Roizman 1985). R325-βTK⁺ contains a 500-bp deletion in the *Bam*N fragment within the coding sequences of α gene 22, and the 700-bp deletion in the TK sequences was repaired (Post and Roizman 1981). (Ba) *Bam*HI; (Bg) *Bgl*II; (S) *Sac*I; (P) *Pvu*II; (Bs) *Bst*EII. P_{27} refers to the promoter-regulatory region of the α27 gene which was linked to *tk* in R3104. N, P, Q, Z, and D′ refer to the *Bam*HI fragments containing the ICP 4 promoters (N and Z), the *tk* gene (Q), the γ_2 promoters (D′), and the terminal fragment into which the α27-TK gene was inserted in R3104 (P).

mutant in which we cannot detect deletions, we suspect that the mutants arise through random deletions but have selective advantages in the cells in which they are selected. In contrast, all other constructs tested were stable. R316 virus is the first and sole example of a recombinant whose stability is cell-dependent. The instability could not be related to the α-regulated TK since all other α-TK viruses tested were stable, or to the presence of *cis*-acting sites or genes in the *Bam*HI N fragment since the recombinant R315, carrying the *Bam*HI N fragment at the same site but in an inverted orientation was stable.

Another system for studying the genetic stability and purity of virus stocks is serial passaging of the virus through mouse brain. Operationally, a measured amount of virus is inoculated intracerebrally. At a specific interval (2–4 days) the amount of infectious virus in the brain is measured, the progeny is amplified in cell culture, and the virus is again inoculated into the brain. A plot of the ratio of recovered to input virus as a function of virus passage in the brain is a good indicator of genetic stability as well as of the presence of trace amounts of contaminating parental virus. As illustrated in Table 1, the R325-βTK^+ virus carrying a restored TK (Fig. 2), yielded a constant ratio of recovered to input virus whereas ts^+ HSV-1 × HSV-2 recombinants appeared to be genetically unstable. Thus while the restriction endonuclease patterns of one recombinant virus (C7D) after passage 9 showed evidence of a contaminant, the pattern of the other (A5C), showed no evidence of contamination and therefore represented virus carrying back mutations or compensatory mutations that permitted it to grow to a higher titer. (B. Meignier and B. Roizman, unpubl.).

To date, few recombinants have been tested in this fashion. Among the viruses to remain unchanged after nine serial passages through mouse brain was the recombinant I358 in which 15 kbp comprising the internal inverted repeats were replaced by an *α-tk* gene (B. Meignier and B. Roizman, unpubl.).

Table 1
Latency of R325-βTK^+ in the Mouse

Site of inoculation	Virus	Dose (pfu)	Number of positive ganglia / Number of eyes/ears infected	% latency
Eye	HSV-1(F)	10^7	5/5	100
	R325-βTK^+	10^7	5/10	50
	R325-βTK^+	$10^{7.3}$	10/20	50
Ear	HSV-1(F)	10^7	10/10	100
	R325-βTK^+	10^7	0/10	0

BALB/c mice were inoculated in either one eye or one ear with approximately 10^7 pfu of virus as described (Meignier et al. 1983). Ipsilateral trigeminal ganglia or pooled cervical dorsal root ganglia were assayed for latent virus. Latency percentage is given as the number of animals from which virus could be recovered divided by the number of animals infected.

Virulence

As a general rule, any genetically stable mutation, insertion, or deletion in the virus genome that reduces the ability of the virus to multiply in experimental animal systems also reduces the capacity of the virus to ascend to and multiply in the CNS. Of the two operational measures of virulence, penetration of the virus into the CNS, and ability to multiply and destroy the CNS, the latter is easier to measure. Loss of ability to grow in the CNS may reflect the loss of either the overall ability to grow in experimental animal tissues or loss of an organ-specific function essential for CNS invasion but not for multiplication in other organs. HSV-1 TK^- mutants are examples of viruses with reduced capacity to grow in experimental animal tissues. Other examples include the R325 and I358 recombinants described above. Whereas the pfu/LD_{50} ratio for the wild-type parent HSV-1(F) is $10^{1.5}$, those for the R325 and I358 recombinants are $10^{7.0}$ and $10^{6.0}$, respectively. A CNS tissue-specific mutation has also been described (Thompson and Stevens 1983; Thompson et al. 1983).

Latency

To establish latency in an experimental animal, HSV must infect sensory nerve endings and ascend to the nucleus of the neurons. Operationally, the detection of latent virus requires incubation of ganglionic tissue to induce virus multiplication, and therefore the virus must be capable of at least limited multiplication in neurons. The rich and varied literature on HSV latency in the mouse suggests that for a given mouse strain and portal of entry, the amount of virus required for establishment of latency may vary from one virus strain to another. For a given virus strain and mouse genotype, the efficiency of establishment of latency may vary according to the portal of entry and may be virus dosage-dependent. One, not totally accurate way to describe the system is that for a given virus, mouse, and site of inoculation, the minimal amount of virus required to establish latency is fixed and can be attained either by virus multiplication at the portal of entry or by inoculation of that amount of virus at that site.

The basis for these conclusions rests on the observation that mutants like TK^- strains with reduced capacity to multiply, fail to establish latency when given by certain routes. Another example, illustrated in Table 1, is that of R325 which is capable of establishing latency when given by the eye route but not when inoculated into the ear of BALB/c mice.

It has been suggested that TK activity is required for establishment of latency. Studies on the αTK recombinants cited earlier in the text and especially on the deletion mutants generated during the growth of R316 virus in the mouse (Table 2) show little correlation between TK activity and ability to establish latency and suggest that the inability to demonstrate latency in some instances might be due to other mutations introduced during the selection and which affect viral multiplica-

Table 2
Frequency of Establishment of Latency in Trigeminal Ganglia of BALB/c Mice and Properties of Parent and Progeny Viruses

Virus	No. positive ganglia[a] total eyes infected	Latency (%)	Max TK[b] (%F)	TK[c] regulation
HSV-1(F)	20/20[d,f]	100	100[b]	β
R314	10/19[f]	53	ND	α
M314-1A	8/10	80	ND	α
M314-5A	4/10	40	ND	α
M314-10A	8/10	80	ND	α
R3104	7/20[f]	35	ND	α
M3104-1A	5/10	50	ND	α
M3104-3A	7/10	70	ND	α
M3104-7A	5/10	50	ND	α
R316	6/25[e,f]	24	57	α
M316-2	9/15[f]	60	2	not measurable
M316-6	15/15[f]	100	23	non-α
M316-10	14/15[f]	93	2	not measurable
M316-8D	0/19[f]	0	42	α
R3112	9/20[f]	45	35	γ

[a]Ipsilateral to the inoculated eye
[b]Measured TK activity at 8 hrs postinfection in replicate cell cultures infected at the same multiplicity
[c]Data from Sears et al. (1985b). The natural *tk* gene is regulated as a β gene (Honess and Roizman 1974).
[d]The procedure for isolation of latent virus was previously described (Meignier et al. 1983).
[e]Two separate experiments: 3 out of 10 and 10 out of 15 ganglia
[f]Virus inoculated in one eye only
(ND) Not done

tion either at the site of inoculation or in tissue culture after explantation of the ganglion.

To date, all recombinant viruses generated in these studies were able to establish latency in the mouse when given by the eye route, but the efficiency varied, the least efficient being the recombinant virus I358.

A key question that remains unresolved is the identity of genes whose expression is required for establishment of latency. Excluded from the list of such genes would be those that are required for multiplication at the portal of entry for infection of nerve endings and for the synthesis of structural proteins involved in the transport of the virus to the nucleus of the neuron. In the case of viruses remaining latent in cells undergoing rapid turnover (e.g., Epstein-Barr virus in B lymphocytes), it could be predicted that the viral genes expressed in the latently infected cells would preclude integration of the viral genome into the host chromosomes and would ensure replication of the viral episomes by host enzymes, and may also coincidentally stimulate the host cell to multiply. In the case of HSV, the need for such functions is less obvious since the neurons do not turn over. The need to preclude integration of the viral chromosome is also less apparent since the presence of the internal inverted repeats of the terminal sequences would permit the regeneration of terminal sequences lost as a consequence of viral integration. Another complication is that viral gene expression is tightly regulated; HSV-1 genes form several groups designated as α, β, and γ whose expression is coordinately regulated and sequentially ordered in a cascade fashion (Honess and Roizman 1974). The onset of expression of β genes is likely to result in the synthesis of viral enzymes leading to virus multiplication and destruction of the host. There is little evidence that viral gene expression in latently infected cells is restricted to one or more α genes. The recent evidence indicating that α genes are induced by a structural component of the virus (Batterson and Roizman 1983), that all α gene regulatory domains contain a specific recognition sequence for that structural component (Kristie and Roizman 1984), and that this sequence resembles sequences found in the regulatory domains of a variety of eukaryotic genes (Kristie et al. 1983), suggests the possibility that the inducer protein does not reach the nucleus of the neuron and that no viral genes are expressed during the HSV latency. Induction of virus multiplication would then result from the induction of a host factor capable of recognizing the α gene induction sequence and substituting for the viral structural protein that normally induces α gene expression in lytic infections.

CONCLUSIONS

The technology for construction of HSV genomes carrying insertions or deletions at specific sites is available. The recombinants vary in their genetic stability, virulence, and ability to establish latency in model systems. Most of the viable recombinants generated to date exhibit a restricted cell host range and are therefore invaluable tools for the study not only of gene function but also of the biology of the virus in animal model systems.

ACKNOWLEDGMENTS

These studies were aided by grants to the University of Chicago from the National Cancer Institute, United Public Health (5 ROI PHS CA 08494 and 5 ROI PHS CA 19264), the American Cancer Society (ACS MV 2T), and Institute Merieux. A.E.S. is a postdoctoral trainee, on the 5 T32 CA 09241 Training Grant, and M.A. is a Damon Runyon-Walter Winchell Cancer Fund fellow (DRG 017).

REFERENCES

Batterson, W. and B. Roizman. 1983. Characterization of the herpes simplex-virion associated factor responsible for induction of α genes. *J. Virol* **46**: 371.

Centifanto-Fitzgerald, Y.M., T. Yamaguchi, H.E. Kaufman, M. Tognon, and B. Roizman. 1982. Ocular disease pattern induced by herpes simplex virus is genetically determined by a specific region of viral DNA. *J. Exp. Med.* **155**: 475.

Gibson, M.G. and P.G. Spear. 1983. Insertion mutants of herpes simplex virus have a duplication of the glycoprotein D gene and express two different forms of glycoprotein D. *J. Virol.* **48**: 396.

Honess, R.W. and B. Roizman. 1974. Regulation of herpesvirus macromolecular synthesis. I. Cascade regulation of the synthesis of three groups of viral proteins. *J. Virol.* **14**: 8.

Hubenthal-Voss, J. and B. Roizman. 1985. Herpes simplex virus 1 reiterated S component sequences (C_1) situated between the 'a' sequence and α4 gene are not essential for virus replication. *J. Virol.* **54**: 509.

Kristie, T.M. and B. Roizman. 1984. Separation of sequences defining basal expression from those conferring α gene recognition within the regulatory domains of herpes simplex virus 1 α genes. *Proc. Natl. Acad. Sci. U.S.A.* **81**: 4065.

Kristie, T.M., W. Batterson, S. Mackem, and B. Roizman. 1983. The regulatory elements in the domains of α genes of herpes simplex virus 1 (HSV-1). *Proc. Cold Spring Harbor conference on enhancer and eukaryotic gene expression.* (ed. Y. Gluzman and T. Shenk), p. 141. Cold Spring Harbor Laboratory, Cold Spring Harbor, New York.

Mackem, S. and B. Roizman. 1982. Structural features of the α gene 4,0, and 27 promoter-regulatory sequences which confer α regulation on chimeric thymidine kinase genes. *J. Virol.* **44**: 939.

Meignier, B., B. Norrild, and B. Roizman. 1983. Colonization of murine ganglia by a superinfecting strain of herpes simplex virus. *Infect. Immun.* **41**: 702.

Mocarski, E.S., L.E. Post, and B. Roizman. 1980. Molecular engineering of the herpes simplex virus genome: Insertion of a second L-S junction into the genome causes additional genome inversions. *Cell* **22**: 243.

Paoletti, E., B.R. Lipinskas, C. Samsonoff, S. Mercer, and D. Panicalli. 1984. Construction of live vaccines using genetically engineered poxviruses: Biological activity of vaccinia virus recombinants expressing the hepatitis B virus surface antigen and the herpes simplex virus glycoprotein D. *Proc. Natl. Acad. Sci. U.S.A.* **81**: 193.

Poffenberger, K.L., E. Tabares, and B. Roizman. 1983. Characterization of a viable, noninverting herpes simplex virus 1 genome derived by insertion and deletion of sequences at the junction of components L and S. *Proc. Natl. Acad. Sci. U.S.A.* **80**: 2690.

Post, L.E. and B. Roizman. 1981. A generalized technique for deletion of specific genes in large genomes: α Gene 22 of herpes simplex virus 1 is not essential for growth. *Cell* **25**: 227.

Post, L.E., S. Mackem, and B. Roizman. 1981. Regulation of α genes of herpes simplex virus: Expression of chimeric genes produced by fusion of thymidine kinase with α gene promoters. *Cell* **24**: 555.

Sears, A.E. and B. Roizman. 1985. Cell-specific selection of mutants of a herpes simplex virus recombinant carrying deletions. *Virology* **145**: 176.

Sears, A.E., I.W. Halliburton, B. Meignier, S. Silver, and B. Roizman. 1985a. Herpes simplex virus mutant deleted in the $\alpha 22$ gene: Growth and gene expression in permissive and restrictive cells, and establishment of latency in mice. *J. Virol.* **55**: 338.

Sears, A.E., B. Meignier, and B. Roizman. 1985b. Establishment of latency in mice by herpes simplex virus 1 recombinants carrying insertions affecting the regulation of the thymidine kinase gene. *J. Virol.* **55**: 410.

Shih, M.-F., M. Arsenakis, P. Tiollais, and B. Roizman. 1984. Expression of hepatitis B virus Σ gene by herpes simplex virus type 1 vectors carrying α- and β-regulated gene chimeras. *Proc. Natl. Acad. Sci. U.S.A.* **81**: 5867.

Silver, S.S. and B. Roizman. 1985. γ_2-thymidine kinase chimeras are identically transcribed but regulated as γ_2 genes in herpes simplex virus genomes and as β genes in cell genomes. *Mol. Cell. Biol.* **5**: 518.

Smith, G.L., M. Mackett, and B. Moss. 1983. Infectious vaccinia virus recombinants that express hepatitis B virus surface antigen. *Nature (Lond)* **302**: 490.

Thompson, R.L. and J.G. Stevens. 1983. Biological characterization of a herpes simplex virus intertypic recombinant which is completely and specifically non-virulent. *Virology* **131**: 171.

Thompson, R.L., E.K. Wagner, and J.G. Stevens. 1983. Physical location of a herpes simplex virus type 1 gene function(s) specifically associated with a 10 million-fold increase in HSV neurovirulence. *Virology* **131**: 180.

COMMENTS

MARTIN: Does HSV contain dispensable genes?

ROIZMAN: I don't think there is such a thing as a dispensable gene. There are tissue-specific genes which can be deleted if the virus is grown in culture. I estimate that we can delete about 15% of the genome; but that doesn't mean that the virus carrying these deletions will grow in animals.

MARTIN: But, in terms of being able to replicate in tissue culture and make a particular gene product?

ROIZMAN: We have at the moment about 25 kb of space.

MERIGAN: Can you tell us more about the virus you grew out of the one ganglia? Was it an authentic S mutant?

ROIZMAN: Yes, it was identical to its parent virus. I should have shown one more point which I did not have time to discuss. There is a very good test for stability. The test simply involves serial passage in the brain of mice. You put virus into the mouse brain, isolate the virus in cell culture, and put the virus back into the mouse brain. This process is repeated nine times. This creates an enormous selective pressure for viruses capable of growth in the central nervous system. None of the deletion viruses has ever shown an increase in virulence, which indicates that we cannot compensate for loss of a gene by a mutation at a different rate.

HOWLEY: Do you know if the α-TIF is working directly or indirectly on the *cis*-acting sequence and do you know whether α-TIF is virally encoded?

ROIZMAN: Yes, it is a viral gene factor. It has been sequenced. The problem is that we don't know whether it binds DNA or not. The only thing we know is that CMV has its own *trans*-acting factor. There is a rumor that the Epstein-Barr virus has a similar structural component. The HSV α-TIF does not act directly although definitive evidence is not available.

WATSON: Do you turn on any of the heat shock genes?

ROIZMAN: Yes, the heat shock genes are turned on, but I'm not sure that it is regulated by this factor.

WATSON: How many of them? All of them?

ROIZMAN: It varies depending on the cell line. In different cell lines we see different heat shock proteins.

AHMED: It's probably not an easy experiment, but can you look in latently infected cells and see which viral proteins are being made?

ROIZMAN: Very few cells carry latent virus. So far only one report indicated the presence of the α proteins in neurons of rabbit ganglia carrying latent virus. Other attempts have not been successful.

KILBOURNE: Why isn't your induction of latency either all or none? Is that the threshold for detection of reactivation?

ROIZMAN: I don't think it's all or none. I can talk about situations–for example, people who have both genital and oral lesions very frequently will have recurrences of both. But my impression is that not all latent viruses are induced all at once. The failure to induce all viruses at once may reflect differences in the amounts of inducer that reach the neurons or differential response of neurons to the inducer.

MOSS: Do you believe it's mainly a problem in reactivation of the virus? Can you say that a virus doesn't become latent because of its attenuation?

ROIZMAN: The conditions of the experiment are such that the neuron is overwhelmed with virus, so I don't think multiplication is a crucial issue. For example, the α-22 deletion virus does not grow in mice, yet it establishes latency. But if you inject enough virus into the eye, you can get latent virus which can be detected by cocultivation. The failure to detect virus does not exclude the possibility that latent, nonreactivable virus is present in neurons.

Human Host Responses to Genetically Altered Viruses

ROBERT MERRITT CHANOCK AND BRIAN ROBERT MURPHY
Laboratory of Infectious Diseases
National Institute of Allergy and Infectious Diseases
National Institutes of Health
Bethesda, Maryland 20205

OVERVIEW

Humans are constantly being exposed in nature to viruses that have undergone genetic alteration. Fortunately, these alterations rarely lead to an increase in viral transmission or virulence. The changes that occur in human viruses under natural conditions are, for the most part, beyond our control. The only altered viruses that are under our control are (1) those that are constructed for study in the laboratory and (2) live attenuated virus mutants that are administered to large segments of the human population for prevention of disease. Properties that are shared by the safe, effective, attenuated vaccine viruses are reviewed to provide a frame of reference that may prove useful in evaluating future candidate live vaccine virus strains or other genetically altered viruses that are proposed for introduction into the ecosystem.

INTRODUCTION

Virus virulence is a composite property that is polygenic because the efficient functioning of every, or almost every gene product is required for expression of the full manifestations of disease. The other factor in the equation of disease is host response to infection and this area is less well understood than the contribution made by viral genes. Generally, the role of host response to infection only becomes evident when there are obvious deficiencies of immunity. Because our understanding of the role of host factors in severe viral diseases that occur in immunocompetent individuals is rudimentary, this paper will focus on viruses and the manner in which genetic alteration affects virulence of these micro-organisms for humans.

Many animal viruses have a high mutation rate, particularly RNA viruses which lack a mechanism for correction of transcriptional errors made by viral polymerases. For example, the error rate for the polymerase of influenza A virus, Sendai virus, or vesicular stomatitis virus is estimated to be $10^{-5.5}$ per nucleotide per replication (Portner et al. 1980; Holland et al. 1982). This means that virus populations are not as monolithic or as extremely homogeneous as previously thought. Instead, viruses consist of heterogeneous populations of virions that fluctuate about a norm that constitutes the consensus sequence. Mutants that deviate from the consensus sequence are constantly being produced and tested for survival in nature. Often new consensus sequences (genotypes) emerge and continue to evolve while co-circulating with prior genotypes. This has been demonstrated for rotaviruses (analyzed by gel electrophoresis of viral RNAs), polioviruses (analyzed by oligonucleotide fingerprinting), and influenza A viruses (studied by nucleotide sequence analysis) (Nottay et al. 1981; Kendal et al. 1984; Kapikian and Chanock 1985). In the case of the human rotaviruses, and most other human viruses as well, the consequences of such genetic drift are usually not obvious in terms of altered virulence or transmissibility. The human influenza A viruses represent a special situation because drift in the major epitopes of the hemagglutinin confers an advantage of increased transmissibility, but there is no evidence that virulence is affected (Kendal et al. 1984).

Viruses that possess a segmented genome can also undergo change by gene reassortment during dual infection with a related virus. For example, reassortment of genes of the human influenza A H3N2 and H1N1 viruses has occurred under natural conditions, presumably in man (Young and Palese 1979). To date there is no evidence that this form of gene exchange has produced influenza A viruses with increased virulence for humans. Reassortment of influenza A virus genes also occurs in animals and birds and this is thought to be one of the mechanisms responsible for sudden emergence of antigenically novel, human pandemic influenza A viruses (Young and Palese 1979).

Variation in virulence among strains belonging to the same serotype also occurs but this form of polymorphism has not been studied systematically. Almost 30 years ago Sabin observed that poliovirus strains recovered from healthy individuals exhibited significant variation in neurovirulence as measured by intracerebral inoculation of monkeys (Sabin 1956). In fact, the parent of the live vaccine type 2 poliovirus was initially identified in this manner. Also, several unique rotavirus strains with distinct genotype (as defined by gel electrophoresis of viral RNAs) were recently observed to be attenuated in young infants (Banatvala et al. 1978; Bishop et al. 1983). The molecular basis for these and most other examples of virulence polymorphism are not understood. Nonetheless, it is important to recognize virulence polymorphism when it occurs because failure to do so can easily confound the evaluation of host factors that affect severity of disease. It should be noted, however, that it is difficult to study virulence of human viruses in the laboratory because infection of susceptible experimental animals rarely induces an illness that closely resembles human disease. There are a few situations in which experimental animals develop a response to infection that resembles that of man, but, unfortunately these animal model systems have rarely been used to evaluate naturally occurring human viruses for virulence in a systematic manner.

In only one instance has a convincing molecular explanation been offered for polymorphism of virus virulence under natural conditions (Rott 1979). Newcastle disease viruses (NDV) vary in virulence for their natural hosts, chickens, and other avian species. The NDV fusion (F) glycoprotein must undergo posttranslational cleavage by a host cell protease for the virus to become infectious. Naturally occurring avirulent strains of NDV possess an F glycoprotein that is not readily cleaved by host cell proteases, whereas the F glycoprotein of naturally occurring virulent strains of NDV is readily cleaved. Ease of cleavage is presumably responsible for the pantropism of virulent NDV strains that disseminate throughout the infected host. In contrast, avirulent NDV strains, which are not readily cleaved, remain localized to the respiratory tract.

Viruses can also suddenly enter the human chain of transmission and develop sufficient virulence to cause disease. A recent example is enterovirus 70 which causes disease in the eyes and central nervous system (CNS) of infected individuals (Kono 1975). The molecular basis for this type of change is not understood.

Very infrequently a virus can suddenly develop a marked increase in virulence. The most recent instance of this sort of abrupt change is unique because laboratory studies quickly provided a reasonable explanation for the genetic events responsible for the dramatic shift in virulence. During 1983 an avian influenza A virus (H5N2) developed a sudden and dramatic increase in virulence during transmission among chickens in Pennsylvania and New Jersey (Kawaska et al. 1984). Initially, infection was associated with a low mortality ($< 1\%$), but the situation changed abruptly and a mortality rate of $\sim 80\%$ was observed. The change from a temperate to a highly virulent virus was associated with a mutation in the hemagglutinin glycoprotein that resulted in loss of a glycosylation site. Loss of carbohydrate presumably rendered the nearby cleavage site of the hemagglutinin more accessible to the action of host cell proteases. This change probably enhanced infectivity of the virus and extended the range of tissues in which virus could replicate and cause disease because cleavage of the hemagglutinin is required for initiation of influenza infection. A similar dramatic increase in virulence of a recognized human virus has not been documented. Perhaps the marked increase in mortality that occurred during the second wave of the 1918 influenza A virus pandemic may represent a human counterpart of the avian influenza A (H5N2) experience of 1983 (Shope 1944). Unfortunately this possibility cannot be investigated directly because influenza A virus strains are not available from the era of the 1918 pandemic.

It is clear that we are frequently exposed to altered viruses that arise under natural conditions, but most alterations do not appear to enhance virulence. Currently, the changes which occur in human viruses in nature are, for the most part, beyond our control because effective measures for prevention or treatment of infection are only available for a limited number of viruses. These measures, which include effective vaccines, chemoprophylaxis, chemotherapy, and/or vector control, could be developed for many other viruses, but in most instances the cost/benefit ratio has been viewed as unfavorable by public health authorities or pharmaceutical manufacturers.

In the near future it is unlikely that we will be in a position to affect most genetic alterations that occur in human or animal viruses under natural conditions. However, we must be careful not to add to the existing burden of genetically altered viruses by experimentally constructing infectious viruses that have the potential for enhanced transmission and increased virulence in man or lower animals and by allowing these viruses to enter the human and/or animal chain of transmission. Examples of constructs that should be carefully controlled, or perhaps in some instances forbidden, include fully infectious influenza or parainfluenza viruses that possess an attachment or fusion glycoprotein that is more readily cleaved by host cell proteases than the corresponding moeity of contemporary wild-type viruses. We should also be concerned about fully infectious viruses in which foreign enhancer or promoter sequences are inserted since this could result in an alteration in organ tropism or host range.

Recombination of viral genomes or reassortment of viral genes should also be approached with caution if both virus parents are fully virulent or if one of the parents exhibits an unusually high degree of virulence. On the other hand, the possibility for generation of viruses with enhanced virulence diminishes markedly if the previously cited admonitions are observed and if one or both of the parents are significantly attenuated for humans (or their natural animal host) (Murphy et al. 1984a,b). Thus far only two virus reassortants have been described which are more virulent than either parental virus (Rott et al. 1984). However, two of the admonitions cited above apply to the construction of these influenza A virus reassortants. First, a virus which was used as a parent for both reassortants had a readily cleavable hemagglutinin and, as a consequence, this parental virus was highly virulent for its natural host. Second, this hemagglutinin was transferred to both reassortants thereby permitting these viruses to expand their tissue tropism and cause disease in a site not ordinarily affected by influenza A viruses. For these reasons the preparation of similar reassortants should be discouraged.

Although the distribution, properties, and clinical consequences of naturally occurring, genetically altered human viruses are not well understood, there is considerable information about another type of altered virus, namely, attenuated mutants developed for use in live virus vaccines. Widespread distribution and utilization of the licensed attenuated virus vaccines has been recommended, or in some instances, mandated by public health authorities. As a consequence each of the licensed live attenuated virus vaccines, with one exception, has been administered to more than 10^8 individuals. In addition, six experimental live attenuated virus vaccines have been evaluated thoroughly in small-to-moderate size groups of susceptible persons. The cumulative experience with the licensed and experimental vaccines is of some interest for several reasons. First, the vaccine strains represent a form of genetically altered virus over which we have control. Second, the response of humans to each vaccine strain has been well characterized. Third, each of the vaccines has proved to be quite safe. Fourth, the properties that are shared by these safe vaccine viruses may provide a standard of reference for evaluating future candidate live vaccine virus strains or other genetically altered viruses that are being considered for introduction into the ecosystem.

RESULTS

Origin of Live Attenuated Virus Vaccines

Most live virus vaccine strains were derived from a human virus (Table 1). Exceptions to this rule are the simian and bovine rotaviruses and the avian influenza A virus donor, which is used as a source of attenuating genes that are transferred into avian-human influenza A reassortant viruses by gene reassortment during mixed infection with a human influenza A virus. The origin of vaccinia virus is not known at this time; it does not appear to have been derived directly from either variola or cowpox virus.

Procedures Used to Attenuate Live Virus Vaccine Strains

Except for vaccinia virus whose origin is shrouded in mystery, the licensed vaccine strains were attenuated during serial propagation in cell culture (Table 2). Satisfactory attenuation and antigenicity of measles, mumps, and rubella vaccine viruses were ascertained exclusively by clinical evaluation in susceptible individuals.

Table 1
Derivation of Attenuated Viruses Used in Live Vaccines

Status of vaccine	Vaccine virus	Derivation of virus
Licensed	Vaccinia	Not known
	Yellow fever, polio, measles, rubella, mumps	Human virus
Experimental	Cytomegalo (CMV), respiratory syncytial (RSV), rotavirus	Human virus
	Rotavirus	Animal virus–bovine or simian
	Influenza A	Reassortant virus with six viral nonsurface protein ("internal") genes derived from human *ca* mutant or avian wild-type influenza A donor virus
	Influenza B	Reassortant virus with six viral "internal" genes derived from human *ca* mutant influenza B donor virus

Table 2
Procedures Used to Attenuate Live Virus Vaccine Strains

Virus	Passage in	Additional selective procedure
Vaccinia	Calf skin	None
Yellow fever (17D), mumps	Chick embryo cell culture and eggs	Identification of mutant (17D) with decreased viscerotropism in monkeys
Measles	Chick embryo cell culture	Selection of cold adapted (*ca*) mutant by passage at 32°C
Rubella, CMV	Human embryo lung diploid fibroblast culture	None
Varicella	Human embryo lung fibroblast and guinea pig embryo cell cultures	None
Polio–types 1, 2, and 3	Primary simian kidney cell culture	Identification of plaque progeny with reduced neurovirulence for monkey spinal cord
Respiratory syncytial	Primary bovine kidney cell culture	*ts* mutation (*ts*-1 or *ts*-2) induced by 5FU
Human rotavirus	Primary simian kidney cell culture	None
Simian rotavirus	Primary simian kidney and fetal simian diploid cell cultures	None
Bovine rotavirus	Primary bovine and simian cell cultures	Selection of *ca* mutant at 30°C
ca influenza A donor, *ca* influenza B donor[a]	Primary chick kidney cell culture	Selection at 25°C of *ca* mutant with reduced virulence for ferrets
Avian influenza A donor[a]	Primary chick kidney cell culture	Identification of donor virus which is restricted in replication in simian respiratory tract

[a]Six nonsurface protein ("internal") genes of donor virus transferred into reassortant vaccine virus by gene reassortment

Measles virus required further passages at low temperature (32°C) in order to select additional mutations that were required for satisfactory attenuation. In the case of yellow fever virus and the polioviruses, studies in experimental animals provided a signal that mutants with the desired phenotype and level of attenuation had been selected by passage of virus in cell culture. The mutant of yellow fever virus (17D) that was identified in this manner had decreased tropism for the liver of experimentally infected monkeys (Theiler 1951). Later clinical studies indicated that the mutant was satisfactorily attenuated and antigenic in humans (Smith 1951). Similarly, mutants of poliovirus with reduced neurovirulence were initially identified by direct inoculation into the spinal cord of monkeys (Sabin 1956). Subsequently these poliovirus vaccine strains were shown to be satisfactorily attenuated and antigenic for susceptible individuals.

Attenuation of candidate vaccine strains of cytomegalovirus (CMV), varicella virus, and rotavirus for susceptible individuals was established directly during clinical trials because experimental animals that could provide information predictive of human response were not available (Plotkin 1984; Weibel et al. 1984; Kapikian et al. 1985; Vesikari et al. 1985). In contrast, reduced growth of *ts* mutants of RS virus in the lower respiratory tract of hamsters provided a basis for predicting, correctly in this instance, that the mutants would be attenuated in humans (Wright et al. 1982). Similarly, decreased replication of cold adapted (*ca*) influenza A virus reassortants in the respiratory tract of ferrets and restriction of avian-human influenza A virus reassortants in the lower respiratory tract of monkeys provided a basis for predicting that these viruses would be satisfactorily attenuated in susceptible humans (Murphy et al. 1984a,b).

Molecular Basis for Attenuation

The molecular basis for attenuation of most live vaccine viruses has not been established (Table 3). At present the most information concerning nucleotide sequence changes associated with attenuation has been obtained for poliovirus types 1 and 3. During its attenuation the type 1 vaccine virus sustained 56 base changes that resulted in 21 amino acid substitutions; these changes were distributed over the entire viral genome (Omata et al. 1985). Tests in monkeys for neurovirulence of recombinant viruses constructed in vitro from infectious cDNAs of parental wild-type virus and vaccine virus indicated that mutations responsible for attenuation of type 1 vaccine virus did not cluster in one region of the genome (Omata et al. 1985). Instead these mutations appeared to be located throughout the viral genome.

During its passage in tissue culture and during the selection of clonal populations of virus with minimal neurovirulence, the type 3 poliovirus vaccine strain sustained fewer mutations than the type 1 vaccine strain. Only ten nucleotide substitutions occurred and this resulted in only three amino acid substitutions (Almond et al.

Table 3
Basis of Attenuation of Live Vaccine Viruses

Virus		Basis of attenuation
Vaccinia, yellow fever (17D), measles, rubella, mumps, varicella		Not known. Host range mutations have not been mapped. Deletion, insertion or rearrangement not detected in varicella vaccine genome
Polio		Type 1; 56 nucleotide changes coding for 21 amino acid substitutions distributed over entire genome. Type 3; ten nucleotide changes coding for three amino acid substitutions; base changes at positions 472 (noncoding), 2034 (VP3), and 3333 (VP1) probably most important in attenuation
Influenza	Type A *ca* donor	Limited number of mutations in "internal" genes of *ca* donor virus; mutations in PA gene probably most important in attenuation
	Type B *ca* donor	Not known
	Type A avian donor	Host range restriction associated with extensive sequence divergence of avian influenza A donor virus NP and M genes from corresponding genes of human influenza A viruses
Respiratory syncytial		*ts*-1 mutation not mapped. *ts*-2 mutation affects viral fusion glycoprotein
Rotavirus	Human	Not known; probably mutations affecting VP_3 capsid protein
	Bovine and simian	Host range restriction associated with extensive sequence divergence of genes from corresponding genes of human rotaviruses

1985; Evans et al. 1985). Sequence analysis of type 3 "revertant" viruses recovered from patients with vaccine virus-associated paralytic disease and tests for neurovirulence of type 3 wild-type and vaccine virus recombinants suggest that base substitutions in only three positions play a significant role in attenuation. These changes are at position 472, which is in the 5′ noncoding region; position 2034, which is located in the VP3 capsid protein region; and position 3333, which is located in the VP1 capsid protein region (Almond et al. 1985; Evans et al. 1985). Mutation of nucleotide 472 to wild-type sequence can occur early during replication of vaccine virus in the intestinal tract and this reversion has been detected in every type 3 poliovirus vaccine isolate that exhibits increased neurovirulence examined thus far (Almond et al. 1985; Evans et al. 1985).

Segregation analysis of the nonsurface protein ("internal") genes of the avian influenza A donor virus was performed by selecting single avian influenza gene reassortants (i.e., a single avian influenza A gene on a background of seven human influenza A genes) and determining which avian influenza gene or genes restrict growth of avian-human influenza A reassortant viruses in monkeys (Murphy et al. 1984a,b). These studies indicated that the avian influenza NP and M genes play a major role in restricting replication of avian-human influenza virus reassortants in the simian respiratory tract. This form of host range restriction is associated with significant nucleotide sequence divergence of the avian influenza M and NP genes from the corresponding genes of human influenza A viruses (Buckler-White et al. 1985). In the case of the avian influenza M gene, the available sequence comparisons indicate that there are ten sites in the M1 and M2 proteins which exhibit a host species-specific dimorphism (Buckler-White et al. 1985). At each of these sites a given amino acid is specific for avian influenza A viruses, while another amino acid is specific for human influenza A viruses. Six of these amino acid differences result in a change in hydrophobicity or charge.

Segregation analysis of the "internal" genes of the *ca* influenza A donor virus suggests that the PA gene plays a major role in attenuation of *ca* reassortant influenza A viruses for humans (M. Snyder et al., unpubl.). This gene codes for one of the proteins of the virus polymerase complex. The nature and extent of the mutations in the *ca* PA gene are not known at this time. Sequence analysis of other "internal" *ca* genes, which is in progress, indicates that very few mutations occurred during growth and selection of the donor virus at low temperature (N.J. Cox and A.P. Kendal, pers. comm.).

The host range restriction and attenuation of bovine and simian rotaviruses for humans is associated with extensive sequence divergence of each of the 11 animal rotavirus genes from the corresponding genes of human rotaviruses (Kapikian et al. 1985). The location and extent of the sequence divergences that specify attenuation for humans is not known.

Finally, the *ts* mutations responsible for attenuation of the *ts*-1 mutant of RS virus for humans have not been mapped. However, the mutations of the *ts*-2 mutant of RS virus appear to affect the viral surface glycoprotein responsible for entry of virus into the host cell and for cell fusion; the location of this lesion on the fusion protein gene remains to be determined (Wright et al. 1982).

Pattern of Infection of Attenuated Viruses

Most vaccine viruses are highly restricted in their capacity to grow in humans (Table 4). For example, the measles, mumps, and CMV vaccine strains cannot be recovered from sites at which wild-type virus replicates to high titer. Yellow fever (17D) vaccine virus also produces a restricted infection in humans during which viremia is reduced significantly. Similarly, replication of rubella vaccine virus is markedly reduced in the oropharynx of vaccinees (Center for Disease Control 1984). The influenza A and B virus reassortants are also restricted in their replication in the respiratory tract; duration of infection is shortened and the level of virus replication is reduced $\sim 10^{-3}$ (Murphy et al. 1984a,b). Intestinal infection with the bovine and simian rotaviruses is also restricted; susceptible infants shed these viruses in their feces in greatly reduced amounts compared to human rotaviruses. Finally the *ts*-1

Table 4
Pattern of Infection Produced by Live Vaccine Viruses

Vaccine	Virus recoverable	Sites, amount (and frequency)
Measles, mumps, CMV	No	Oropharynx, respiratory tract, blood (CMV),[a] genital tract (CMV)[a]
Yellow Fever (17D)	Yes	Blood–low level (50%)
Rubella	Yes	Oropharynx–low level; blood (70%)
Vaccinia	Yes	Vaccination site. Transient viremia but virus rarely in throat
Polio	Yes	Stools–high titer. Alimentary infection continues for several weeks during which mutations that confer increased neurovirulence (as measured in monkeys) occur and are selected for. Viremia
Varicella	Yes	Skin lesions of immunocompromised vaccinees
Influenza A and B	Yes	Respiratory tract–short duration and reduced amount, $\sim 10^{-3}$
Respiratory syncytial	Yes	Respiratory tract. *ts*-1 mutant–high titer. *ts*-2 mutant greatly reduced amount
Rotavirus	Yes	Stools–Simian and bovine rotaviruses shed in greatly reduced amount

[a]Blood and genital tract secretions tested only for CMV

mutant of RS virus does not appear to be restricted significantly in the upper respiratory tract of susceptible individuals, whereas the RS virus *ts*-2 mutant is markedly restricted. Study of the *ts* mutants has been terminated because of genetic instability (*ts*-1) or overattenuation (*ts*-2).

The soon to be licensed varicella virus vaccine is also markedly restricted in its replication. Thus far this vaccine virus has been recovered only from the skin lesions of immunocompromised vaccinees (Weibel et al. 1984; Gershon et al. 1985). This observation illustrates the importance of host immune mechanisms in bringing about resolution of infection by vaccine viruses. Thus, the desired degree of restriction and attenuation occurs only in the context of a normal host immune response. In the case of varicella virus vaccine, the consequences of an inadequate host immune response to the vaccinee are not serious. Mild or moderate varicella occurs but illness is not serious or life-threatening.

The consequences of an inadequate immune response are more serious during vaccination with vaccinia virus. In immunocompetent individuals, infection with vaccinia virus is restricted in extent compared to variola. Vaccinia virus replicates to high titer at the site of vaccination and transient viremia occurs (Blattner et al. 1964). However, virus is usually not detected in the throat and significant spread of virus to other areas of the skin is rare (Connor et al. 1977). The most serious circumstance in which vaccinia virus spreads centrifugally is progressive vaccinia (or vaccinia necrosum), a devastating and inexorable disease. This disease occurs in vaccinees who have a profound underlying immunodeficiency, especially in cell-mediated immune response (Fulginiti et al. 1968). Fortunately, failure to resolve infection at the site of inoculation is extremely rare; it occurs approximately once in 10^6 vaccinations (Lane et al. 1969).

Of the various live virus vaccine strains, the polioviruses appear to be least restricted in replication in humans. Virus is shed from the intestinal tract in moderately high titer and for a relatively long interval, usually several weeks to a month or more. Viremia also occurs. In view of the fact that infection of humans with the poliovirus vaccine strains is asymptomatic, except in very rare instances (approximately one in $10^{6.9}$ vaccinations), it would appear that these viruses are markedly restricted in the target organ, namely the spinal cord, whereas there is considerably less restriction in the intestinal tract. This is not surprising because the poliovirus vaccine strains were selected by identifying plaque progeny of tissue culture passaged virus that exhibited the lowest possible level of neurovirulence following direct intraspinal inoculation of monkeys (Sabin et al. 1954; Sabin 1956, 1957). At the same time candidate vaccine viruses with low neurovirulence were screened to identify strains that grew well in the intestinal tract of the chimpanzee. In this manner the dual properties of extensive growth in the intestinal tract and lowest possible level of neurovirulence were selected for in the most permissive experimental systems available—the chimpanzee's intestinal tract and the monkey's spinal cord.

An experimental model for selective restriction of virus in the CNS has been described by Spriggs and his colleagues (Spriggs et al. 1983). They observed that reovirus mutants selected for resistance to neutralizing monoclonal antibodies exhibited an altered and restricted pattern of infection in the brain, whereas virus replication in the intestines and other viscera was not modified. These mutants presumably sustained a mutation in the outer capsid attachment protein (S_1) because the monoclonal antibodies used for selection were directed against this protein. Although the phenotype of Spriggs' mutants resembles that of the poliovirus vaccine strains, the genetic changes responsible for attenuation of the latter viruses are more complex because attenuating mutations in the type 1 vaccine strain appear to be scattered throughout the genome (Omata et al. 1985). Also one of the major attenuating mutations in the type 3 vaccine strain appears to be in the 5′ noncoding region (Almond et al. 1985; Evans et al. 1985).

Clinical Response

Two of the licensed live virus vaccines, polio and mumps, characteristically produce an asymptomatic infection, while the remaining licensed live virus vaccine strains retain a low level of residual virulence. The signs and symptoms that develop following infection by the latter vaccine viruses represent a mild form of the signs and symptoms induced by wild-type virus. Thus, attenuated vaccine viruses do not exhibit an altered pathogenesis and there have been no surprises or unanticipated reactions. The most common reactions produced by the vaccines with residual low virulence (yellow fever, measles, rubella, and vaccinia) are minor and have not interferred with acceptance and widespread use of these products (Table 5). These common reactions have not been a cause for concern. Instead, concern has focused on the very rare reactions associated with two of the vaccines, vaccinia and poliovirus.

Soon after eradication of variola virus from the Americas, at a time when this virus was still endemic in several African and Asian countries, public health authorities in the U.S. were sufficiently concerned about the rare complications of vaccinia that they recommended termination of mandatory vaccination. Two of the three severe reactions to vaccinia occur only in individuals with a pre-existing disease condition, eczema in the case of eczema vaccinatum and severe immunodeficiency in the case of progressive vaccinia (vaccinia necrosum) (Lane et al. 1969, 1970). Thus, these two reactions can be prevented by withholding vaccine from individuals with these conditions and from contacts who might transmit vaccinia to individuals with one of these conditions.

Paralytic disease caused by poliovirus vaccine is also extremely rare, i.e., one in $10^{6.9}$ vaccinations (Nkowane and Wassilak 1984). A small proportion of these vaccine-associated illnesses (~ 15%) occurs in infants who are at risk because of a profound immunodeficiency. It should be noted that most vaccine-associated paralytic disease occurs in individuals who do not have obvious signs of immuno-

Table 5
Clinical Response to Vaccination

Virus	Clinical response
Mumps, CMV	Asymptomatic
Polio	Asymptomatic; very rarely paralytic disease (one in $10^{6.9}$ vaccinations)
Vaccinia	Local lesion at vaccination site. Fever (10%), myalgias, and headache (2-5%). Very rarely eczema vaccinatum (~ 1 in 10^5), encephalitis (~ 1 in $10^{5.5}$), or progressive vaccinia (~ 1 in 10^6, primarily in immunodeficient individuals)
Yellow fever (17D)	Fever (10%), myalgia and headache (2-5%)
Measles	Fever (5-15%), mild rash (5%). Encephalitis rare (~ 1 in 10^6); relationship to vaccination not clear
Rubella	Transient arthritis in adult females (10-15%) and less commonly in children
Varicella	Papular rash (2%) in healthy vaccinees. Papulo-vesicular rash in leukemia patients—6% prior to chemotherapy and 37% after chemotherapy. Incidence of zoster not increased within 2-3 year observation period
Influenza A and B	Mild upper respiratory symptoms (< 10%) and/or transient fever (< 10%). Symptoms are dose-dependent and rarely occur when 10-32 HID_{50} given to susceptible individuals
Respiratory syncytial	Upper respiratory symptoms and mild otitis (*ts*-1) or asymptomatic infection (*ts*-2) in seronegative children
Rotavirus	Human and bovine rotavirus infections asymptomatic in susceptible adults and infants, respectively. Simian rotavirus infection asymptomatic in susceptible infants except for transient fever after high dose of virus

deficiency. In these instances it appears that vaccine virus regains sufficient neurovirulence to induce paralysis.

During replication in the intestinal tract, poliovirus vaccine strains, particularly type 3, commonly undergo a drift toward greater neurovirulence as measured by direct intraspinal inoculation of monkeys (Evans et al. 1985). In most instances this type of change is probably not significant because it occurs with high frequency, whereas vaccine-associated disease is very rare, approximately one in $10^{6.9}$ vaccinations. Rarity of disease may reflect the low frequency with which the appropriate constellation of mutations necessary for full expression of neurovirulence develops or it may reflect the effectiveness of vaccine-induced immunity in preventing disease by fully neurovirulent mutants that arise late in infection. Possibly both factors are important.

Twenty-two type 2 or type 3 "revertants" isolated from vaccine-associated paralytic illnesses have been studied by oligonucleotide mapping and/or sequence analysis (Kew et al. 1981; Nottay et al. 1981; Kew and Nottay 1984; Almond et al. 1985; Evans et al. 1985). Many of the "revertant" viruses were estimated by oligonucleotide mapping to have undergone 50–100 mutations, while one "revertant" type 3 virus sustained only seven mutations as indicated by sequence analysis. The pattern of mutation for each isolate was unique, suggesting that only mutations in the appropriate region of the vaccine virus genome are required for restoration of neurovirulence and that the majority of mutations that develop are silent with respect to neurovirulence.

Spread of Infection with Resulting Disease

The available evidence indicates that ten of the 15 licensed or experimental vaccine strains are not transmitted from infected vaccinees to intimate contacts (Table 6). Three of the vaccine viruses (measles, mumps, and CMV) do not appear to be shed by vaccinees (Table 4) and hence infection cannot be transmitted. The yellow fever vaccine strain (17D) induces a low level viremia, but the virus is not transmitted by insect vectors which are competent to transmit wild-type virus (Whitman 1951). Other vaccine virus strains (rubella, influenza A [*ca* reassortant and avian-human reassortant] and B [*ca* reassortant], and bovine and simian rotaviruses) apparently do not spread because they are shed by vaccinees in markedly reduced amounts compared to wild-type virus. In two instances, *ca* influenza A reassortants and avian-human influenza A reassortants, infectivity for susceptible individuals is known to be reduced significantly ($\sim 10^{-2}$); and this undoubtedly contributes to lack of transmission (Murphy et al. 1984a,b).

Varicella vaccine virus does not spread from immunocompetent vaccinees to contacts (Weibel 1984). However, infection can be transmitted from immunocompromised vaccinees with leukemia who develop skin lesions from which vaccine virus can be recovered (Gershon et al. 1985). Limited experience to date indicates that the risk to contacts is not significant because they undergo an asymptomatic infection or develop mild varicella and there is no evidence that tertiary spread occurs.

Spread of the *ts*-1 mutant of RSV was observed on several occasions but contact infection was asymptomatic and isolates recovered from an infected contact retained the ts phenotype.

Only two vaccines pose a threat to contacts who acquire infection from vaccinees, i.e., vaccinia and poliovirus. Poliovirus vaccine strains spread to contacts with considerably greater frequency than does vaccinia (Fox et al. 1960). In a prospective study of the three poliovirus vaccine strains in families during 1958–1959, it was observed that spread of infection to susceptible contacts occurred with a frequency of 8% in upper economic families and 51% in lower economic families

Table 6
Spread of Infection to Contacts and Its Consequences

Vaccine	Spread to contacts	Conditions which favor spread or which are required for spread	Consequences of contact infection
Yellow fever,[a] measles, rubella, mumps, CMV, influenza A[b] and B, bovine and simian rotaviruses	No		
Vaccinia	Yes–un-common	Intimate contact; source of infection is vaccination site. Most at risk are persons with eczema	Eczema vaccinatum with involvement of areas of eczematous skin. Persons without eczema most commonly develop vaccinia lesions on the face.
Varicella	Yes	Spread from immunocompromised vaccinees but not from normal vaccinees	Asymptomatic infection or mild varicella
Polio–types 1, 2, 3	Yes–very common (8–51% in families)	Close contact within family or with persons in immediate environment. Low socioeconomic status favors spread.	Rarely paralytic disease in family contacts (one in $10^{6.7}$ vaccinations) or nonhousehold contacts (one in $10^{7.2}$ vaccinations)
Respiratory syncytial	Yes	Limited information. Spread of *ts*-1 mutant in immediate environment but not *ts*-2 mutant	Asymptomatic. Retention of *ts* phenotype

[a] 17D yellow fever virus not transmitted by competent insect vector
[b] Influenza A virus replication reduced ~ 10^{-3} and infectivity for susceptible individuals reduced ~ 10^{-2}

(Fox et al. 1960). Significant spread also occurred to susceptible contacts in the immediate community, 33% in one locality and 70% in another locality. Comparison of intrafamilial spread of vaccine strains with that observed when wild-type polioviruses entered similar households indicated that the vaccine viruses had a reduced capacity for transmission. For example, wild-type polioviruses infected 86% of susceptible upper economic family members and 93% of susceptible lower economic family members. As mentioned above, the comparable values for vaccine viruses were 8% and 51%. It would appear that the better hygiene of the upper economic families seriously impeded spread of vaccine strains but not wild-type viruses. Also spread of vaccine viruses ceased at a time when a large proportion of susceptibles were still available. In contrast, wild-type polioviruses exhausted most of the susceptibles in a comparable population.

Surveillance of unvaccinated individuals following large-scale administration of vaccine over a short interval in the United States (Cincinnati), Mexico, or Czechoslovakia indicated that polioviruses disappeared from the community within several months after the campaign although the seasonal conditions were optimal for the spread of other enteroviruses (Sabin et al. 1960, 1961; Zacek et al. 1962). These observations, together with the results of the family studies, suggest that the live poliovirus vaccine strains are capable of only one or two cycles of transmission and hence are unable to persist in the community. Reduced transmissibility of the type 1 vaccine strain is probably due in part to reduced infectivity by the oral route. A virus in the lineage of the type 1 vaccine very close to the final vaccine strain was demonstrated to be approximately 1000 times less infectious than its wild-type parent when tested by the oral route in monkeys (Sabin et al. 1954).

Although spread of vaccine strains to close susceptible contacts occurs with appreciable frequency under optimal circumstances, i.e., crowding and poor sanitation, the development of vaccine-associated paralytic disease by contacts is a very rare event, approximately one in $10^{6.7}$ vaccinations for family contacts and one in $10^{7.2}$ vaccinations for nonhousehold contacts. Nonetheless, such vaccine-associated illness in contacts accounts for approximately 55% of all vaccine-related disease (Nkowane and Wassilak 1984). One factor that probably acts to diminish the incidence of disease in contacts is the high frequency of immunity among contacts resulting from prior vaccination or prior infection with wild-type virus. On the other hand, it is likely that the risk of paralytic disease may be greater in susceptible contacts than in vaccinees because mutations in poliovirus are progressive during infection of the intestinal tract and this probably favors the emergence of vaccine virus mutants with progressively greater neurovirulence late during infection (Nottay et al. 1981; Kew et al. 1981; Kew and Nottay 1984). Many of the most neurovirulent mutants would thus arise at a time when vaccinees had developed some immunity and hence were resistant to disease, whereas susceptible contacts would remain unprotected.

Capacity for extensive growth in the intestinal tract is a property that was selected for in poliovirus vaccine strains because it is required for satisfactory antigenicity and induction of effective resistance to disease. It is paradoxical that this property also plays a critical role in the only serious problem associated with the vaccine, namely the emergence of mutants with increased neurovirulence that cause disease with very low frequency (one in $10^{6.7}$-$10^{7.2}$) in vaccinees and their contacts. It is hoped that more stable attenuating mutations, perhaps in the form of deletions or insertions, can be introduced into poliovirus vaccine strains by appropriate manipulation of infectious cDNA copies of the viral genome.

Vaccinia virus can spread from vaccinees to contacts. Although the precise frequency with which this occurs has not been established, it appears to be a rare event somewhere in the range of one in 10^4-10^5 vaccinations (Lane et al. 1969). Transmission of vaccinia virus to contacts requires intimate contact and usually occurs between siblings who sleep in the same bed or between parent and child (Neff et al. 1967). Clearly, vaccinia virus can not sustain itself in humans because this virus has not been recovered from humans since termination of routine vaccination in the United States and other countries. The most serious consequence of contact infection is eczema vaccinatum, which occurs primarily in individuals with eczema. Approximately one-half the cases of eczema vaccinatum occur in contacts of vaccinees; this type of contact infection happens approximately once in $10^{5.4}$ vaccinations. Eczema vaccinatum acquired by contact is more severe than eczema vaccinatum that develops after vaccination. In fact, fatal eczema vaccinatum occurs only in individuals who are infected by contact and this form of fatal disease accounts for 20% of all mortality attributable to vaccination with vaccinia. Other forms of contact infection are almost always trivial and do not pose a threat to the health of the affected individual.

DISCUSSION

The licensed and experimental live attenuated vaccine viruses do not pose a threat to the human population for several reasons. First, these virus strains cause few if any symptoms and those that occur are not of concern except in very rare instances (one in 10^5-10^7 for poliovirus and vaccinia). Second, except for the poliovirus strains, vaccine viruses exhibit a global restriction of replication in humans and this appears to be a stable property. Poliovirus vaccine strains, which are not significantly restricted in replication in the intestines, are, however, markedly restricted in the target organ, the spinal cord. Third, without exception, vaccine viruses are not transmissible or are unable to sustain transmission. Fourth, restriction of replication of vaccine viruses and/or failure of these strains to transmit or to sustain transmission make it unlikely that restoration of virus virulence and transmissibility could occur by the usual genetic mechanisms that correct previous mutations, i.e., reversion of attenuating mutations, "suppression" of attenuating mutations by second site mutations, recombination and/or gene reassortment. Clearly, restriction of replication and transmission of vaccine viruses should markedly limit the opportunity for such genetic events to occur. Fifth, a number of observations indicate that virulence is a polygenic property and thus an appropriate mutation or sequence divergence in almost any gene or region of the genome can cause or contribute to attenuation. Indeed, the mechanism of attenuation of the three vaccine strains studied in greatest detail, i.e., types 1 and 3 poliovirus and the avian-human influenza A virus reassortants, appears to be multigenic. Although the genetic basis of attenuation of animal rotaviruses for humans has not been established formally, it is likely to be multigenic because there is significant sequence divergence between each of the 11 genes of bovine or simian rotavirus and the corresponding genes of human rotaviruses. Also, one or more mutations have been identified in each of the "internal" genes of the *ca* influenza A donor virus. The multigenic nature of attenuation of some (and possibly all) live vaccine viruses provides an additional constraint on restoration of virulence because correction or modification of all of the attenuating mutations in a vaccine strain would be required and this complete series of independent genetic events is unlikely to occur when both virus replication and transmission are restricted.

REFERENCES

Almond, J.W., G.D. Westrop, A.J. Cann, G. Stanway, D.M.A. Evans, P.D. Minor, and G.C. Schild. 1985. Attenuation and reversion to neurovirulence of the Sabin poliovirus type-3 vaccine. In *Vaccines 85: Molecular and chemical basis of resistance to parasitic, bacterial, and viral diseases* (ed. R.A. Lerner, R.M. Chanock, and F. Brown), p. 271. Cold Spring Harbor Laboratory, Cold Spring Harbor, New York.

Banatvala, J.E., I.L. Chrystie, and B.M. Totterdell. 1978. Rotaviral infections in human neonates. *J. Am. Vet. Med. Assoc.* **173**: 527.

Bishop, R.F., G.L. Barnes, E. Cipriani, and J.S. Lund. 1983. Clinical immunity after neonatal rotavirus infections: A prospective longitudinal study in young children. *N. Engl. J. Med.* **309**: 72.

Blattner, R.J., J.O. Norman, F.M. Heys, and I. Aksu. 1964. Antibody response to cutaneous inoculation with vaccinia virus: Viremia and viruria in vaccinated children. *J. Pediatr.* **64**: 839.

Buckler-White, A.J., B.R. Murphy, and C.W. Naeve. 1985. Characterization of the genes involved in host-range restriction of avian influenza-A viruses in primates. In *Vaccines 85: Molecular and chemical basis of resistance to parasitic, bacterial and viral diseases* (ed. R.A. Lerner, R. Chanock, and F. Brown), p. 345. Cold Spring Harbor Laboratory, Cold Spring Harbor, New York.

Center for Disease Control. 1984. U.S. Dept. of Health and Human Services, Public Health Service. Recommendation of the immunization practices advisory committee (ACIP): Rubella prevention. *Morbidity and Mortality Weekly Report* (*MMWR*) **33**: 301.

Connor, J.D., K. McIntoch, J.D. Cherry, A.S. Benenson, D.W. Alling, U.T. Rolfe, J.E. Schanberger, and M.J. Mattheis. 1977. Primary subcutaneous vaccination. *J. Infect. Dis.* **135**: 167.

Evans, D.M.A., G. Dunn, P.D. Minor, G.C. Schild, A.J. Cann, G. Stanway, J.W. Almond, K. Currey, and J.V. Maizel Jr. 1985. Increased neurovirulence associated with a single nucleotide change in a noncoding region of the Sabin type 3 poliovaccine genome. *Nature* **314**: 548.

Fox, J.P., H.M. Gelfand, D.R. LeBlanc, L. Potash, D.I. Clemmer, and D. Lapenta. 1960. The spread of vaccine strains of poliovirus in the household and in the community in southern Louisiana. In *Poliomyelitis: Papers and discussions presented at the Fifth International Poliomyelitis Conference, Copenhagen, Denmark, July 26-28, 1960* (edited for International Poliomyelitis Congress), p. 368. J.B. Lippincott, Philadelphia.

Fulginiti, V.A., C.H. Kempe, W.E. Hathaway, D.S. Pearlman, O.F. Sieber, J.J. Eller, J.J. Joyner, and A. Robinson. 1968. Progressive vaccinia in immunologically deficient individuals. *Immunologic deficiency diseases of man* (ed. D. Bergsma), vol. 4, p.129. The National Foundation, New York.

Gershon, A.A., S.P. Steinberg, L. Gelb, G. Galsson, S. Borkowsky, P. LaRussa, A. Ferrarri, et al. 1985. A multicenter trial of live attenuated varicella vaccine in children with leukemia in remission. *Post. Grad. Med. J.* (in press).

Holland, J., K. Spindler, F. Horodyski, E. Grabau, S. Nichol, and S. VandePol. 1982. Rapid evolution of RNA genomes. *Science* **215**: 1577.

Kapikian, A.Z. and R.M. Chanock. 1985. Rotaviruses. In *Virology* (ed. B.N. Fields), p. 863. Raven Press, New York.

Kapikian, A.Z., K. Midthun, Y. Hoshino, J. Flores, R.G. Wyatt, R.I. Glass, J. Askaa, O. Nakagomi, T. Nakagomi, R.M. Chanock, M.M. Levine, M.L. Clements, R. Dolin, P.F. Wright, R.B. Belshe, E.L. Anderson and L. Potash. 1985. Rhesus rotavirus: A candidate vaccine for prevention of human rotavirus disease. In *Vaccines 85: Molecular and chemical basis of resistance to parasitic, bacterial, and viral diseases* (ed. R.A. Lerner, R.M. Chanock, and F. Brown), p. 357. Cold Spring Harbor Laboratory, Cold Spring Harbor, New York.

Kawaska, Y., C.W. Naeve, and R.G. Webster. 1984. Is virulence of H5N2 influenza viruses in chickens associated with loss of carbohydrate from the hemagglutinin? *Virology* **139**: 303.

Kendal, A.P., N.J. Cox, S. Nakajima, K. Nakajima, L. Raymond, A. Caton, G. Brownlee, and R.G. Webster. 1984. Structures in influenza A/USSR/90/77 hemagglutinin associated with epidemiologic and antigenic changes. In *Modern approaches to vaccines: Molecular and chemical basis of virus virulence and immunogenicity* (ed. R.M. Chanock and R.A. Lerner), p. 151. Cold Spring Harbor Laboratory, Cold Spring Harbor, New York.

Kew, O.M. and B.K. Nottay. 1984. Evolution of the oral polio vaccine strains in humans occurs by both mutation and intramolecular recombination. In *Modern approaches to vaccines: Molecular and chemical basis of virus virulence and immunogenicity* (ed. R.M. Chanock and R.A. Lerner), p. 357. Cold Spring Harbor Laboratory, Cold Spring Harbor, New York.

Kew, O.M., B.K. Nottay, M.H. Hatch, J.H. Nakano, and J.F. Obijeski. 1981. Multiple genetic changes can occur in the oral poliovaccines upon replication in humans. *J. Gen. Virol.* **56**: 337.

Kono, R. 1975. Appollo 11 disease or acute hemorrhagic conjunctivitis: A pandemic of a new enterovirus infection of the eyes. *Am. J. Epidemiol.* **101**: 383.

Lane, J.M., F.L. Ruben, J.M. Neff, and J.D. Millar. 1969. Complications of smallpox vaccination. 1968. National surveillance in the United States. *New Engl. J. Med.* **281**: 1201.

Lane, J.M., F.L. Ruben, E. Abrutyn, and J.D. Millar. 1970. Deaths attributable to smallpox vaccination, 1959 to 1966, and 1968. *J. Am. Med. Assoc.* **212**: 441.

Murphy, B.R., M.L. Clements, H.F. Maassab, A.J. Buckler-White, S.-F. Tian, W.T. London, and R.M. Chanock. 1984a. The basis of attenuation of virulence of influenza virus for man. In *The molecular virology and epidemiology of influenza* (ed. Sir C. Stuart-Harris and C.W. Potter), p. 211. Academic Press, London.

Murphy, B.R., A.J. Buckler-White, S. Tian, R.M. Chanock, M.L. Clements, H.F. Maassab, and W.T. London. 1984b. Attenuation of wild-type influenza-A viruses for man by genetic reassortment with attenuated donor viruses. In *Modern approaches to vaccines: Molecular and chemical basis of virus virulence and immunogenicity* (ed. R.M. Chanock and R.A. Lerner), p. 329. Cold Spring Harbor Laboratory, Cold Spring Harbor, New York.

Neff, J.M., R.H. Levine, J.M. Lane, E.A. Ager, H. Moore, B.J. Rosenstein, J.D. Millar, and D.A. Henderson. 1967. Complications of smallpox vaccination, United States 1963: II. Results obtained by four statewide surveys. *Pediatrics* **39**: 916.

Nkowane, B. and S.G.F. Wassilak. 1984. Update on paralytic poliomyelitis. In *19th Immunization Conference Proceedings, Boston, Mass., May 1984*, p. 67. U.S. Dept. of Health and Human Services, Public Health Service, Centers for Disease Control, Atlanta, Georgia.

Nottay, B.K., O.M. Kew, M.H. Hatch, J.T. Heyward, and J.F. Obijeski. 1981. Molecular variation of type 1 vaccine-related and wild polioviruses during replication in humans. *Virology* **108**: 405.

Omata, T., M. Kohara, A. Abe, H. Itoh, T. Komatsu, M. Arita, B.L. Semler, E. Wimmer, S. Kuge, A. Kameda, and A. Nomoto. 1985. Construction of recombinant viruses between Mahoney and Sabin strains of type-1 poliovirus and their biological characteristics. In *Vaccines 85: Molecular and chemical basis of resistance to parasitic, bacterial, and viral diseases* (ed. R.A. Lerner, R.M. Chanock and F. Brown), p. 279. Cold Spring Harbor Laboratory, Cold Spring Harbor, New York.

Plotkin, S.A. 1984. Vaccination with live attenuated human cytomegalovirus. In *Modern approaches to vaccines: Molecular and chemical basis of virus virulence and immunogenicity* (ed. R.M. Chanock and R.A. Lerner), p. 307. Cold Spring Harbor Laboratory, Cold Spring Harbor, New York.

Portner, A., R.G. Webster, and W.J. Bean. 1980. Similar frequences of antigenic variants in Sendai, vesicular stomatitis, and influenza A viruses. *Virology* **104**: 235.

Rott, R. 1979. Molecular basis of infectivity and pathogenicity of myxovirus. *Arch. Virol.* **59**: 285.

Rott, R., C. Scholtissek, and H.-D. Klenk. 1984. Alterations in pathogenicity of influenza virus through reassortment. In *Modern approaches to vaccines: Molecular and chemical basis of virus virulence and immunogenicity* (ed. R.M. Chanock and R.A. Lerner), p. 345. Cold Spring Harbor Laboratory, Cold Spring Harbor, New York.

Sabin, A.B. 1956. Present status of attenuated live-virus poliomyelitis vaccine. *J. Am. Med. Assoc.* **162**: 1589.

———. 1957. Properties and behavior of orally administered attenuated poliovirus vaccine. *J. Am. Med. Assoc.* **164**: 1217.

Sabin, A.B., W.A. Hennessen, and J. Winsser. 1954. Studies on variants of poliomyelitis virus: I. Experimental segregation and properties of avirulent variants of three immunologic types. *J. Exp. Med.* **99**: 551.

Sabin, A.B., R.H. Michaels, I. Spigland, W. Pelon, J.S. Rhim, and R.E. Wehr. 1961. Community-wide use of oral poliovirus vaccine. *Am. J. Dis. Child.* **101**: 38.

Sabin, A.B., M. Ramos-Alvarez, J. Alvarez-Amezquita, W. Pelon, R.H. Michaels, I. Spigland, M.A. Koch, J.M. Barnes, and J.S. Rhim. 1960. Live, orally given poliovirus vaccine: Effects of rapid mass immunization on population under conditions of massive enteric infection with other viruses. *J. Am. Med. Assoc.* **173**: 1521.

Shope, R.E. 1944. Old, intermediate, and contemporary contributions to our knowledge of pandemic influenza. *Medicine* **23**: 415.

Smith, H.H. 1951. Controlling yellow fever. In *Yellow fever* (ed. G.K. Strode et al.), p. 539, McGraw-Hill, New York.

Spriggs, D.R., R.T. Bronson, and B.N. Fields. 1983. Hemagglutinin variants of reovirus type 3 have altered central nervous system tropism. *Science* **220**: 505.

Theiler, M. 1951. The virus. In *Yellow fever* (ed. G.K. Strode et al.), p. 39. McGraw-Hill, New York.

Vesikari, T., E. Isolauri, F.E. Andre, E. d'Hondt, A. Delem, G. Zissis, G. Beards, and T.H. Flewett. 1985. Protection of infants against human rotavirus diarrhea by the RIT 4237 live attenuated bovine rotavirus vaccine. In *Vaccines 85: Molecular and chemical basis of resistance to parasitic, bacterial, and viral diseases* (ed. R.A. Lerner, R.M. Chanock and F. Brown), p. 369. Cold Spring Harbor Laboratory, Cold Spring Harbor, New York.

Young, J.F. and P. Palese. 1979. Evolution of human influenza A viruses in nature: Recombination contributes to genetic variation of H1N1 strains. *Proc. Natl. Acad. Sci. U.S.A.* **76**: 6547.

Weibel, R.E., B.J. Neff, B.J. Kuter, H.A. Guess, C.A. Rothenberger, A.J. Fitzgerald, K.A. Connor, A.A. McLean, M.R. Hilleman, E.B. Buynak, and E.M. Scolnick. 1984. Live attenuated varicella virus vaccine: Efficacy trial in healthy children. *New Engl. J. Med.* **310**: 1409.

Whitman, L. 1951. The arthropod vectors of yellow fever. In *Yellow fever* (ed. G.K. Strode et al.), p. 229. McGraw-Hill, New York.

Wright, P.F., R.B. Belshe, H.W. Kim, L.P. Van Voris, and R.M. Chanock. 1982. Administration of a highly attenuated, live respiratory syncytial virus vaccine to adults and children. *Infect. Immun.* **37**: 397.

Zacek, K., E. Adam, V. Adamova, V. Burian, D. Rezacova, E. Skridlovska, N. Vaneckova, and V. Vonka. 1962. Mass oral (Sabin) poliomyelitis vaccination. *Br. Med. J.* **1**: 5285.

COMMENTS

KILBOURNE: Without wishing for a moment to beg the question on attenuation, there is another way of looking at these data, that is, in none of these viruses has an absolute barrier been established against replication in the target tissue. Even with respect to rubella, which is one of the safest of vaccines, there is an occasional transmission and even a very rare vertical transmission. So, I think we could shift the distribution curve of disease severity over markedly, and I think a lot of this depends on the circumstances of administration. Smallpox virus was given years ago as a so-called attenuated agent. It wasn't attenuated at all. It was probably the route of inoculation or delivery that spared some people from smallpox.

CHANOCK: You mean variolation?

KILBOURNE: Yes, variolation.

CHANOCK: With regard to rubella, the rubella vaccine virus does pass the placenta and infects the fetus. Rubella virus has been recovered from the products of conception of women who were inadvertently vaccinated and who later underwent surgical abortion. However, there have been a number of women who were inadvertently vaccinated during early pregnancy and who later gave birth to a normal infant. In no instance was congenital rubella identified.

KILBOURNE: You don't believe that report of cataracts?

CHANOCK: The most recent summary from CDC, published about 9 months ago, states that rubella vaccine administered during early pregnancy has not been associated with the congenital rubella syndrome.

KILBOURNE: Yes, but you're also talking about primary vaccination causing occasional arthralgia in women.

CHANOCK: The percentage of women who develop transient arthritis is 10–15%, but it is very mild.

KILBOURNE: My only point is there hasn't been any absolute prevention of cytopathic effect.

CHANOCK: What I said is that the signs and symptoms that develop following the infection with vaccine virus are considerably less severe than those that occur during infection with wild-type virus. In most instances there are no signs or symptoms. Symptoms are not associated with oral poliovirus vaccine except once in 10^7 vaccinations. With mumps vaccine there are no signs or symptoms. With measles and rubella vaccines there are mild symptoms.

MERIGAN: But the host does make a difference.

CHANOCK: The host makes a difference.

MERIGAN: You can shift it more toward virulence in certain hosts, but still they're not dangerous so far except in vaccinia.

CHANOCK: This is true except for vaccinia virus, which can produce progressive vaccinia in individuals with profound immunodeficiency. In this condition the virus continues to spread and ultimately kills the host. Also immunodeficiency plays a role in a small proportion of poliovirus vaccine-associated paralytic illness.

MERIGAN: Is this due to B cell immunodeficiency?

CHANOCK: B cell or combined immunodeficiency. These conditions account for 15% of poliovirus vaccine-associated disease.

Session 5: Viral Vectors

Use of Vaccina Virus Vectors for the Development of Live Vaccines

BERNARD MOSS
Laboratory of Viral Diseases
National Institute of Allergy and Infectious Diseases
National Institutes of Health
Bethesda, Maryland 20205

OVERVIEW

Vaccinia virus was originally used for immunoprophylaxis of smallpox but has now been developed into a vector for the expression of genes from other microorganisms. Genes, unrelated to those of vaccinia virus, may be placed under control of vaccinia transcriptional regulatory signals and stably integrated into the vaccinia virus genome. It is important to note that infectivity for tissue culture cells and animals is retained. Vaccinia expression vectors have proven particularly useful for laboratory investigations involving the synthesis and processing of expressed proteins and the analysis of antibody and cell-mediated immune responses. Animals vaccinated with some recombinant vaccinia viruses have been protected upon subsequent challenge with the virus from which the donor gene was obtained. Serious consideration, therefore, is being given to the use of recombinant vaccinia viruses as vaccines for medical and veterinary purposes. Information obtained during smallpox eradication campaigns should be useful for assessing the impact of the recombinant vaccines on the individual and the environment. The possibility that recombinant viruses possess unique properties, however, needs to be thoroughly investigated.

INTRODUCTION

Global implementation of Edward Jenner's recommendation regarding vaccination has led to the eradication of smallpox. With the elimination of this disease and the absence of known reservoirs of variola virus, there is no further need to continue smallpox vaccination. Several years ago, however, the feasibility of genetically engineering vaccinia virus so that it could express genes of other organisms was demonstrated (Mackett et al. 1982; Panicali and Paoletti 1982). Since then, genes from a variety of different viruses have been inserted into the vaccinia genome and in many cases animals have been protectively immunized against the virus that donated the gene. Although the technology for construction of efficient vaccinia virus vectors is still under development, it seems likely that vaccination for medical or veterinary purposes will soon be attempted. For this reason, attention needs to be given to the advantages and disadvantages of this approach and to its potential effects on the individual recipient and the environment.

RESULTS

Construction of Vaccinia Virus Vectors

Vaccinia virus, a member of the poxvirus family, is a large DNA virus that replicates in the cytoplasm of infected cells (Moss 1985). The genome consists of a linear duplex molecule of about 185,000 bp with ends that are covalently linked as hairpins. The presence in the infectious virus particle of a DNA-dependent RNA polymerase and enzymes that modify mRNA is a characteristic feature of poxviruses that contributes to their ability to replicate in the cytoplasm. Expression of vaccinia virus genes is transcriptionally regulated in a temporal manner. Recent studies indicate that the promoter signals regulating early and late gene expression are located just before the RNA start sites (Mackett et al. 1984; Weir and Moss 1984; Bertholet et al. 1985; Cochran et al. 1985). Moreover, the vaccinia transcription system specifically recognizes vaccinia promoters (Puckett and Moss 1983).

To use vaccinia virus as a vector, it is necessary to insert an isolated gene or cDNA into the vaccinia virus genome, arrange to have it expressed efficiently, and retain vaccinia virus infectivity. Our strategy (Mackett et al. 1984) has been to construct chimeric genes by ligating together a foreign protein coding sequence and a vaccinia transcriptional promoter. The next step, insertion of the chimeric gene into the vaccinia genome is technically difficult to carry out in vitro because of the large size of the DNA. Instead, we construct a plasmid in which the chimeric gene is flanked by DNA that surrounds the proposed site of insertion in the vaccinia genome. When cells are infected with vaccinia virus and transfected with such a plasmid, insertion occurs by homologous recombination. If the site of insertion is properly chosen, i.e., no essential gene function is interrupted, then infectivity will be retained. Since the proportion of recombinants is only about 1×10^{-3}, a procedure for their selection is desirable. By flanking the chimeric gene with segments of the vaccinia thymidine kinase (*tk*) gene, the *tk* locus will be the site of insertion and consequently all recombinants will have a TK^- phenotype. The latter includes the ability to form plaques on TK^- cells in the presence of 5-bromodeoxyuridine (BUdR). In this manner, recombinants can be selected from TK^+ parental virus. TK^- recombinants can be distinguished from the occasional (1×10^{-4}) spontaneous TK^- mutants by DNA hybridization or binding of specific antibody to plaques formed by the recombinant virus.

Virus from positive plaques is then repeatedly plaque-purified prior to preparation of virus stocks. Analysis of the recombinant virus DNA has invariably indicated correct insertion of the chimeric gene.

Expression of Cloned Genes

The time and level of expression of the chimeric gene depends on the vaccinia virus promoter used (Mackett et al. 1984; Weir and Moss 1984). Estimates of 10^8 polypeptide molecules synthesized per cell was made for hepatitis B virus surface antigen (HBsAg) by a vaccinia virus recombinant (Smith et al. 1983a,b). The screening of promoters for still higher expression is in progress. The proteins expressed by vaccinia virus recombinants appear to be synthesized and processed in a correct manner. For example, vaccinia recombinants containing the influenza hemagglutinin (HA) (Smith et al. 1983c), and the vesicular stomatitis virus (VSV) G protein (Mackett et al. 1985) genes synthesized polypeptides that comigrated with authentic products and were glycosylated and inserted into the plasma membrane. Similarly HBsAg made by recombinant vaccinia virus contains glycosylated and unglycosylated polypeptides of correct size that form characteristic 22 nm particles (Smith et al. 1983b).

Immunization of Experimental Animals

The production of specific antibody has been demonstrated by inoculating experimental animals with vaccinia virus recombinants that express HBsAg (Smith et al. 1983b; Paoletti et al. 1984); influenza virus HA (Panicali et al. 1983; Smith et al. 1983c), VSV G (Mackett et al. 1985), rabies G (Wiktor et al. 1984), and herpes simplex virus glycoprotein D (gD) (Cremer et al. 1985). In addition, priming of a cytotoxic T cell response could be shown with mice that were inoculated with vaccinia virus recombinants which express influenza virus type A HA (Bennink et al. 1984) and nucleoprotein (Yewdell et al. 1985). With many of the recombinants listed above, vaccination protected experimental animals against challenge (Table 1).

Altered Properties of Recombinant Vaccinia Virus

The possibility that expression of new genes, e.g., rabies G protein or HBsAg, might alter the tissue tropism or host range has been considered. Although this has not been critically analyzed, recombinants appeared to have reduced virulence as judged by intraperitoneal or intracranial inoculations of mice (Buller et al. 1985). This reduction in virus virulence has been attributed to the TK^- phenotype, which is a consequence of the site used for gene insertion. The local vaccination response in chimpanzees (Moss et al. 1984) and other primates (unpubl. data) also appeared milder with TK^- recombinants than with wild-type TK^+ virus. In addition, deletion of other portions of the vaccinia genomes have been associated with decreased virulence.

Table 1
Protective Immunization with Recombinant Vaccinia Virus

Gene[a]	Experimental animal	Inoculation site	Challenge Virus	Challenge Site	Protection[b]	Reference
Influ HA	Hamster	ID	Influ	LR	High	Smith et al. (1983c)
HSV-1 gD	Mouse	IP	HSV-1	IP	High	Paoletti et al. (1984)
		ID,SC,IP	HSV-1	IP	HIgh	Cremer et al. (1985)
		ID	HSV-2	IP	High	Cremer et al. (1985)
		ID	HSV-1	ID[c]	2/3	Cremer et al. (1985)
HBsAg	Chimp	ID	HBV	ID	High	Moss et al. (1984)
Rabies G	Mouse	ID	Rabies	IC	High	Wiktor et al. (1984)
VSV G	Mouse	ID	VSV	IV	High	Mackett et al. (1985)
	Cow	ID	VSV	ID[d]	2/3	Mackett et al. (1985)

(ID) Intradermal; (IP) intraperitoneal; (SC) subcutaneous; (IV) intravenous; (LR) lower respiratory
[a]Gene inserted into vaccinia virus
[b]High, greater than 95%
[c]HSV-1 was inoculated by abrasion of lip, and latent trigeminal ganglia infection was measured.
[d]VSV was injected intradermalingually and formation of vesicles in 48 hrs was measured.

DISCUSSION

In this section, we will consider possible advantages and disadvantages of recombinant vaccinia virus vaccines. On the positive side, experience with smallpox vaccine suggests that production costs (in animals' skin or tissue culture cells) will be relatively low, refrigeration will not be required, and administration with a bifurcated needle or jet injector will be simple. The large capacity of the vaccinia genome for additional DNA (Smith and Moss 1983) makes it technically feasible to construct recombinants that express multiple genes from one or more microorganisms.

It is important to consider potential harmful effects of recombinant vaccinia virus on the individual recipient, the possibility of virus dissemination, and the likelihood of spread of recombinant genetic material. The experience obtained during the extensive use of vaccinia virus as a smallpox vaccine can form the basis for such a discussion.

The formation of a vesicular skin lesion was, of course, a desired consequence of a primary smallpox vaccination and occurred in most recipients. In some cases, secondary lesions also formed, perhaps due to scratching. In about 10%, fever and myalgia accompanied a primary vaccination. Inoculation of immunodeficient individuals was contraindicated since it could lead to a progressive infection with a fatal outcome. Vaccination of infants with eczema, or individuals in contact with such infants, also was contraindicated because of the tendency of vaccinia virus to spread in the skin lesions. Because of its unpredictable occurrence and serious consequence, encephalitis was probably the side effect that was of most concern. Encephalitis occurred almost exclusively after primary infection and its frequency varied greatly in different parts of the world, apparently because of vaccine strain differences. The New York City Board of Health strain of virus was associated with one of the lowest frequencies of encephalitis with only three cases per 1,000,000 primary vaccinations (Lane et al. 1969).

Will a recombinant vaccinia virus have a higher or lower incidence of side-effects? Experiments with laboratory animals carried out thus far suggest that TK^- recombinants are less pathogenic than the parental virus. Deletion of other vaccinia virus genes could also reduce virulence. Thus, the prospects are for a safer vaccine. Nevertheless, further experiments need to be carried out to ensure that certain properties of vaccinia virus, such as tissue tropism, are not altered because of expression of genes of other microorganisms.

Dissemination of vaccinia virus occurs infrequently and requires close physical contact during the vesicular stage of infection. Third person spread is unusual and transmission cannot be sustained. No known animal reservoirs of vaccinia have developed despite its widespread use.

The third consideration, spread of genetic material, could theoretically occur by recombination with other members of the poxvirus family. Orthopoxviruses share considerable DNA sequence homology and can undergo recombination readily

(Mackett and Archard 1979). The members of the orthopoxvirus genus include vaccinia, cowpox, variola, ectromelia, and monkeypox. As previously mentioned, there appears to be no natural reservoir of vaccinia, and variola has been eradicated. Ectromelia is found in laboratory mice; however, its occurrence in wild mice has not been documented. Monkeypox infects primates, including man, infrequently in parts of Africa particularly Zaire; the usual host is unknown but some small mammal is suspected. Similarly, cowpox is suspected of having a small wild rodent as natural host (Baxby 1977).

DNA sequence homology has not been demonstrated between vaccinia virus and other poxviruses. Nevertheless, a suspected recombinant between Shope fibroma virus and vaccinia virus, which presumably arose in the laboratory, has been described (Berkowitz and Pogo 1985).

CONCLUSION

Current studies have indicated that immunization with vaccinia virus recombinants can protect experimental animals against challenge by a variety of different viruses. Moreover, the technology for using vaccinia virus as an expression vector is improving. It seems likely that in the near future, there will be requests for clinical testing of a vaccine intended for human use or field testing of one developed for veterinary purposes. Such testing should be preceded by laboratory tests to determine that the recombinant viruses are not more virulent, and preferably are less virulent, than strains approved for smallpox vaccination. In each case, the likely benefits versus the anticipated side effects must be carefully evaluated.

REFERENCES

Baxby, D. 1977. Is cowpox misnamed? A review of ten human cases. *Br. Med. J.* **1**: 1379.

Bennink, J.R., J.W. Yewdell, G.L. Smith, C. Moller, and B. Moss. 1984. Recombinant vaccinia virus primes and stimulates influenza virus HA-specific CTL. *Nature* **311**: 578.

Berkowitz, E.M. and B.G.-T. Pogo. 1985. Molecular characterization of two strains of Shope fibroma virus. *Virology* **142**: 437.

Bertholet, C., R. Drillien, and R. Wittek. 1985. One hundred base pairs of 5′ flanking sequence of a vaccinia virus late gene are sufficient to temporally regulate late transcription. *Proc. Natl. Acad. Sci. U.S.A.* **82**: 2096.

Buller, R.M.L., G.L. Smith, B. Moss, K. Cremer, and A.L. Notkins. 1985. Infectious vaccinia virus TK⁻ recombinants that express foreign genes are less virulent than wild-type virus in mice. In *Vaccines 85* (ed. R.A. Lerner, R.M. Chanock, and F. Brown), p. 163. Cold Spring Harbor Laboratory, Cold Spring Harbor, New York.

Cochran, M.A., C. Puckett, and B. Moss. 1985. *In vitro* mutagenesis of the promoter region for a vaccinia virus gene; evidence for tandem early and late regulatory signals. *J. Virol.* **54**: 30.

Cremer, K.J., M. Mackett, C. Wohlenberg, A.L. Notkins, and B. Moss. 1985. Vaccinia virus recombinant expressing herpes simplex type 1 glycoprotein D prevents latent herpes in mice. *Science* **228**: 737.

Lane, J.M., F.L. Ruben, J.M. Neff, and J.D. Millar. 1969. Complications of smallpox vaccination, 1968. National surveillance in the United States. *N. Engl. J. Med.* **281**: 1201.

Mackett, M. and L.C. Archard. 1979. Conservation and variation in the orthopoxvirus genome structure. *J. Gen. Virol.* **45**: 683.

Mackett, M., G.L. Smith, and B. Moss. 1982. Vaccinia virus: A selectable eukaryotic cloning and expression vector. *Proc. Natl. Acad. Sci. U.S.A.* **79**: 7415.

——— . 1984. A general method for the production and selection of infectious vaccinia virus recombinants expressing foreign genes. *J. Virol.* **49**: 857.

Mackett, M., T.Y. Yilma, J.A. Rose, and B. Moss. 1985. Vaccinia virus recombinants express vesicular stomatitis virus genes and protectively immunize mice and cattle. *Science* **227**: 433.

Moss, B., G.L. Smith, J.L. Gerin, and R. Purcell. 1984. Live recombinant vaccinia virus protects chimpanzees against hepatitis B. *Nature* **311**: 67.

Moss, B. 1985. Replication of poxviruses. In *Virology* (ed. B.N. Fields, D.M. Knipe, R.M. Chanock, et al.), p. 685. Raven Press, New York.

Panicali, D. and E. Paoletti. 1982. Construction of poxviruses as cloning vectors: Insertion of the thymidine kinase gene of herpes simplex virus into the DNA of infectious vaccinia virus. *Proc. Natl. Acad. Sci. U.S.A.* **79**: 4927.

Panicali, D., S.W. Davis, R.L. Weinberg, and E. Paoletti. 1983. Construction of live vaccines using genetically engineered poxviruses: Biological activity of recombinant vaccinia virus expressing influenza virus hemagglutinin. *Proc. Natl. Acad. Sci. U.S.A.* **80**: 5364.

Paoletti, E., B.R. Lipinskas, C. Samsonoff, S. Mercer, and D. Panicali. 1984. Construction of live vaccines using genetically engineered poxviruses: Biological activity of vaccine virus recombinants expressing the hepatitis B surface antigen and the herpes simplex virus glycoprotein D. *Proc. Natl. Acad. Sci. U.S.A.* **81**: 193.

Puckett, C. and B. Moss. 1983. Selective transcription of vaccinia virus genes in template dependent soluble extracts of infected cells. *Cell* **35**: 441.

Smith, G.L. and B. Moss. 1983. Infectious poxvirus vectors have capacity for at least 25,000 base pairs of foreign DNA. *Gene* **25**: 21.

Smith, G.L., M. Mackett, and B. Moss. 1983a. Expression of hepatitis B virus surface antigen by infectious vaccinia virus recombinants. In *UCLA symposia molecular cellular biology* (ed. M. Rosenberg and D. Hamer), vol. 8, p. 543. A. Liss, New York.

——— . 1983b. Infectious vaccinia virus recombinants that express hepatitis B virus surface antigen. *Nature* **302**: 490.

Smith, G.L., B.R. Murphy, and B. Moss. 1983c. Construction and characterization of an infectious vaccinia virus recombinant that expresses the influenza hemagglutinin gene and induces resistance to influenza virus infection in hamsters. *Proc. Natl. Acad. Sci. U.S.A.* **80**: 7155.

Weir, J.P. and B. Moss. 1984. Regulation of expression and nucleotide sequence of a late vaccinia virus gene. *J. Virol.* **51**: 662.

Wiktor, T.J., R.I. Mcfarlan, K.J. Reagan, B. Dietzschold, P.J. Curtis, N.H. Wunner, M.P. Kieny, R. Lathe, J.-P. Lecocq, M. Mackett, B. Moss, and H. Koprowski. 1984. Protection from rabies by a vaccinia virus recombinant containing the rabies virus glycoprotein gene. *Proc. Natl. Acad. Sci. U.S.A.* **81**: 7194.

Yewdell, J.W., J.R. Bennink, G.L. Smith, and B. Moss. 1985. Influenza A virus nucleoprotein is a major target antigen for cross-reactive anti-influenza A virus cytotoxic T lymphocytes. *Proc. Natl. Acad. Sci. U.S.A.* **82**: 1785.

COMMENTS

WATSON: How fast do we lose our immunity to vaccinia. I remember that we were revaccinated every 5 years or something in the old days.

MOSS: During the smallpox era, travelers were vaccinated every 3-5 years. A previously vaccinated individual had either an accelerated response, indicating partial immunity, or an immune response, indicating complete immunity. With time, it is possible to pass from an immune to a partially immune state.

WATSON: So it's only children who develop the immune response to your recombinant DNA-modified vaccines.

MOSS: Children or previously unvaccinated adults will have the best immune response. If a vaccine is sufficiently immunogenic, however, it might be effective despite partial immunity.

MILLER: Does a higher inoculum enable you to overcome the problems of a partially immune population? Have you tried that in your animal studies?

MOSS: A higher inoculum might help, but we haven't tried that.

SHOPE: Bernie [Moss], there are entomopox viruses, i.e., viruses of insects, in the pox virus family. Has anyone tried to adapt some of these techniques to those viruses? This might be of great interest to EPA, I would think.

MOSS: The entomopox viruses may be very useful because they infect grasshoppers whereas baculoviruses don't. I don't know any laboratory that is currently working with entomopox viruses as vectors.

SHOPE: Is there any reason why this technology shouldn't be applicable to those viruses?

MOSS: No. We don't know how closely they're related to vaccinia virus in terms of transcription signals. One might have to do considerable basic work in order to understand what is necessary to express genes in entomopox virus. One might also think of using fowlpox virus as a vector for chicken vaccines since it is currently used as a veterinary vaccine.

KAMELY: How easy is it to create these recombinant viruses?

MOSS: It takes several weeks.

KAMELY: What is the selective pressure? What is the probability that a hazardous virus could appear during the creation of a recombinant virus?

MOSS: We make these recombinant viruses by taking a foreign gene and flanking it with vaccinia DNA so that the desired recombination event occurs. It's a very specific process.

FIELDS: Bernie [Moss], can you say something about the natural hosts of vaccinia and how the virus is naturally spread, and then how the inserts or the mutation in the *tk* gene would affect the ability of the virus to spread in a natural situation?

MOSS: The precise origin of vaccinia virus is obscure and no natural host is known. During the smallpox vaccination era, spread was infrequent and required physical contact. Despite the extensive use of the vaccine, there is currently no known natural reservoir of vaccinia virus. Since TK^- recombinants grow more poorly than unmodified vaccinia virus, they should have an even lower probability of spreading.

FIELDS: So, in terms of the issues that have been raised here—release, survival—even if all of the above occurred, the potential harm is all markedly reduced in terms of environmental impact?

MOSS: I would think so. But there is one possibility that shouldn't be overlooked. If we put an envelope protein of another virus into vaccinia virus, it is theoretically possible to make a pseudotype with altered properties. We're investigating this possibility now. Thus far we haven't seen any evidence of increased pathogenicity.

AHMED: There have been reports that the majority of the cross-reactive cytotoxic T cells (CTL) recognize the nuclear protein of influenza virus. Have you tested the vaccinia recombinants containing influenza NP in protection experiments?

MOSS: Yes.

AHMED: Is the protection broader than with the vaccinia recombinants containing flu HA?

MOSS: Right now we're having difficulty showing good protection by vaccinating with the nucleoprotein recombinant. What we would expect, however, would be not protection but accelerated recovery from infection. That's what we're looking at. We do see protection with the hemagglutinin but that is sub-type specific.

KILBOURNE: I just wanted to point out that the CDC has recently reported that ordinary vaccinia has been spread among little girls at slumber parties. The virus was inadvertantly given to the daughter of a military person. The military is still immunized for protection against possible germ warfare. So, this girl was inadvertantly immunized and then spread it to susceptible little girls under conditions of intimate contact.

Bovine Papillomavirus Vectors

PETER M. HOWLEY
Laboratory of Tumor Virus Biology
National Cancer Institute
Bethesda, Maryland 20205

OVERVIEW

The bovine papillomavirus (BPV-1) is able to transform a variety of mammalian cells in vitro, and the DNA persists as an extrachromosomal multicopy plasmid in the transformed cells. This property was the basis of the development of BPV-1 DNA into a vector for introducing foreign genes into mammalian cells. BPV-1-based plasmids replicate on the average of once per cell cycle and the plasmids appear to be faithfully partitioned to daughter cells. No expression of late RNAs is detected in these transformed cells and no virus particles are produced. The use of BVP-1-based vectors does not involve the use of infectious virus particles and there is no production of infectious virus particles in the stably transformed cells. At the current stage of development of this vector system, therefore, the use of BPV-1-based vectors poses little, if any, threat to the environment.

INTRODUCTION

The papillomaviruses are a group of small DNA viruses that have remained refractory to standard virologic study primarily because they have never been successfully propagated in tissue culture. Papillomaviruses are widespread in nature infecting a variety of higher vertebrates, including man. The proliferative lesions induced by most papillomaviruses are limited to the squamous epithelium, although some papillomaviruses are able to induce fibropapillomas containing both proliferative fibroblastic and squamous epithelial components. There is a subgroup of papillomaviruses, however, that are capable of transforming certain rodent cells in vitro. Transformation by these viruses has provided a biologic system for studying the latent, nonproductive infection of cells by papillomaviruses. The most extensively studied transforming papillomavirus is the bovine papillomavirus type 1 (BPV-1) which has served as the prototype for unraveling the molecular biology and genetics of this group of viruses.

Bovine papillomavirus type 1 causes benign fibropapillomas in cattle and also induces fibrosarcomas in heterologous hosts. The virus or its molecularly cloned DNA can transform certain mouse cells in vitro and the DNA persists as an extrachromosomal multicopy plasmid in these transformed cells (Howley et al. 1980; Law et al. 1981). A specific 69% subgenomic fragment (BPV_{69T}) bounded by the unique *Hin*dIII and *Bam*HI sites is sufficient to induce transformation in vitro (Lowy et al.

1980), and recombinants containing this DNA segment can be maintained as free plasmids in transformed cells (Law et al. 1981). Cell lines established from these transformed foci have a fully transformed phenotype in that they are anchorage independent and are tumorigenic in nude mice (Sarver et al. 1984). Furthermore, BPV-1 transformed cells do not produce virus particles, detectable levels of late messenger RNA, or late capsid proteins (Howley 1983).

The genomic organization of BPV-1 has been established from the DNA sequence (Chen et al. 1982) and from the transcription data available for BPV-1 transformed cells (Heilman et al. 1982) as well as for productively infected bovine papillomas (Amtmann and Sauer 1982; Engel et al. 1983). All of the RNAs expressed in transformed cells are transcribed from a single strand and map to the 69% subgenomic fragment sufficient for inducing transformation (Heilman et al. 1982). This region of the genome contains eight open reading frames (ORFs) located on the same strand (the coding strand) (Chen et al. 1982) as well as important regulatory sequences. A schematic representation of the genomic organization of BPV-1 is depicted in Figure 1. The 69% segment sufficient for transformation and plasmid maintenance is indicated by the heavy dark line. This region contains a 1000 base noncoding region (ncr) 5′ to the 8 ORFs that are actively transcribed in productively infected cells. This noncoding region contains important regulatory sequences: a transcriptional control element (Sarver et al. 1984); a DNAase I hypersensitivity site (Rosl et al. 1983); the initiation site for plasmid DNA replication in rodent cells (Waldeck et al. 1984); a plasmid maintenance sequence (Lusky and Botchan 1984); and an enhancer that is *trans*-activated by a specific BPV-1 early gene product (Spalholz et al. 1985).

BPV-1 transformed cells contain multiple, different, viral-specific transcripts varying in size from approximately 1000–4000 bases (Heilman et al. 1982). Because of the low abundance of these transcripts, precise mapping of these species had been difficult. Recent sequence analysis of full length cDNA copies of the RNAs in transformed cells, however, (Yang et al. 1985), and electron microscopic analysis of these RNAs (Stenlund et al. 1985) have revealed that multiple species of the RNAs are generated by differential splicing. Apparently, two different viral promoters are functional in transformed cells since the different mRNA species can be grouped into two sets of species based on their 5′ ends. One set has 5′ ends at approximately base 89, just downstream from a TATAAA sequence and base 58. Another set of RNAs has a 5′ end mapping to approximately base 2440. All of the viral RNAs detected in transformed cells appear to be polyadenylated at the same site (base 4203), immediately downstream from a polyadenylation recognition sequence AATAAA (Yang et al. 1985).

Functional studies from a number of laboratories have now identified regions of the BPV-1 genome that influence the expression of viral transforming functions and viral plasmid maintenance functions. The E6 ORF appears to encode a protein that is by itself sufficient for inducing transformation in mouse cells (Schiller et al.

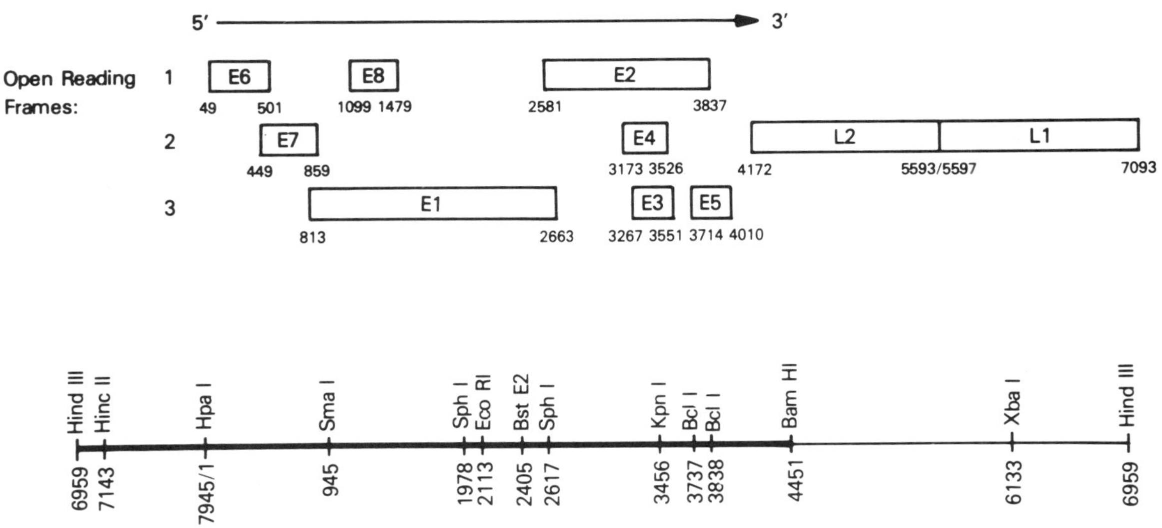

Figure 1
Genomic organization of BPV-1 DNA. The full length molecule (7945 bp) of BPV-1 opened at the unique *Hin*dIII site (base 6959) is marked off with restriction sites in bases noted at the bottom of the figure. The transforming segment from *Hin*dIII to *Bam*HI site is indicated by the heavy bar. The region transcribed in transformed cells and the direction of transcription are indicated by the arrow at the top of the figure (Heilman et al. 1982). Open bars represent potential coding regions for the BPV-1 proteins in each of the ORFs (Chen et al. 1982). ORFs within the transforming region have been designated E1-E8. The L1 and L2 ORFs are expressed only in productively infected cells of a bovine fibropapilloma (Engel et al. 1983). Numbers beneath the ORFs represent the first and last bases of the ORF. Reprinted, with permission, from Sarver et al. (1984).

1984; Yang et al. 1985). The integrity of this region is required for the full transformed phenotype as measured by anchorage independence and tumorigenicity of BPV-1 transformants (Sarver et al. 1984). An independent transforming function can also be mapped to the 3′ ORFs (Nakabayashi et al. 1983; Sarver et al. 1984; Yang et al. 1985). Specific deletion mutagenesis within this region indicates the importance of the E5 ORF for this transformation (D. DiMaio, unpubl. data; D. Groff and W. Lancaster, unpubl. data; M.S. Rabson and P. Howley, unpubl. data; and J. Schiller and D. Lowy, unpubl. data). The E1 gene has been implicated in DNA replication and plasmid maintenance function for BPV-1 base plasmids in transformed cells. Mutations mapping in the E1 ORF have a minimal effect on transformation efficiency but result in the integration of the viral DNA into the host chromosome (Sarver et al. 1984; Lusky and Botchan 1985). Furthermore, the E1 gene product provided in *trans* allows BPV-1 plasmids with mutations in E1 to remain extrachromosomal (Lusky and Botchan 1985). cDNA analysis has indicated that the E6 and E7 ORFs can be spliced together in frame at the level of RNA (Yang et al. 1985), and genetic studies suggest that the gene product of these fused exons may be involved in the copy number control of BPV-1 plasmids in transformed cells (Lusky and Botchan 1985).

Previous genetic studies have implicated the E2 gene product as having a role in transformation and in plasmid maintenance (Sarver et al. 1984). In studies using the authentic viral promoter elements to direct transcription of the transforming region, deletions affecting the E2 ORF and premature termination codons located in the E2 ORF dramatically decrease the efficiency of focus formation in C127 cells (Sarver et al. 1984; Lusky and Botchan 1985; D. DiMaio, pers. comm.). In addition, the transformants induced by these E2 mutants contain integrated viral DNA (Sarver et al. 1984; D. DiMaio, pers. comm.; M.S. Rabson and P. Howley, unpubl. data) indicating a role for the E2 gene product in plasmid maintenance. Recent studies indicate that the E2 gene products are required in *trans* for the activation of an ncr enhancer element (Spalholz et al. 1985). Thus it seems likely that E2 may have an indirect effect on the viral functions of plasmid maintenance and transformation through the activation of this ncr enhancer element. A direct role for E2 in transformation or in plasmid maintenance, however, cannot be ruled out at this time.

RESULTS AND DISCUSSION

Plasmid Shuttle Vectors

The stable introduction of foreign DNA into mammalian cells has proved to be a critical technique for the study of the function and of the regulated expression of specific genes. The development of a stable extrachromosomal nuclear plasmid vector affords several advantages. First, the plasmid DNA can be readily separated from host chromosomal DNA thus facilitating rescue of the DNA in bacterial cells if the plasmid contains a bacterial origin of DNA replication and a bacterial selective marker. Second, based on the physical property of being separated from the host chromosome, the minichromosomes with their associated proteins can potentially be separated from the host chromosomes by physical means. Third, a plasmid vector system offers the possibility of uniform expression for foreign genes introduced into cells as part of stable plasmids since each plasmid will be maintained in a uniform sequence and chromatin environment. Finally, the potential problem of integration of transferred DNA into inactive or improperly regulated portions of the host chromosome can be eliminated.

The initial development of BPV-1 as a plasmid vector in mammalian cells utilized the 69% transforming segment of the genome to introduce a 1.6-kbp rat preproinsulin gene into mouse C127 cells (Sarver et al. 1981). This hybrid DNA was maintained stably as an extrachromosomal plasmid in mouse cells where it was transcriptionally active. Mouse cells selected by virtue of their transformed phenotype produced and secreted rat proinsulin (Sarver et al. 1981). Because of the *cis*-inhibition imparted by pBR322 sequences on BPV-1 DNA mediated transformation (Sarver et al. 1982), further vectors had to be developed that could transform mouse cells efficiently in the presence of prokaryotic sequences and replicate stably both in mammalian cells as well as *E. coli*. One such shuttle vector consists of the full BPV-1 genome cloned into a deletion derivative of pBR322 (deleted of specific pBR322 sequences inhibitory to SV40 DNA replication in monkey cells) (Lusky and Botchan 1981); it can serve as a dual host replicon and exist as a stable plasmid in either bacterial or mouse cells (Sarver et al. 1982). A variety of other shuttle vectors have also been described (DiMaio et al. 1982; Kushner et al. 1982; Karin et al. 1983). Each of these shuttle vectors consists of the 69% transforming fragment of BPV-1, a prokaryotic DNA replication origin and selective marker, and a DNA segment, such as the human beta-globin gene or the 31% late region of BPV-1, which by some yet undefined mechanism can stabilize these plasmids in mammalian cells and can enhance their ability to transform susceptible mammalian cells.

Studies from a number of laboratories have indicated that certain genes can be properly regulated when maintained in BPV-1-based vectors. The human β-interferon gene that has been introduced into mouse cells with BPV-1 vectors can be induced by viral infection, polyriboinosinic acid-polyribocytidylic acid, and by cyclohexamide treatment (Zinn et al. 1982; Maroteaux et al. 1983; Mitrani-Rosenbaum et al. 1983). The mouse and human metallomethionine genes introduced with BPV-1 vectors can be induced by heavy metals but not by corticosteroids when maintained extrachromosomally (Karin et al. 1983; Pavlakis and Hamer 1983). Finally, the mouse mammary tumor virus (MMTV) LTR, which contains a steroid inducible promoter, is properly regulated when introduced on an extrachromosomal BPV-1-based vector (Ostrowski et al. 1983).

BPV-1 Functions Effect the Expression of Linked Genes

Recent studies have indicated that BPV-1 functions can have a significant effect on the expression of foreign gene segments. The BPV-1 genome is not an inert vector but contains elements which in *cis* can affect the expression of a linked gene. Also, BPV-1 contains a gene or genes that may, apparently, in *trans* also effect regulated gene expression. The effect of gene position on the expression of the rat preproinsulin gene has recently been examined (N. Sarver, in prep). Using the prototype BPV-1 vector developed in our laboratory, containing the full BPV-1 genome cloned into pML2 at its unique *Bam*HI site (Sarver et al. 1982), the rat preproinsulin gene was inserted at either of the two BPV-1 pML2 junction. This molecule is shown in Figure 2 and the two possible cloning sites are depicted as *A* and *B*. In

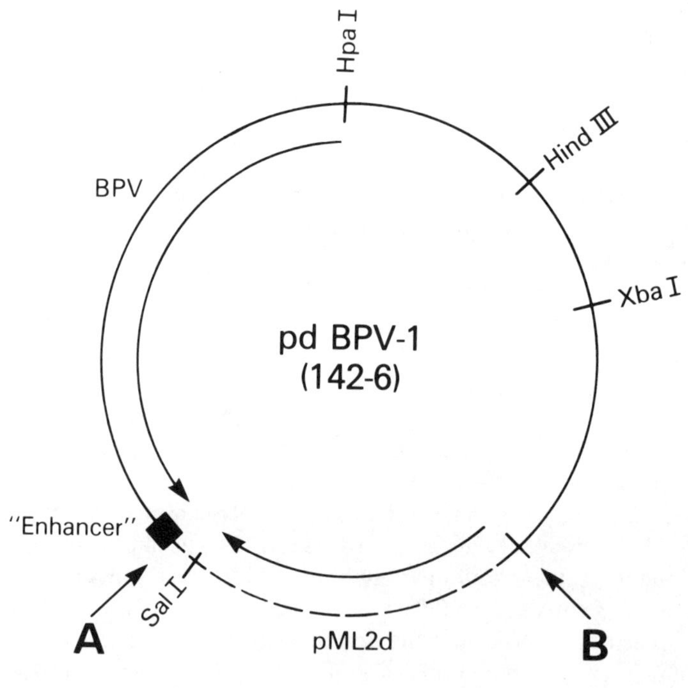

Figure 2
Physical map of the pML2/BPV-1 hybrid plasmid shuttle vector p142-6 (Sarver et al. 1982). The transcriptional direction of the BPV-1 genome is counterclockwise on this map (Heilman et al. 1982) and the location of the BPV-1 distal enhancer is indicated (Lusky et al. 1983). Transcription of the pML2 β-lactamase gene proceeds clockwise on this diagram. The positions of insertions of the rat preproinsulin gene either adjacent to the BPV-1 distal enhancer, or in a position blocked from its effect, are indicated as *A* and *B*, respectively. Reprinted, with permission, from Howley (1985).

position *A*, the insert is placed immediately adjacent to a BPV-1 enhancer element located at the 3′ end of the transforming region that has been shown by Lusky et al. (1983) to contain a sequence which can function as a transcriptional enhancer in a position- and orientation-independent manner. In cell lines transformed by BPV-1/rI_1/pML2 hybrids with the rat preproinsulin gene adjacent to the viral distal enhancer, the rat preproinsulin gene was expressed. Lines containing the BPV-1/rI_1/pML2 hybrid DNA with the rat preproinsulin gene at position B, however, did not express the rat preproinsulin gene. Insertion of a defined enhancer fragment from either BPV-1, SV40, or the Moloney sarcoma virus (MSV), adjacent to the rat preproinsulin gene in position *B* resulted in the activation and expression of the rat preproinsulin gene. Thus, the expression of a gene devoid of a transcriptional enhancer or containing a tissue-specific enhancer, requires positioning of the gene in a nonblocked position with respect to the BPV-1 distal enhancer for expression. Sequences within the BPV-1 genome (presumably the BPV-1 promoters) and sequences within pBR322 or pML2 can effectively block this enhancer effect.

BPV-1 also apparently encodes gene products that can act on foreign promoter elements to affect their expression. This was first shown by Spalholz et al. (1985) who demonstrated that the BPV-1 E2 gene product can *trans*-activate a BPV-1 transcriptional regulatory element located in the ncr. In these studies it was also noted that plasmids containing the intact SV40 early promoter are about five times more active in BPV-1-transformed cells, suggesting that BPV-1 encodes a function that can *trans*-activate other promoters. Furthermore, plasmids bearing the mouse metallomethionine promoter directing the expression of the neomycin resistance gene from Tn 5 are able to induce drug-resistant colonies at a much higher frequency in BPV-1 transformed cells, again suggesting that BPV-1 encodes a function that can act in *trans* to activate the metallothionein promoter (Sarver et al. 1984). Thus, BPV-1 cannot be thought of as an inert plasmid for the insertion of foreign genes for studying their regulated gene expression. Investigators must be cognizant of the BPV-1 viral functions and their effects on inserted genes for expression.

Expansion of Host Range for BPV-1 Vectors

A number of problems still exist regarding papillomavirus vectors. To a large extent these problems result from the paucity of our knowledge concerning the biology and molecular biology of this virus group. The dependence on transformation as a selective phenotype presents a problem to investigators interested in studying the expression of other transforming genes in BPV-1 vectors, as well as to investigators interested in studying the expression of specialized genes whose differentiated functions could potentially be affected by BPV-1 transforming functions. Other genes providing other selective markers have been introduced into BPV-1 vectors to provide an alternative selection. The amino glycoside phosphotransferase gene from Tn 5 has been expressed in a number of different mammalian cell transcriptional cassettes and has provided an alternative selection for several laboratories (Law et al. 1983; Matthias et al. 1983; Lusky and Botchan 1984). The use of herpes thymidine kinase (*tk*) gene (Lusky et al. 1983) or the *E. coli* xanthine-quanine phosphoribosyltransferase gene for selection has thus far been less successful (Law et al. 1982). By using the neomycin resistance gene and the aminoglycoside G418, it should now be possible to assess whether or not the range of cells capable of supporting BPV-1 plasmid replication can be expanded beyond those capable of manifesting the neoplastic transformed phenotype. Initial studies using mouse lymphocytes, mouse keratinocytes, and mouse hepatocytes have not been promising. It may be that in order to establish BPV-1-based vectors in these specialized cells, one may need to introduce tissue-specific, transcriptional regulatory elements, such as enhancers, which are capable of functioning in such specialized cells.

As we begin to understand the viral functions involved in viral transformation and viral plasmid maintenance, it may be possible to design plasmids that are capable of expressing the plasmid replication and maintenance functions but that are impaired in expressing the transformation functions. To this end, Lusky and Botchan have described a replication-competent deletion derivative of BPV-1 which does not readily transform mouse cells (Lusky and Botchan 1984), although cells harboring these plasmids may have altered growth properties. An alternative approach to developing nontransforming papillomavirus-based vectors may be to use papillomaviruses, such as the human papillomaviruses, which do not induce cellular transformation in tissue culture. Whether the genomes of these nontransforming papillomaviruses are able to establish themselves as plasmids that can faithfully partition at mitosis, however, is not yet established.

SUMMARY

The bovine papillomavirus remains extrachromosomal as a multicopy plasmid in transformed rodent cells. This property has been used to develop BPV-1-based vectors. These vectors have been useful in introducing a variety of foreign genes into mammalian cells for expression and for proper posttranslational modifications. They have also been useful in studying the regulated expression of inducible genes, including the interferon genes and the metallothionein genes. This vector system does not use virus particles for efficient infection of cells and the transformed cells do not produce infectious virus particles. It is, therefore, unlikely that this vector system or the transformed cells generated by transfection with these vectors pose a significant hazard to the environment.

ACKNOWLEDGMENT

I am grateful to Nan Freas for the preparation of this manuscript.

REFERENCES

Amtmann, E. and G. Sauer. 1982. Bovine papillomavirus transcription: Polyadenylated RNA species and assessment of the direction of transcription. *J. Virol.* **43**: 59.

Chen, E.Y., P.M. Howley, A.D. Levinson, and P.H. Seeburg. 1982. The primary structure and genetic organization of the bovine papillomavirus (BPV) type 1. *Nature* **299**: 529.

DiMaio, D., R. Treisman, and T. Maniatis. 1982. A bovine papillomavirus vector which propagates as an episome in both mouse and bacterial cells. *Proc. Natl. Acad. Sci. U.S.A.* **79**: 4030.

Engel, L.W., C.A. Heilman, and P.M. Howley. 1983. Transcriptional organization of the bovine papillomavirus type 1. *J. Virol.* **47**: 516.

Heilman, C.A., L. Engel, D.R. Lowy, and P.M. Howley. 1982. Virus specific transcription in bovine papillomavirus transformed mouse cells. *Virology* **119**: 22.

Howley, P.M. 1983. The molecular biology of papillomavirus transformation. *Am. J. Pathol.* **113**: 414.

——— . 1985. Papillomavirus vectors. In *Microbiology–1985* (ed. L. Lieve), p. 224. American Society for Microbiology, Washington, D.C.

Howley, P.M., M.-F. Law, C.A. Heilman, L.W. Engel, M.C. Alonso, W.D. Lancaster, M.A. Israel, and D.R. Lowy. 1980. Molecular characterization of papillomavirus genomes. In *Viruses in naturally occurring cancers* (ed. M. Essex, G. Todaro, and H. zur Hausen), p. 233. Cold Spring Harbor Laboratory, Cold Spring Harbor, New York.

Karin, M., G. Cathala, and M.C. Nguyen-Huu. 1983. Expression and regulation of a mouse metallothionein gene carried on an autonomously replicating shuttle vector. *Proc. Natl. Acad. Sci. U.S.A.* **80**: 4040.

Kushner, P.J., B.B. Levinson, and H.M. Goodman. 1982. A plasmid that replicates in both mouse and *E. coli* cells. *J. Mol. Appl. Genet.* **1**: 527.

Law, M.-F., J.C. Byrne, and P.M. Howley. 1983. A stable bovine papillomavirus hybrid plasmid that expresses a dominant selective trait. *Mol. Cell. Biol.* **3**: 2110.

Law, M.-F., D.R. Lowy, I. Dvoretzky, and P.M. Howley. 1981. Mouse cells transformed by bovine papillomavirus contain only extrachromosomal viral DNA sequences. *Proc. Natl. Acad. Sci. U.S.A.* **78**: 2727.

Law, M.-F., B. Howard, N. Sarver, and P.M. Howley. 1982. Expression of selective traits in mouse cells transformed with a BPV DNA derived hybrid molecule containing *E. coli* gpt. In *Viral vectors* (ed. Y. Gluzman), p. 79. Cold Spring Harbor Laboratory, Cold Spring Harbor, New York.

Lowy, D.R., I. Dvoretzky, R. Shober, M.-F. Law, L. Engel, and P.M. Howley. 1980. *In vitro* tumorigenic transformation by a defined sub-genomic fragment of bovine papilloma virus DNA. *Nature* **287**: 72.

Lusky, M. and M. Botchan. 1981. Inhibition of SV40 replication in simian cells by specific pBR322 DNA sequences. *Nature* **293**: 79.

——. 1984. Characterization of the bovine papilloma virus plasmid maintenance sequences. *Cell* **36**: 391.

——. 1985. Genetic analysis of the bovine papillomavirus type 1 transacting replication factors. *J. Virol.* **53**: 955.

Lusky, M., L. Berg, H. Weiher, and M. Botchan. 1983. The bovine papillomavirus contains an activator of gene expression at the distal end of the transcriptional unit. *Mol. Cell. Biol.* **3**: 1108.

Maroteaux, L., L. Chen, S. Mitrani-Rosenbaum, P.M. Howley, and M. Revel. 1983. Cyclohexamide induces expression of the human interferon beta-$_1$ gene in mouse cells transformed by bovine papillomavirus-interferon beta-$_1$ recombinants. *J. Virol.* **47**: 89.

Matthias, D.D., H.U. Bernard, A. Scott, T. Hashimoto-Gotoh, and G. Schutz. 1983. A bovine papillomavirus vector with a dominant resistance marker replicates extrachromosomally in mouse and *E. coli* cells. *Eur. Mol. Biol. Organ. J.* **2**: 1487.

Mitrani-Rosenbaum, S., L. Maroteaux, Y. Mory, M. Revel, and P.M. Howley. 1983. Inducible expression of the human interferon beta-$_1$ gene linked to a bovine papillomavirus DNA vector and maintained extrachromosomally in mouse cells. *Mol. Cell. Biol.* **3**: 233.

Nakabayashi, Y., S.K. Chattopadhyah, and D.R. Lowy. 1983. The transforming function of bovine papillomavirus DNA. *Proc. Nat. Acad. Sci. U.S.A.* **80**: 5832.

Ostrowski, M.C., H. Richard-Foy, R.G. Wolford, D.S. Berard, and G.L. Hager. 1983. Glucocorticoid regulation of transcription at an amplified episomal promoter. *Mol. Cell. Biol.* **3**: 2048.

Pavlakis, G.N. and D.H. Hamer. 1983. Regulation of a metallothionein-growth hormone hybrid gene in bovine papilloma virus. *Proc. Natl. Acad. Sci. U.S.A.* **80**: 397.

Rosl, F., W. Waldeck, and G. Sauer. 1983. Isolation of episomal bovine papillomavirus chromatin and identification of a DNase I-hypersensitive region. *J. Virol.* **46**: 567.

Sarver, N., J.C. Byrne, and P.M. Howley. 1982. Transformation and replication in mouse cells of a bovine papillomavirus/pML2 plasmid vector that can be rescued in bacteria. *Proc. Natl. Acad. Sci. U.S.A.* **79**: 7147.

Sarver, N., P. Gruss, M.F. Law, G. Khoury, and P.M. Howley. 1981. Bovine papilloma virus DNA–A novel eukaryotic cloning vector. *Mol. Cell. Biol.* **1**: 486.

Sarver, N., M.S. Rabson, Y.-C. Yang, J.C. Byrne, and P.M. Howley. 1984. Localization and analysis of bovine papillomavirus type 1 transforming functions. *J. Virol.* **52**: 377.

Schiller, J.T., W.C. Vass, and D.R. Lowy. 1984. Identification of a second transforming region in bovine papillomavirus DNA. *Proc. Natl. Acad. Sci. U.S.A.* **81**: 7880.

Spalholz, B.A., Y.-C. Yang, and P.M. Howley. 1985. Transactivation of a bovine papillomavirus transcriptional regulatory element by the E2 gene product. *Cell* **42**: 183.

Stenlund, A., J. Zabielski, H. Ahola, J. Moreno-Lopez, and U. Pettersson. 1985. The messenger RNAs from the transforming region of bovine papilloma virus type I. *J. Mol. Biol.* **182**: 541.

Waldeck, W., F. Rosl, and H. Zentgraf. 1984. Origin of replication in episomal bovine papilloma virus type 1 DNA isolated from transformed cells. *Eur. Mol. Biol. Organ. J.* **3**: 2173.

Yang, Y.-C., H. Okayama, and P.M. Howley. 1985. Bovine papillomavirus contains multiple transforming genes. *Proc. Natl. Acad. Sci. U.S.A.* **82**: 1030.

Zinn, K., P. Mellon, M. Ptashne, and T. Maniatis. 1982. Regulated expression of an extra chromosomal human beta-interferon gene in mouse cells. *Proc. Natl. Acad. Sci. U.S.A.* **79**: 4897.

The Use of Adenovirus Recombinants to Study Viral Gene Expression

JOHN LOGAN* AND STEPHEN PILDER
Department of Molecular Biology
Princeton University
Princeton, New Jersey 08544

OVERVIEW

Human adenoviruses that contain specific alterations at defined sites can be constructed. These viruses can be used to ellucidate the function of adenovirus gene products. This technology can also be used to produce adenoviruses that contain and express foreign genes. We will describe the construction and analysis of an adenovirus variant, *dl*338, which contains a mutation in the E1b-55K coding region. The protein appears to function as a posttranscriptional regulator of adenovirus late mRNA accumulation. We will also describe the construction of an adenovirus expression vector.

INTRODUCTION

Adenoviruses have been isolated from a wide variety of hosts. Human adenoviruses, especially type 2 and type 5, are the most extensively studied at the molecular level (Tooze 1981; Ginsberg 1984). They contain a linear double-stranded DNA genome of approximately 36,000 base pairs. The lytic cycle can be divided into an early and a late phase which is separated by the onset of viral DNA replication. At early times postinfection a limited amount of the viral chromosome is expressed as mRNA. These are known as the early regions E1a, E1b, E2a, E2b, E3, and E4. Viral DNA replication commences at between 8 to 10 hours postinfection and is followed by a dramatic change in the expression of the viral chromosome. The major late promoter is activated and this accounts for the majority of mRNA produced in the infected cells at late times postinfection. Host cell protein synthesis is switched off, resulting in almost exclusive production of viral polypeptides. By 24 to 48 hours postinfection, virions are assembled, the cells lyse, and the infection cycle is completed.

The E1b region is located between 4.5 and 11 map units on the adenovirus chromosome. It encodes three polypeptides of molecular weight 17K, 21K, and 55K (Fig. 1) at early times postinfection. The 21K and 55K polypeptides are syn-

Present address: American Cyanimid, P.O. Box 400, Princeton, New Jersey 08540.

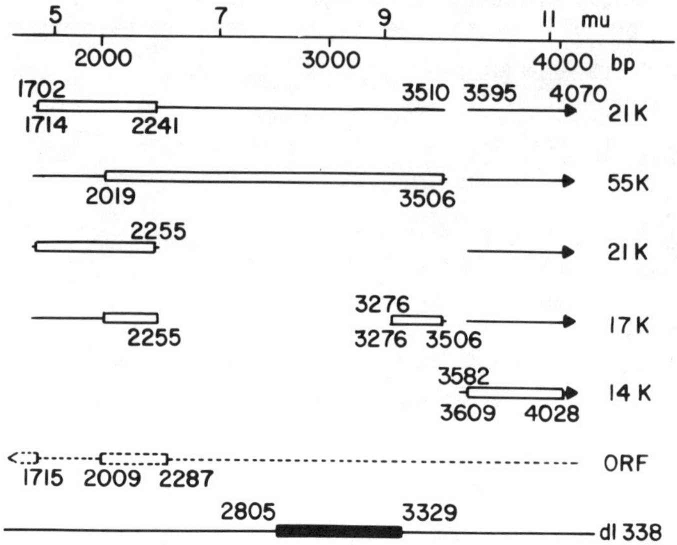

Figure 1
Physical map of E1b mRNAs, coding regions and the *dl*338 mutation. The top of the figure positions the map in terms of map units and nucleotide sequence position relative to the left end of the viral chromosome (Bos et al. 1981). mRNAs are designated by lines, introns by spaces, and coding regions by open rectangles. Open reading frames (ORFs) encoded by the opposite DNA strand are designated by open rectangles outlined with a broken line. The segment deleted in *dl*338 is represented by a solid rectangle. The nucleotide sequence numbers of the base pairs present bracketing the deletion are indicated.

thesized from different reading frames whereas the 17K polypeptide is produced from a mRNA species which splices out the internal portion of the 55K open reading frame and therefore has the same amino and carboxy terminus as the 55K polypeptide (Bos et al. 1981; Anderson et al. 1984). Here we describe a mutation in the E1b region which specifically removes the 55 and 17K polypeptides but leaves the 21K species intact. The mutation is constructed in a plasmid and then rebuilt into virus using overlap recombination. (Chinnadurai et al. 1979). The phenotype of the virus reveals that the function of the E1b-55K open reading frame is to regulate posttranscriptionally the accumulation of adenovirus late mRNAs.

A similar approach in the construction of adenovirus recombinants can be used to generate viruses that express foreign genes. In this paper we will describe a plasmid in which the E1a and E1b regions are replaced with a cassette that includes the major late promoter, tripartite leader sequence, gene of interest, splice and polyadenylation signals. This cassette is then reconstructed back into virus by overlap recombination. Therefore, the construction of recombinant adenoviruses can be used to study functions involved in the regulation of viral gene expression and also to express foreign genes at high levels in human cells.

RESULTS

Role of the E1b-55K Polypeptide during Lytic Growth

To study the function of the E1b-55K polypeptide a deletion mutant, *dl*338, was constructed that removed the sequences between nucleotides 2805 to 3329 (Fig. 1). This disrupts the 55K coding region and also removes the 3′ splice site for the newly identified 17K coding region. The 17K polypeptide is completely contained within the 55K coding region. The phenotype can be attributed to the 55K open reading frame but cannot be specifically localized to the 55K polypeptide; however, other mutations that only affect the 55K polypeptide have recently been described (Babiss and Ginsberg 1984). These mutations have similar phenotypes to *dl*338, and therefore it is likely that the 17K polypeptide has no discernible independent function. This alteration does not affect the 21K polypeptide or the polypeptide, protein IX (Fig. 1). The mutated plasmid sequences were rebuit into intact virus by overlap recombination (Chinnadurai et al. 1979; Ho et al. 1982) and were propagated on 293 cells, which are a human embryonic kidney cell line containing and expressing the adenovirus E1a and E1b regions (Graham et al. 1977). This cell line is therefore capable of complementing any defect caused by the lack of the E1b-55K polypeptide. When *dl*338 is assayed for growth in Hela cells, the virus grows 50–100-fold more slowly than wild-type. Analysis of the defect reveals that *dl*338 expresses its early regions normally and replicates its DNA normally. However, when expression of late proteins are analyzed by ^{35}S-methionine labeling of Hela cells at various times postinfection (Fig. 2), *dl*338 is defective for expression of its late polypeptides. All polypeptides from the late transcription unit are reduced between three to tenfold. Analysis of mRNA levels reveals a comparable reduction. Transcription rates are normal, which suggest a defect in either transport or stability of the late mRNAs or both. Indeed more detailed studies using continuous labeling and pulse-chase protocols have indicated that this is the case (S. Pilder, J. Logan, and T. Shenk, in prep.). It appears, therefore, that the function of the adenovirus E1b-55K polypeptide is to regulate posttranscriptionally the accumulation of late viral mRNAs.

Expression of Foreign Genes in Adenovirus

Adenovirus recombinants can be constructed to express foreign genes. An illustration of a helper-independent conditionally defective system is shown in Figure 3. In this plasmid the E1a and E1b regions are replaced by sequences which are required for high-level expression of genes in the adenovirus system. The gene of interest is placed under the control of the major late promoter which is the most active adenovirus transcriptional control region at late times postinfection. It is constructed such that the 5′ noncoding region contains the tripartite leader sequence. This sequence has been shown to be necessary for efficient translation of mRNAs at late times postinfection in adenovirus infected cells. (Thummel et al.

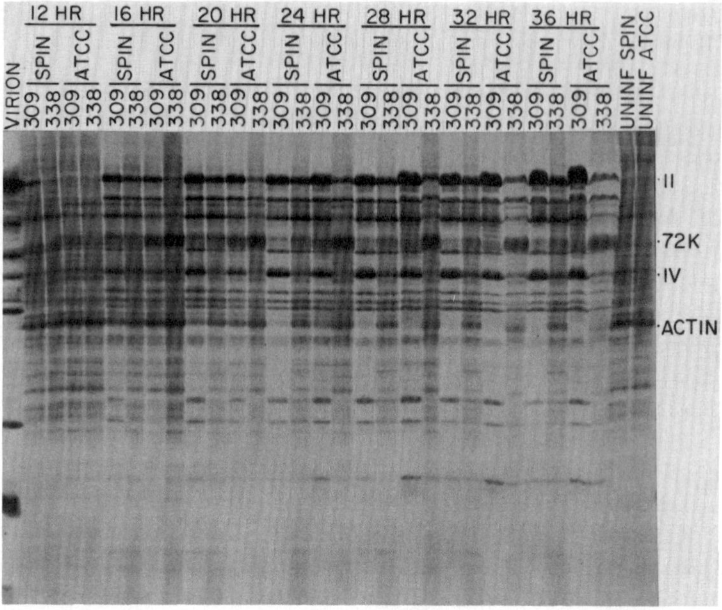

Figure 2

Electrophoretic analysis of polypeptides synthesized in HeLa cells infected with mutant and wild-type viruses. At the times after infection indicated, cultures were labeled for 30 min with S-methionine. Cell extracts were prepared and aliquots subjected to electrophoresis in a 12.5% polyacrylamide gel containing SDS. The bands corresponding to cellular actin and several viral polypeptides are designated.

1983; Logan and Shenk 1984; Berkner and Sharp 1985). Therefore, the gene of interest is expressed efficiently at both the levels of transcription and translation. The plasmid is constructed back into the virus by overlap recombination and propagated in 293 cells. Since the virus will lack the E1a and E1b regions, it will only be able to grow in 293 cells. Polypeptides have been expressed using this system, e.g., polyoma and SV40 T-antigens (J. Logan, unpubl. data) at levels comparable to those of adenovirus late proteins.

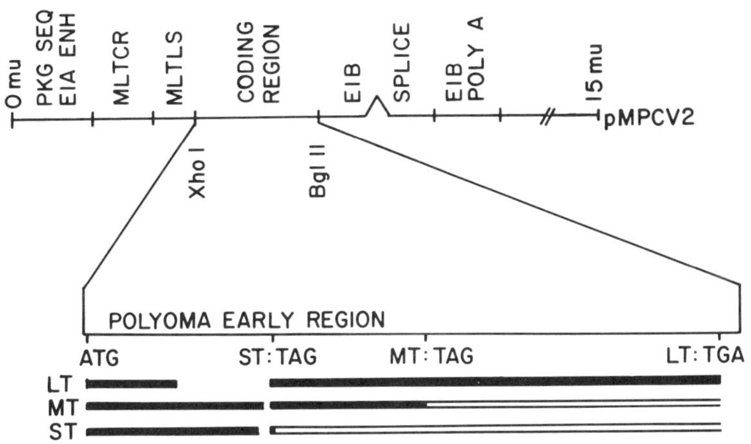

Figure 3
Structure of the plasmid pMPCV2 used to express foreign genes in adenovirus. Sequences of the E1a and E1b regions between nucleotides 354 and 2805 are removed and replaced with sequences that contain the major late promoter (MLTCR) and the tripartite leader sequence (MLTLS). The E1b region between nucleotides 2805 and 3328 can now be replaced with the gene of interest. The gene of interest illustrated is the polyoma early region. Sequences downstream contain a splice site and a polyadenylation signal. This plasmid can be reconstructed into virus by overlap recombination using the downstream homology.

DISCUSSION

The recombinants described in this paper were constructed to answer two questions. First, what is the function of the adenovirus E1b-55K polypeptide. Second, can one use adenovirus as an expression vector. It is clear that the E1b product is required for efficient expression of adenovirus encoded polypeptides at late times postinfection (Fig. 2) (Babiss and Ginsberg 1984; Logan et al. 1984). It apparently acts posttranscriptionally either to mediate transport of specifically adenovirus late mRNA or to stabilize that RNA once it exits into the cytoplasm. How it does this and what it recognizes is still unclear, but it does demonstrate the importance of posttranscriptional events to gene expression and also identifies a gene product that can regulate these events.

Adenovirus can be used as an expression vector. The basic system outlined in this paper uses a conditional defective helper-independent system that achieves high-level expression using adenovirus transcriptional and translational signals. These recombinants are less viable than wild-type virus and would not be able to grow without a helper virus being present. It is conceivable that the adenovirus expression system could be made nondefective and helper-independent. In this context it is interesting to speculate that the recombinant viruses could be used in a manner analogous to that of vaccinia virus, i.e., as a vehicle for the introduction of protective antigens into the host system via a viral vector. Irrespective of the viability of the previous proposal, it is clear that recombinant adenoviruses can be used both to answer basic questions concerning gene regulation and also as a vehicle for the expression of foreign genes.

REFERENCES

Anderson, C.W., R.C. Schmitt, J.E. Smart and J.B. Lewis. 1984. Early region 1B of adenovirus 2 encodes two coterminal proteins of 495 and 155 amino acid residues. *J. Virol.* **50**: 387.

Babiss, L.E. and H.S. Ginsberg. 1984. Adenovirus type 5 early region 1b gene product is required for efficient shut off of host protein synthesis. *J. Virol.* **50**: 202.

Berkner, K. and P. Sharp. 1985. Effect of the tripartite leader or synthesis of a non-viral protein in an adenovirus 5 recombinant. *Nucl. Acid Res.* **13**: 841.

Bos, J.L., L.J. Polder, R. Bernards, P.I. Schrier, P.J. van den Elsen, A.J. van der Eb, H. van Ormondt. 1981. The 2.2 kb E1B mRNA of human Ad12 and Ad5 codes for two tumor antigens starting at different AUG triplets. *Cell* **27**: 121.

Chinnadurai, G., S. Chinnadurai, J. Brusca. 1979. Physical mapping of a large-plaque mutation of adenovirus type 2. *J. Virol.* **32**: 623.

Ginsberg, H.S. 1984. *The adenoviruses.* Plenum Press, New York.

Graham, F.L., J. Smiley, W.C. Russell, and R. Nairn. 1977. Characteristics of a human cell line transformed by DNA from human adenovirus type 5. *J. Gen. Virol.* **36**: 59.

Logan, J. and T. Shenk. 1984. The adenovirus tripartite leader sequence enhances translation of mRNAs late after infection. *Proc. Nat. Acad. Sci. U.S.A.* **81**: 3655.

Logan, J., S. Pilder, and T. Shenk. 1984. Functional analysis of adenovirus type 5 early region 1B. *Cancer Cells* **2**: 527.

Thummel, C., R. Tjian, S-L. Hu and T. Grodzicker. 1983. Translational control of SV40 T-antigen expressed from the adenovirus late promoter. *Cell* **33**: 455.

Tooze, J. 1981. In *Molecular biology of tumor viruses*, 2nd ed., revised, part 2, DNA tumor viruses. Cold Spring Harbor Laboratory, Cold Spring Harbor, New York.

Genetic Engineering of the Genome of the *Autographa californica* Nuclear Polyhedrosis Virus

MAX DUANNE SUMMERS AND GALE EUGENE SMITH
Department of Entomology and
The Texas Agricultural Experiment Station
Texas A&M University
College Station, Texas 77843

OVERVIEW

The polyhedrin gene in the baculovirus genome of *Autographa californica* has been genetically engineered to develop a helper-independent expression vector for the high-level expression of foreign procaryotic and eucaryotic genes in insect cells. The recombinant proteins produced in insect cells with the baculovirus vector are biologically active and appear to undergo posttranslational processing to produce recombinant products very similar to the authentic proteins. This system may be very useful for the study of the structure and function of foreign genes and their products in invertebrate cells and for the high-level expression of recombinant proteins of human, animal, and agricultural importance.

The genetic engineering of baculovirus genomes to produce more efficacious microbial pesticides is one practical application in the development of a new generation of pest control agents. Genetically engineered baculoviruses expressing highly specific and effective insecticidal products under the regulation of the polyhedrin or other baculovirus promoters could be an excellent test system for the evaluation of the fate and persistence of a genetically engineered agent after introduction into the environment. This, combined with the use of baculovirus mutants containing transposable elements, could provide safe models to evaluate the potential of genetically modified viruses to infect nontarget organisms, transfer genetic material to other microbial agents or to susceptible and nonsusceptible hosts, and to determine whether such foreign genetic sequences would be inheritable if transferred to organisms.

INTRODUCTION

Baculoviruses are insect pathogenic viruses which, until recently, were studied mostly for their potential use as viral insecticides for insects of agricultural importance. This emphasis is because certain baculoviruses are highly virulent for pest insects and some of the most promising have been commercially developed and are used as biological pesticides (Miltenburger and Krieg 1984). Baculoviruses have the property of being very stable and are able to persist for longer times in the environment as compared to other animal viruses. This unusual biological stability is the result of a unique association of the infectious virus particles and a viral occlusion which is a crystalline assembly of a viral encoded structural protein called polyhedrin. Late in viral replication, baculovirus particles become embedded in a protein occlusion composed of the polyhedrin protein. This relationship between the virus and viral occlusion is the structural property which provides the protection and stability that facilitates horizontal transmission among susceptible insects in the environment.

Of the 450–500 species of known baculoviruses, practically all encode for a polyhedrin protein. The viral occlusion is a paracrystalline assembly of a polyhedrin monomer which, for most viruses, has an average molecular weight of 28,000–30,000 (Summers and Smith 1978). Baculoviruses are unique among animal viruses, not only in the protective function of the viral occlusion in the viral life cycle but also because the polyhedrin gene is the most highly expressed eucaryotic virus gene known. The polyhedrin protein can accumulate to greater than 1 mg/ml of infected cultured insect cells (70–75% of the total cellular protein) or can comprise up to 25% of the total protein of an infected insect. Although very highly expressed, neither the polyhedrin gene nor its protein is essential for viral infection or replication in cultured insect cells or insects, thus making the polyhedrin gene an ideal target for genetic manipulation.

The genetic engineering of the baculovirus polyhedrin gene to express high levels of biologically active recombinant human β-interferon was first reported by Smith et al. (1983a,b). The genetic engineering of baculoviruses now has the following additional potential uses in the areas of human and animal health and agriculture: (1) the engineering of baculovirus genes and the use of a recombinant baculovirus as a helper-independent expression vector for the high-level expression of genes encoding for products useful in basic research and in the therapy and diagnosis of disease, (2) the potential for the genetic engineering of baculoviruses as more effective viral pesticides, or (3) as vectors for the transfer of foreign genes into insect cells or chromosomes.

The most extensively studied baculovirus is the *Autographa californica* nuclear polyhedrosis virus (Ac*M*NPV). Studies of the physical and functional organization of the Ac*M*NPV genome have resulted in the mapping, cloning, and sequencing of the Ac*M*NPV polyhedrin gene and its regulatory sequences (Iddekinge et al. 1983; Smith et al. 1983a). Not only does the polyhedrin gene exhibit a strong promoter, but expression can continue late in infection well beyond the point of repression of nearly all other baculovirus and host genes. For our initial report of engineering the polyhedrin gene to express a foreign gene, the Ac*M*NPV polyhedrin promoter was cloned into a bacterial plasmid. The cloned gene was altered by in vitro mutagenesis to construct the appropriate transfer vectors that would allow the fusion of any foreign gene to polyhedrin transcriptional regulatory sequences to produce a polyhedrin promoter-foreign gene chimera (Smith et al. 1983b). During co-transfection of insect cells with the transfer vector and wild-type Ac*M*NPV DNA, a hybrid gene

Table 1
AcMNPV Expression Vectors

Gene	Source	Gene size (bp)	Protein size (MW)	Modifications	Expression[a] (mg/l)
Polyhedrin[2]	Genomic	1050	29K	Phosphorylated	1200
Human fibroblast[1] interferon	Genomic	780	17K/21K	Glycosylated, signal peptide cleaved, secreted	1–5
Fusion product:[1] polyhedrin/IfN	Genomic	780	29K	–	100–500
Chloramphenicol[2] acetyl transferase	Genomic	770	27K	–	100
β-galactosidase:					
(polyhedrin[2,8] promoter)	Genomic	3000	110K	–	600
(10K promoter)[2]	Genomic	3000	110K	–	600
Simian rotavirus[3] VP-6	cDNA	1397	41K	–	50
Human reovirus SI[4]	cDNA	1430	14.5K	–	1–5
Human interleukin-2[5]	cDNA	1000	15.5	Signal peptide cleaved, secreted	10–20
Influenza-A[6]	cDNA	2300	85K	–	1–5
Influenza-PB2[6]	cDNA	2300	85K	–	1–5
Human *c-myc*[7]	cDNA	–	64K	Phosphorylated	1

[a] Relative estimates
[1] Smith et al. (1983b)
[2] G.E. Smith and M.D. Summers, unpubl. data
[3] Research collaboration with Dr. M. Estes (Baylor College of Medicine)
[4] Research collaboration with Dr. W.K. Joklik (Duke University)
[5] G.E. Smith et al., in prep.
[6] Research collaboration with Dr. R. Krug (Sloan Kettering Memorial Hospital)
[7] Miyamoto et al., in prep.
[8] Pennock et al. (1984)

is substituted for the wild-type polyhedrin gene in the viral genome by homologous recombination. A recombinant baculovirus is identified in infected insect cell monolayers by screening for the occlusion negative phenotype. The recombinant virus is plaque-purified, and upon infection of insect cells the foreign gene is expressed as a baculovirus gene under the transcriptional control of the polyhedrin promoter. Our initial success with the expression of several human and viral genes of medical importance (Table 1) demonstrates the successful development of Ac*M*NPV as a new eucaryotic expression vector system, and it shows some of the advantages of the baculovirus expression vector when compared to other procaryotic and eucaryotic expression systems. Not only is the quantity of recombinant products expressed from the polyhedrin promoter equal to or better than that reported for bacterial, yeast, and vertebrate expression systems, but our preliminary analyses of the recombinant proteins of selected vertebrate genes have shown that the posttranslational processing and biological properties of recombinant vertebrate proteins produced in insect cells may be very similar to those documented to occur with the natural proteins.

In agriculture, genetically engineered baculoviruses have potential applications in two major areas. First, the engineering of more effective viral pesticides through the introduction of genes encoding for products which may more quickly and effectively debilitate a pest insect (Kirschbaum 1985). Second, preliminary studies of baculovirus mutants containing insertions of host cell DNA have revealed DNA sequences with properties similar to those of transposable elements (Fraser et al. 1983, 1985). Little is reported of the transposition and molecular biology of these elements in baculoviruses, but their potential use as vectors for foreign genes into insect cell chromosomes is clear.

RESULTS

Ac*M*NPV Expression Vector

Recombinant transfer vectors designed to fuse foreign genes to the polyhedrin promoter were initially developed by cloning the Ac*M*NPV *Eco*RI-I restriction enzyme fragment, which contains the polyhedrin gene, into a bacterial plasmid (Smith et al. 1983a). The recombinant plasmid was engineered to contain a natural, unique *Bam*HI site at + 175 relative to the polyhedrin translation start signals (Smith et al. 1983b). This plasmid was cleaved with *Bam*HI and digested with Bal 31 exonuclease to remove either all or portions of coding sequences for the aminoterminal polyhedrin amino acids. As a result, transfer vectors were constructed to allow fusion of a foreign gene with its translation start and stop signals intact to the polyhedrin promoter either before or after the polyhedrin translation site. Vectors pAc360 (fusion of ten aminoterminal polyhedrin amino acids) and pAc101 (fusion of 58 aminoterminal polyhedrin amino acids) were designed to produce a fusion protein. Vectors pAc373 and pAc380 permitted insertion of a gene into the polyhedrin 5′ noncoding leader sequence so as to produce a nonfused recombinant product. Recombinant viruses containing the hybrid gene were selected after cotransfection with a mixture of Ac*M*NPV and the transfer vector DNAs by screening insect cell monolayers for the presence of occlusion negative plaques (Smith et al. 1983a,b). This procedure is now routine and only 2–4 weeks are needed to prepare a transfer vector containing a hybrid gene and select for the recombinant Ac*M*NPV. In these vectors the polyhedrin 3′ nontranslated leader sequence and polyadenylation site were left intact to allow for the correct processing of the recombinant mRNA.

Some of the procaryotic and eucaryotic genes cloned into Ac*M*NPV and expressed under the transcriptional control of the polyhedrin promoter using primarily the pAc373 vector are listed in Table 1. A summary is given of the relative levels of expression, size of the protein, source of the gene, and known posttranslational modifications. The sizes of the recombinant proteins are as predicted and each, where possible, is identified by either immunological, enzymatic, or biological assays.

Preliminary transcription and translation studies comparing levels of expression of fused and nonfused recombinant proteins (Table 1); suggest that, in some cases, a greater than 100-fold difference may be due to the translational efficiency of the recombinant mRNA in *Spodoptera frugiperda* cells. The importance of the eucaryotic gene sequences flanking the polyhedrin AUG in ribosome binding and translation initiation (Kozak 1984) may be reflected in the higher levels of expression of some foreign genes fused to the aminoterminal amino acids of polyhedrin (Table 1) (Smith et al. 1983b). To test this hypothesis vector constructs are needed to allow fusion of foreign genes in phase immediately following the polyhedrin AUG. The potential for achieving increased levels of expression is obvious by using transfer vectors engineered to fuse more precisely a foreign gene to the polyhedrin 5′ and 3′ noncoding translational and regulatory signals.

Co- and posttranslational processing of recombinant proteins have been shown to occur in baculovirus-infected *Spodoptera frugiperda* cells. Recombinant human β-interferon is glycosylated by an N-linked, mannose containing glycan. We have not studied the structure of the carbohydrate in the insect cell-derived β-interferon since there are no reports on the structure of authentic β-interferon glycosylation. The nature of glycosylation of recombinant proteins in insect cells is an important question. Studies of the glycosylated Ac*M*NPV protein 64K (B. Ericson and M.D. Summers, in prep.) show that the 64K protein follows the classical route of processing (from rough endoplasmic reticulum to golgi to cell surface). Data from sugar incorporation (D-mannose, n-acetylglucosamine and N-acetyl galactosamine), tunicamycin sensitivity and glycosidase (endonuclease H, endonuclease D, α-mannosidase) digestion experiments strongly suggest that 64K contains at least two glycans of the MAN (4-3) $NAcGluN_2$ type and that conversion of the "high mannose" oligosaccharides into the "complex" type does not occur for the Ac*M*NPV 64K protein.

The efficient secretion of human β-interferon from insect cells indicates that the signal sequence of the recombinant interferon is correctly recognized and cleaved (Smith et al. 1983b) although this was not confirmed by an amino acid analysis of the aminoterminal amino acid sequence. However, a cDNA coding for human interleukin-2 (IL-2) is also expressed at high levels (Smith et al. 1985), the majority of which is secreted into the culture medium during infection. Recombinant IL-2 is biologically active, supports the growth of an IL-2-dependent murine cell line CTLL, and the aminoterminal amino acid sequence of the insect-derived IL-2 is identical to natural IL-2 (Smith et al. 1985).

Transposable Elements

Serial propagation of Ac*M*NPV in *Trichoplusia ni* or *Spodoptera frugiperda* cells results in the production of spontaneous mutants which produce fewer viral occlusions when compared to the wild-type virus. These mutants, termed FP (few polyhedra) mutants, exhibit an altered plaque phenotype. Miller and Miller (1982) reported that an FP mutant of Ac*M*NPV had an insertion of host cell DNA between 86.4 and 86.6 map units of the genome. The insert is homologous to moderately repetitive host DNA and by physical mapping resembled a *Drosophila* copia-like element. Fraser et al. (1983) analyzed several FP mutants of Ac*M*NPV and *Galleria mellonela* (a closely related strain to Ac*M*NPV), most of which had in common the insertion of host cell DNA in a specific region between 35.0 and 37.7 map units of the genome. These host DNA insertions vary in size from 0.8 kbp to 2.8 kbp, and DNA homology and restriction enzyme recognition sites are shared among some of the insertions. Several FP mutant insertion sequences were cloned and hybridized to host chromosomal DNA and are derived from moderately repeated host cell DNA.

One of the host insertion DNA and flanking viral sequences in one FP mutant, GMFP3, (Fraser et al. 1985) was sequenced. The host insertion is 750 bp, with a perfect duplication of a four nucleotide viral sequence, 5′ TTAA 3′, at each end of the host DNA insertion. Immediately adjacent to these direct repeats is a seven-nucleotide perfect inverted repeat, 5′ CCCATTC...GAATGGG 3′. No direct or inverted repeats were detected in the other sequences analyzed. The presence of short inverted terminal repeats is a characteristic of many transposable elements (Calos and Miller 1980), and the duplication of target sequences upon insertion is considered a defining feature of a transposition event. Although they are very preliminary data, these results indicate that the host DNA insertion of GmFP3 is a transposable genetic element of *Trichoplusia ni*, which we call TFP3. Further studies are underway to evaluate the transposition properties of TFP3, any transcription associated with the host DNA sequences, and effects upon the expression of viral genes encoded within this region of the Ac*M*NPV genome for a comparison of those properties with other Ac*M*NPV host DNA insertion mutants.

Virus Pesticides

The success of engineering the polyhedrin promoter for the expression of foreign genes is merely a first step indicative of the potential to genetically alter baculoviruses for pest control. Presently there are no reports of a recombinant viral pesticide, but clearly the technology is available and the potential is excellent for inserting foreign genes, encoding for proteins or peptides, which could be highly disruptive to some aspect of insect behavior or metabolism. As mapping of the molecular organization and function of the baculovirus genome and the temporal regulation of these functions is determined, it should be possible to select Ac*M*NPV promoters, in addition to that of polyhedrin, which would be more useful for expressing genes with potential pesticidal activities. To be successful, however, more knowledge is needed of the regulation of baculovirus gene expression (Lübbert and Doerfler 1984), as well as a better understanding of the critical target sites in the insect for such products.

DISCUSSION

Studies on the molecular biology of *Autographa californica* nuclear polyhedrosis virus have resulted in the development of a baculovirus expression vector system (BEVS) with a significant potential for interdisciplinary applications in agriculture, forestry, and medicine. The BEVS was developed by genetic manipulation of the unique and highly expressed gene for polyhedrin which is found only in baculoviruses. A variety of expression vector systems have been developed using bacteria, yeast, and vertebrate cell culture systems and are currently used for the cloning and production of protein products that normally exist in very small quantities in man, other animals, or plants (Ginsberg and Vogel 1984). The development of these expression vectors has provided very valuable systems for the production of sufficient quantities of a given protein for studies of its structure and function, or for use in therapy or diagnosis or as a vaccine. The BEVS is unique when compared to bacterial, yeast, and vertebrate cell expression systems: (1) Greater than 1.0 mg of polyhedral protein is normally produced per ml of infected cells and several bacterial, human, and other viral genes are expressed at nearly this high level. The BEVS is virtually untested for additional strategies to obtain even higher levels of expression. (2) Our results show that, in contrast to those produced in bacterial or yeast cells, recombinant proteins produced in insect cells may be co- and post-translationally processed in ways very similar to the processing that occurs in vertebrate cells. (3) Bacterial and yeast expression systems also produce foreign proteins in relatively high quantities. However for eucaryotic proteins requiring complex structural modifications, bacterial and yeast systems are limited by the inability to correctly modify the recombinant protein product, especially by glycosylation or secretion; or there are difficulties encountered when purifying and reconstituting a biologically active recombinant protein. (4) The BEVS provides an opportunity to express foreign genes at high levels without the use of transformed cells or systems employing human viral pathogens or oncogenes. Therefore, the BEVS has several notable advantages for the expression of foreign genes in comparison to other expression vector systems.

During the infection process, the polyhedrin protein assembles into massive numbers of highly stable occlusions in which infectious virions are embedded. The viral occlusions are essential for persistence and horizontal transmission of the infectious virus particles among susceptible insects in nature. This unique form of viral transmission, coupled with the possibility of genetic engineering, makes baculoviruses very promising for use in pest control. In addition, baculoviruses are often highly virulent and species-specific. Those registered as viral pesticides are safe for nontarget organisms and have no effect on natural parasites and predators of pest insects; thus baculovirus pesticides could be used in conjunction with existing biological control programs. Finally, because baculoviruses are safe and a natural part of the environment, they neither pollute nor leave the toxic residues which are normally associated with chemical pesticides. All of these factors, together with the development of the BEVS, provide unexplored potential for the genetic manipulation of baculoviruses as a new generation of biological control agents.

Theoretically, a highly insect-specific toxin or bioregulator could be expressed under the temporal control of any identified baculovirus promoter which, when expressed, could rapidly debilitate the insect (Kirschbaum 1985). There are numerous sites for such products within an insect which, if successfully targeted, would increase the pesticidal effectiveness of the baculovirus many-fold. It might be possible to engineer viruses in ways to improve persistence in the environment, to improve virulence, or to expand host range within an acceptable and limited spectrum. With respect to an expanded host range, there are no viral or host factors identified that would allow genetic engineering to proceed on an informed and predictable basis. Basic research on the molecular biology of baculovirus gene function and regulation and the properties of the host insect that play a role in host range and virulence will be essential to develop these potentials and applications.

Finally, the use of transposable elements and viral DNA sequences that can serve as vectors for the transfer of genes into animal and plant cell chromosomes (Rubin and Spradling 1983; Shapiro 1983) and their potential value in human and animal health and plant genetic engineering is clearly recognized in all areas of biotechnology. The acquisition of host DNA sequences by animal viruses is a well-documented phenomenon that has now been demonstrated to occur with Ac*M*NPV. Studies on the host DNA insertion mutants of Ac*M*NPV strongly suggest that transposable elements will soon be characterized so that they may be developed as vectors for the insertion and stable integration of foreign genes in lepidopteran insect chromosomes. Numerous species of baculoviruses are known to infect a variety of families of insects containing important pest species. Therefore, it may be possible to identify additional vectors by further studies. It may be possible to insert deleterious genes into insect chromosomes and insect populations which could by some chemical or environmental "trigger" be very effective in disrupting growth, differentiation, and development (Gehring 1985). Clearly the technology is available, but at the moment we lack sufficient knowledge of the genetic regulation of these processes in insects to safely develop and exploit any opportunity or application for the purposes of insect control.

CONCLUSION

The baculovirus expression vector should be a very safe and convenient system for the study and expression of foreign genes. Recombinant Ac*M*NPV that do not produce viral occlusions are inefficient for transmission of the virus by natural means and any foreign gene is under the stringent regulation of the Ac*M*NPV polyhedrin promoter. Further, baculovirus vectors can be used to study gene expression, protein structure, posttranslational processing of recombinant proteins, and as a tool to study recombinant proteins in insect cells. Such studies are important for the comparison of biochemical pathways and processes at the molecular level between vertebrate and invertebrate cells. Nothing is known of the specificity or mechanisms of productive infection of the insect pathogenic viruses nor of those mechanisms in cultured insect cells, insects, or other arthropod vectors which allow transmission of vertebrate viruses. Genetically engineered baculoviruses might be used to help study these phenomena.

Ac*M*NPV and genetically engineered forms thereof could provide a safe test system for evaluating the application of genetically engineered viruses in the environment. For viral pesticides, the major concern for environmental hazards is the potential persistence or transfer of genetic information to nontarget invertebrates or other organisms with a phylogeny closer to that of arthropods. Ac*M*NPV mutants with a foreign gene fused to the polyhedrin promoter should be very safe for environmental application for studies of fate, persistence, and inadvertent gene transfer. The techniques and molecular probes are available to assess this. There are sufficient data on the hierarchy of Ac*M*NPV gene regulation during infection that it is possible to prepare recombinant Ac*M*NPV mutants with potentially greater ability to express some foreign gene product in a nontarget organism. This could be done by fusing a foreign gene to a highly regulated late, delayed early, or constitutively-expressed immediate early Ac*M*NPV promoter. Such expression vectors could be used to probe for the expression or transfer of the foreign gene in nontarget organisms or tissues of choice. These vectors could provide some recombinant Ac*M*NPV that would be predicted to be very safe (i.e., with a foreign gene under polyhedrin promoter regulation) and others with the potential to express more easily some undesirable effect or activity. In addition, the use of FP baculovirus mutants containing specific genetic markers within the transposable element could very easily complement the above studies for the environmental assessment of gene transfer from genetically engineered viral pesticides.

ACKNOWLEDGMENT

These studies were funded in part by Public Health Service Grant AI 14755 from the National Institutes of Health and by Texas Agricultural Experiment Station project TEXO 6316.

REFERENCES

Calos, M.P. and J.H. Miller. 1980. Transposable elements. *Cell* **20**: 579.

Fraser, M.J., J.S. Brusca, G.E. Smith, and M.D. Summers. 1985. Transposon mediated mutagenesis of a baculovirus. *Virology* (in press).

Frazer, M.S., G.E. Smith, and M.D. Summers. 1983. Acquisition of host cell DNA sequences by baculoviruses: Relationship between host DNA insertions and FP mutants of *Autographa californica* and *Galleria mellonella* neculear polyhedrosis viruses. *J. Virol.* **47**: 287.

Gehring, W.S. 1985. The homeo box: A key to the understanding of development? *Cell* **40**: 3.

Ginsberg, H.S. and H.J. Vogel. (eds.) 1984. *Transfer and expression of eucaryotic genes.* Academic Press, New York.

Iddekinge, B.J.L. Hooft van, G.E. Smith, and M.D. Summers. 1983. Nucleotide sequence of the polyhedrin gene of *Autographa californica* nuclear polyhedrosis virus. *Virology* **131**: 561.

Kirschbaum, J.B. 1985. Potential implication of genetic engineering and other biotechnologies to insect control. *Annu. Rev. Entomol.* **30**: 51.

Kozak, M. 1984. Point mutations close to the AUG initiator codon affect the efficiency of translation of rat preproinsulin *in vivo. Nature* (Lond.). **308**: 241.

Lübbert, H. and W. Doerfler. 1984. Transcription of overlapping sets of RNAs from the genome of *Autographa californica* nuclear polyhedrosis virus: A novel method for mapping RNAs. *J. Virol.* **52**: 255.

Miller, D.W. and L.K. Miller. 1982. A virus mutant with an insertion of a copra-like transposable element. *Nature* (Lond.) **299**: 562.

Miltenburger, H. and A. Krieg. 1984. Bioinsecticides: II: *Baculoviridae. Adv. Biotechnol. Processes* **3**: 291.

Miyamoto, C., G.E. Smith, J. Farrell-Towt, R. Chizzonite, M.D. Summers, and G. Ju. 1985. Production of human c-*myc* protein in insect cells infected with a baculovirus expression vector. *Mol. Cell. Biol.* (in press).

Pennock, G.D., C. Shoemaker, and L.K. Miller. 1984. Strong and regulated expression of *Escherichia coli* β-galactosidase in insect cells with a baculovirus vector. *Mol. Cell. Biol.* **4**: 399.

Rubin, G.M. and C. Spradling. 1983. Vectors for P-mediated gene transfer in *Drosophila. Nucl. Acids Res.* **11**: 6341.
Shapiro, J.A. (ed.) 1983. *Mobile Genetic Elements.* Academic Press, New York.
Smith, G.E., M.J. Fraser, and M.D. Summers. 1983a. Molecular engineering of the *Autographa californica* nuclear polyhedrosis virus genome: Deletion mutations within the polyhedrin gene. *J. Virol.* **46**: 584.
Smith, G.E., M.D. Summers, and M.J. Fraser. 1983b. Production of human beta interferon in insect cells infected with a baculovirus expression vector. *Mol. Cell. Biol.* **3**: 2156.
Smith, G.E., G. Ju, B.L. Ericson, J. Moschera, H.W. Lahm, and M.D. Summers. 1985. Modification and secretion of human interleukin-2 produced in insect cells by a baculovirus expression vector. *Proc. Natl. Acad. Science. U.S.A.* (in press).
Summers, M.D. and G.E. Smith. 1978. Baculovirus structural polypeptide. *Virology* **84**: 390.

COMMENTS

KAWANISHI: What do you think about the ecological aspects of nonspecific interactions of recombinant pesticidal microorganism with nontarget organism?

SUMMERS: The major concern for potential hazard for the release of a genetically engineered virus into the environment is, we feel, not with man or other animals but with other nontarget organisms that are phylogenetically closer to insects. These may be other invertebrates or lower invertebrate forms in the environment. *Autographa californica* NPV has the widest host range reported for any baculovirus, which includes approximately 28 lepidopteran insect species. I haven't seen the host range expanded outside the Lepidoptera. Let me remind you that engineering the polyhedron gene does not alter any part of the virus that controls that host range. I think it would be possible to survey the population of beneficial insects and other nontarget organisms to make a very quick assessment of whether there is any hazard. If there is a possibility of hazard, it would be possible to conduct studies on selected organisms to evaluate the ability of the virus to infect the nontarget organism, its persistence, and whether expression occurs. If studies were conducted in vitro (insect cell cultures) and with selected organisms dependent upon the results, then you proceed very quickly to some very small animal studies.

FIELDS: In a worst case scenario by genetic engineering, can you envision doing anything worse than the baculovirus itself?

SUMMERS: I don't think so if any foreign gene is under the stringent control of the polyhedron promoter. However, if you want to increase the levels of risk, this could be done by cloning genes under the control of genes that are expressed earlier during infection and testing for expression after entry to a non-target cell or organism. We have representative baculovirus genes available now to conduct those kinds of studies.

WATSON: How widely are these viruses used in the United States as pesticides?

SUMMERS: There have been four baculoviruses registered by EPA–these are natural isolates, not recombinant forms. Baculoviruses are not that widely used simply because they are routinely applied like chemical pesticides and it often doesn't work. There are many factors to consider for effective use: the size and age of the insect population, the time of day, and the means of application. There is also an education problem. Farmers like to see insects die immediately after treatment, and the baculovirus takes 3–4 days to kill. However, if it is possible to engineer the virus such that upon infection of the insect gut a product would be produced that is toxic and would quickly debilitate the insect, then you have an effective viral pesticide. I think it is going to be very possible to do those kinds of studies and produce a safe yet effective virus.

WATSON: Who makes these viruses?

SUMMERS: Elkar is commercially available from Sandoz, Inc. in California. The U.S. Forestry Service has developed two other viral products, one for the gypsy moth and one for the tussock moth. Is there another one?

SHOPE: European pine sawfly virus.

SUMMERS: Yes.

MOSS: Is *Autographa californica* used?

SUMMERS: No, but all the safety tests required by EPA have been done. It's just that the lack of commercial success of the other baculoviruses has not facilitated development of *Autographa californica.*

MOSS: Would it be difficult to grow some of the others in tissue culture?

SUMMERS: No. I think there are three or four baculoviruses that could be used.

MILLER: In a sense, Max [Summers], the very impressive efficiency of your system argues against one of your summary statements, and that was that you could put any potent toxin in and not worry much about it because it's stringently regulated. To take some unlikely examples, if you cloned in, say, aflatoxin or acin or ricin or something else with high toxicity, you could potentially have significant amounts out there during an environmental application if your numbers of target insects were large.

SUMMERS: How stable are these toxins in the environment?

MILLER: I don't know.

SHOPE: Aflatoxin probably is stable.

SUMMERS: I agree that it would not be wise to use an aflatoxin or diphtheria toxin or anything like those, but it might be possible to identify some toxin or toxic product to use in a worst-risk scenario and that would be relatively safe to evaluate. We need a system which can be evaluated and tested environmentally. The baculovirus appears to be one of those systems. It has the ability to express foreign genes; it can be used as a viral pesticide; it also may have transposable elements which could be evaluated for their ability to transfer foreign genes to other organisms after release into the environment. These are all technical realities now; they're not theories. So, it may be an ideal system for evaluating the release of a genetically engineered agent into the environment. Superimposed upon this is the safety testing procedures that have been established by the EPA and evaluated with several baculoviruses.

KAMELY: How do the data look when you extrapolate from the insect cultures to the organisms themselves? Do you observe any differences, such as transformation in animal cells and tumorigenicity in animals? Are there parallels on which you can draw? My point is that when you perform tests, you conduct them in vitro and in insect cell cultures. How easily can you extrapolate your results to the environment and be certain that you are measuring correctly?

SUMMERS: Very superficially; the results of in vivo and in vitro experiments do correlate.

KAMELY: On the other hand, is there anything that you are not detecting in vivo but that may be expressed in vitro?

SUMMERS: It is hard to make something out of negative data. We have very sensitive assays that will allow us to detect expression of specific genes at the RNA and the protein levels. Yes, I suppose there are other gene activities that we don't know about yet, but that is going to depend upon more complete studies of the molecular biology of the *Autographa californica* genome. For certain baculovirus genes, for example for the polyhedron gene, we can very easily develop probes and screen for those in practically any system.

SUMMARY DISCUSSION

FIELDS: I would like to emphasize some of the points that have emerged during this meeting. Let me begin by summarizing the five sessions briefly.

The legislative issues speak for themselves. There are going to be more environmental concerns and there will be legislation. There are important regulatory aspects that were spelled out nicely in that session.

The session on environmental aspects of virology was very useful because it illustrated that, in the real world, viruses are released into the environment. There are important features of respiratory spread and spread of viruses in water; yet there are very few insights into the genetic determinants of viruses that really help us understand, in ways that we could accurately predict right now, the effects of specific genetic manipulation in terms of mechanisms. Environmental virology is a field that is very important and potentially has much to learn from approaches using molecular biology.

Some of the major concerns deal with tropisms, the impact of where viruses localize in the host. The session on tropism illustrated examples where some properties are reasonably well understood from events occurring at cell surfaces involving receptors, to intracellular restrictions. Genetic exchanges or genetic manipulations can in fact alter or restrict host range thus providing a way of thinking about these issues with a few model systems. For the vast majority of viruses, we do not have a detailed understanding of the molecular biology of host interactions, but at least there are some principles emerging.

The impact of genetically altered viruses on the host shows that there can be different outcomes in the host, but one of the very important principles, was, I think, summarized by Bob Chanock, who pointed out that every time we manipulate agents away from the parental type in the making of vaccines, the tendency is to reduce the ability of the virus to spread and to reduce its lethality. In terms of environmental impact, when we manipulate agents in the absence of selective pressures in their natural environment, we tend to select against the properties that are of concern (such as increased virulence and survival in the environment. This has certainly been very reassuring from the vaccine point of view, and I think it is an important principle that we will come back to again.

Last, the nondefective vectors used in cloning DNA have raised the concerns that are going to need to be dealt with. In these cases the genes that are manipulated are not within the same organism, but involve inserting foreign genes and have thus led to the same anxiety. This session was an extremely interesting one. A few points were quite helpful. For example, it is very clear that there is little concern with defective vectors. For nondefective vectors, it has been hard to think of realistic scenarios that are of concern, although a few have emerged.

One issue that emerges again and again is that of host range, that is, if one inserts a gene that broadens the host range, it would change the effect on the host. Altering host range is an issue that comes up with environmental and host factors and always will have to be considered in terms of safety, particularly for nondefective agents.

Another issue that emerged involves placing stable toxins into a vector and expressing them very highly. Such an experiment should not be performed without considerable thought about its impact.

Ed Kilbourne gave me an outline from the journal, *Issues*, and it was suggested by Martin Alexander. This outline may be useful for thinking about the possibility of untoward effects of genetic manifestations of a virus in the environment. There are six steps of probability of an event occurring. The steps were examined in the article in relation to the release of bacteria in the environment; and it is worth mentioning these steps and then opening the points for discussion.

The first event is release of a virus into the environment. The probability of that happening accidentally would have its own probability, P1. But here we're thinking of intentional release into the environment. There are thus two possibilities: creating something where there is an accidental release or intentionally releasing the organism. The second event is survival in the environment, which has its own possibility. The third is not only survival, but multiplication in some target in the environment. The fourth is the issue of dissemination once it has come into the environment. The fifth is the issue of transfer of genetic information to another species, such as by transposable elements. The sixth is, even if all of the above were to occur, what might be the expected harm. If, in the end, we were dealing with something that was not harmful to humans or to another host species in the environment, then it really is not so critical—even if all the prior events were satisfied.

SHOPE: Shouldn't you also substitute "benefits" for "harm" because sometimes we want things to spread; we want them to stay there.

FIELDS: Excellent point. I think it's harm versus benefit. I will suggest as a framework for discussion that there are discrete steps, each of which has its own probability. This means that if we think about what we're doing, most of the types of experiments that have been proposed or thought about can be done in ways where the likelihood of the kind of outcome being a bad one is extremely unlikely. I think we might open the discussion around that framework as a start.

KILBOURNE: I must say this has been a very interesting conference and my hopes and fears have waxed and waned as we have continued our discussion. I think, in the first place, that we are dealing with risks that are rather poorly defined, of necessity, and that at least some of us have looked very searchingly at the past for any kind of anticipatory problems. Really, I think we're coming down on a very reassured position here. My own thinking is—and it has changed perhaps a little bit during the sessions–being really more reassured. The lessons about evolution are, as I mentioned the first day, that all the constraints are really against development of maximal virulence; and I think those constraints are still going to be operative even if the new factor in the equation is man tinkering in the laboratory. In terms of background, there was an excellent piece by Francois Jacob in *Science* a few years ago in which he discusses tinkering. He really comes essentially to the same conclusion, that even human tinkering is not probably going to change evolutionary pathways that much. I think what Bob Chanock has done, which I did very superficially, is to look very deeply at the few prototypes we have of genetically altered viruses that are released. That is from the vaccine story. All of that comes out with a very reassuring sort of statement.

CORTESI: I think the crux, from EPA's point of view, is whether a set of procedures and tests can be developed which will make us comfortable in evaluating each of the various stages outlined. In short, we need procedures so that we can confidently make judgments. This conference has been reassuring in that it appears that changes to existing viruses usually make them less robust and that viruses tend to become less virulent as they evolve. The Pennsylvania chicken episode is disquieting, however.

CHANOCK: The abrupt increase in virulence of the avian influenza A virus is the only instance of this sort that has been described. If we attempted to create this type of mutant using recombinant DNA techniques in the laboratory, it is extremely unlikely that we would succeed at this particular time. At present it is not possible to clone a negative strand virus in the form of complementary DNA, manipulate the DNA and produce a mutation similar to that which occurred in the avian influenza virus, and then transfer the mutant DNA into an RNA form and back into the virus. On the other hand, genetic changes are constantly occurring in viruses which have a segmented genome. These viruses are developing new gene constellations as genes from related viruses become mixed during coinfection. Also mutations occur with high frequency in the genome of negative strand RNA viruses. The dangers that we can project for viruses that are manipulated in the laboratory are miniscule compared to what is occurring in nature. This holds true for mutation, gene reassortment, and recombination–all of which are constantly occurring. The products of these genetic events are also constantly being tested for survival value and for effect on virulence. Nonetheless we must maintain surveillance for the rare event which may lead to increased virulence of a virus that has been manipulated in the laboratory. EPA must develop a very strong science base so that the organization is in a position to test for adverse effects of altered viruses that are being considered for release into the environment. I don't believe that many such issues can be settled simply by discussion.

KILBOURNE: Nature is making the worst case all the time, a worse case than we can devise.

CHANOCK: Yes, nature has exceeded our capacity to produce worse-case situations in the laboratory.

SHOPE: Let me say a little bit more about the question of benefits, rather than the word "harm." We heard earlier, at the beginning of the conference, that one of EPA's missions is not only to protect people in the environment but also to protect the insects and the animals and the plants that are in the environment. We haven't heard anybody in this session tell us what the problems are—what are the diseases that are making animals extinct, for instance? I think if we knew what those problems are, then we, or EPA, or any other scientific organization, could put their minds to solving those problems. The other aspect of it is that we are currently flooding the world with chemical pesticides. I think here we have the opportunity of taking some risks, granted, but to substitute for chemical pesticides some things which have very narrow host ranges and which might in the long run be much more beneficial to the environment.

FOWLE: I just want to comment and follow up on some of the discussion. Obviously, there is no such thing as zero risk in the world. I guess Dr. Shope touched on what I was going to say about the chemicals. Basically, in risk analysis, we don't know everything. We have to make a number of assumptions about certain things; we constantly try to refine our tests in order to develop a philosophy, a working hypothesis, on how to assess risks. So, I think this conference helps us. Obviously, this is just a first step in a long process. Much, much more work is going to be needed. For instance, in your presentation, Dr. Fields, you said our understanding of a problem depends in large measure on how we can define it. When you were defining the various processes involved in viral infection for us, you had neglected to include viral release in the follow-up activities. I think that starting to define these issues is very helpful. Being able to define a process, such as Martin Alexander's six points, makes it possible to start thinking about what type of methodologies and approaches should be developed for test purposes. I think the way the conference has been set up, on various tropisms, enhancers, and receptors is a good start for defining the appropriate questions to address. Also, I think Dr. Chanock put it very well, we're living in a sea of viruses.

CHANOCK: Altered viruses.

FOWLE: We can learn some lessons from the past and these lessons may indicate the first things to focus on—to think about such things as what effect it may have if you're going to put a foreign enhancer or promoter into a virus. For such a case, you might, as a first approach, want to think about testing the organism only in a controlled environment.

CHANOCK: That is the critical issue. It is essential that we learn as much as possible from previous experience. We can benefit from the experience that has been gained from the introduction and widespread use of live virus vaccines. With such vaccines, initial studies are performed in the laboratory followed by studies in experimental animals. Then the investigation is extended into small groups of individuals under carefully controlled conditions. In this manner the possibility of spread to the general population is minimized. Both biologic and physical containment should be employed during the early laboratory and volunteer phases of investigation. Subsequently, progressively larger groups of volunteers are evaluated under less stringent conditions of containment. Finally we reach a level of confidence which allows us to recommend general use of the newly developed vaccine strain. I would imagine that this sort of approach will prove useful in the future as new living organisms are tested for safety and acceptability. We cannot settle many of these questions by discussion. For the most part these questions can only be answered by performing the appropriate tests. We can minimize the risk at each step by using the appropriate level of containment. In this way we maintain control of the organism that we are studying.

KAMELY: Yes. But can you be fully in control and assure yourself that the organism will not spread?

CHANOCK: You can control an organism that is capable of spreading by constructing a physical barrier to its escape into the animal or insect population.

FIELDS: It's worth remembering that the issues are very analogous to the issues of recombinant DNA with respect to humans and the only difference is that the laboratory has to be designed as an environmental laboratory, and therefore has different features; but the same principles exist. The systematic approach to the safety issues involving recombinant DNA was very effective. The process that Dr. Chanock and others have been discussing is a systematic way to begin to analyze the environmental impact of genetically altered viruses and microorganisms.

MILLER: I want to make two points. The first is that in Alexander's sequence of events the probabilities are multiplicative for an adverse outcome. What he often neglects to emphasize is that because the probabilities are multiplicative, if the probability of any one of those events approaches zero, the overall probability of an adverse outcome approaches zero as well. Although often the probability of risk at some of the levels will be uncertain, we can make good guesses about others that can enable us to predict the outcome with some assurance. A second point is essentially the one that Bob Chanock just made: ultimately, the way that much of this has to be decided is in field trials or other trials of progressively greater freedom of experimentation. While it is almost a tautology, it is an important consideration in view of some of the discussions earlier in this conference. For example, extrapolating from Steve Schatzow's presentation, it may be that EPA has already interdicted all proposed field trials of genetically engineered pesticides. Monsanto was one that he mentioned. Another is that proposed by Steven Lindow. While we can't discuss the merits of those proposals here, there are many of us, who are EPA's counterparts at NIH, FDA, and USDA, who feel that both the substance and the process by which those decisions were made are flawed. This situation is reminiscent of the moratorium on recombinant DNA research in the 1970s that Philip Leder has characterized as "hysteria masquerading as responsibility." I think it can be argued that we're seeing a repetition of that. I urge you to watch for EPA's performance on these issues. Watch the pages of *Science* and *Nature* and see whether they are appropriately responsive to the scientific data—and make your opinions heard!

MERIGAN: The critical thing is to plan a path of action, a step-wise path of action, and that we don't prohibit, but that we do restrict. Proceeding that way would build that confidence in the scientific world which Bob [Chanock] is calling for. I think that is what didn't happen with the Asilomar meeting. Instead, we went into suspended animation for a year or two, which was a mistake. It was an unnecessary loss in time for everyone. If, instead, we had utilized progressive barrier isolation, which is a classic way of handling infections, then work could still have gone cautiously ahead.

KAMELY: I want to say that I agree with you, that I think we are in the position now to adopt a step-wise approach in a controlled manner and that the hysteria has somewhat subsided. I have learned two important lessons: First, I was surprised to learn of the stability of these viruses in the environment. Stability is one of the first parameters that needs to be examined. For example, it is important to realize that some viruses survive in water as long as 5 months and may cover a distance of 60 miles. Second, we need to examine the host range. Although it is important to know how a genetically altered virus behaves in the environment, it is equally important to know how the environment reacts to the foreign virus. A set of research guidelines should cover both virus action and altered properties as well as host range.

TEICH: The prospect of genetically altered viruses being released into the environment could have substantial public policy implications. I believe that the scientific community has to be especially careful with this issue. I would make four points in that connection: First, it's evident, I believe, that many of the people who are going to be making policy decisions relevant to this area of genetically altered viruses do not understand much of what most of the people in this room do understand. They are congressional staff or members of Congress, and people in administrative agencies who are lawyers or other nonscientists, and they simply do not have the degree of scientific sophistication present here. Second, there are large segments of the American population–the voting population–which are not going to be willing to take the word of scientists on faith. The scientific community cannot merely say, "Listen, this stuff is safe. Trust us." There is just no reason why the public should trust us–for good reasons or for bad. Third, I think that many lay people are aware of the problems of uncontrolled spread of organisms from everyday experience and news coverage of such things as gypsy moths and killer bees; and–after all–some such organisms were introduced (accidentally or intentionally in connection with scientific experiments. So people realize that such phenomena can occur. Finally, I think, particularly with respect to viruses, that there is a potential public concern which arises from the perception (probably an accurate one) that science does not understand them as well as it understands many other organisms. In common everyday experience, what most people know about viruses is that they go to the doctor with a complaint and the doctor says, "It's a virus." When they ask, "What can you do about it?" the doctor generally replies, "Nothing. It will go away." That's the way most people think about viruses. When you put together the notion of the uncontrolled spread of an organism with the fact that people think the scientific community is not as capable of understanding and dealing with viruses as they are dealing with higher organisms, I think there is a potential here for a serious policy concern. I'm not saying that it necessarily has to become an issue, but I think it could and it's something that we in this room ought to be especially sensitive to.

KILBOURNE: I think that's the argument for the step-wise recurrent research theme. When I say I am reassured, it includes the assurance that before Bob Chanock releases a virus into the environment, he's going to go through the steps he has always done. I think we can't emphasize that too strongly. Something has to be put in place for a step-wise progression in all of this.

FIELDS: Another related point in the same regard is that this is a field that has been neither well funded nor publicized as an important research field. I think, as EPA defines what its role will be, it is very important to decide whether this is or is not a priority area. If it is, then they have to constitute an appropriate scientific board and go through the same kind of processes that the NIH does in their concern about human health.

SUMMERS: I think another point that should be emphasized is that I hope EPA would not again rediscover the wheel. You have produced three major symposia in the past and several publications. There was a special AAAS committee that went to great lengths to evaluate the use of a whole range of insect pathogen agents. More recently, there have been several workshops involved with the review and evaluation of various recommended guidelines specifically targeting aspects involving genetically engineered organisms.

You need to assemble those proceedings into some effective working mechanism and take the recommendations as a basis for developing safety evaluation protocols.

KAMELY: Recommendations from a group of experts such as this would help EPA formulate a research agenda and set policies regarding genetically altered viruses in the environment.

SUMMERS: You already have a considerable amount of this information available and have already taken the first step.

Name Index

Subject Index